Free radicals
in biology and medicine

Free radicals
in biology and medicine

BARRY HALLIWELL
Lister Institute Research Fellow in Preventive Medicine,
King's College, London

and

JOHN M. C. GUTTERIDGE
Senior Scientist, National Institute for Biological
Standards and Control, London

CLARENDON PRESS · OXFORD

Oxford University Press, Walton Street, Oxford OX2 6DP

Oxford New York Toronto
Delhi Bombay Calcutta Madras Karachi
Kuala Lumpur Singapore Hong Kong Tokyo
Nairobi Dar es Salaam Cape Town
Melbourne Auckland
and associated companies in
Beirut Berlin Ibadan Nicosia

Oxford is a trade mark of Oxford University Press

Published in the United States
by Oxford University Press, New York

British Library Cataloguing in Publication Data

Halliwell, Barry
Free radicals in biology and medicine.
1. Radicals (Chemistry)   2. Biological chemistry
I. Title   II. Gutteridge, John M. C.
574.19'282   QP527
ISBN 0-19-854137-6

Library of Congress Cataloging in Publication Data
Halliwell, Barry.
Free radicals in biology and medicine.
Bibliography: p.
Includes index.
1. Free radicals (Chemistry)   2. Free radical
reactions.   3. Biology—Research.   4. Medicine—
Research.   I. Gutteridge, John M. C.   II. Title.
[DNLM: 1. Free Radicals.   2. Biology.   3. Medicine.
QD 471 H191f]
QP527.H35   1984   574.19'283   84-19111
ISBN 0-19-854137-6

Printed in Northern Ireland by
The Universities Press (Belfast) Ltd.

Preface

The importance of radical reactions in radiation damage, food preservation, combustion, and in the rubber and paint industry, has been known for many years to people in the respective fields, but it has rarely been appreciated by biologists and clinicians. The interest in radicals shown by the latter groups has been raised recently by the discovery of the importance of radical reactions in normal body chemistry and in the mode of action of many toxins. The discoveries of hypoxic cell sensitizers that potentiate radiation-induced radical damage to cancerous tumours, of the enzyme superoxide dismutase, and of the mechanism of action of such toxins as paraquat and carbon tetrachloride provide major examples of this importance.

Any expanding field attracts the charlatans, such as those who make money out of proposing that consuming radical scavengers will make you live for ever or that taking tablets containing superoxide dismutase will enhance your health and sex life. In evaluating these and other less-obviously silly claims, it is useful to understand the basic chemistry of radical reactions.

This book is aimed mainly at biologists and clinicians. It assumes virtually no knowledge of chemistry and attempts to lead the reader as painlessly as possible into an understanding of what free radicals are, how they are generated, and how they can react. Having established this basis, the role of radical reactions in several biological systems is critically evaluated in the hope that the careful techniques needed to *prove* their importance will become more widely used. We believe that free-radical chemists should also find these latter chapters useful.

London   B.H.
                                                                         J.M.C.G.

Acknowledgements

We are very grateful to the following scientists and publishers who have granted permission to reproduce material or otherwise supplied us with figures:

Dr R. W. Hardy and the American Society of Plant Physiologists (Fig. 1.1); Dr O. R. Brown and Academic Press (Fig. 1.2); Dr H. B. Michaels (Fig. 1.3); Dr J. D. Balentine (Fig. 1.4); Dr G. L. Huber and Springer-Verlag (Fig. 1.5); Dr C. L. Greenstock and Pergamon Press (Fig. 2.1 part A); Professor R. L. Willson (Fig. 2.1 part B); Dr B. Franck and Verlag Chemie GmbH (Fig. 2.4); Professor B. Chance and Academic Press (Fig. 3.3); Professors B. Chance, H. Sies and the American Physiological Society (Figs. 3.1 and 3.5); Dr H. B. Dunford and Elsevier-North Holland (Fig. 3.6); Professor G. Rotilio and Elsevier-North Holland (Fig. 3.7); Dr L. Hurley and Elsevier-North Holland (Fig. 3.9); Professor P. Hochstein and Academic Press (Fig. 4.8); Dr David Hockley (Fig. 4.4); Professor W. Pryor and Ann Arbor Science Publishers (Figs. 4.10 and 6.4); Dr A. A. Noronha-Dutra (Fig. 4.9); Dr A. W. Segal (Fig. 7.4); Professor W. G. Hocking (Fig. 7.5); Dr W. Dawson (Fig. 7.6); Professor R. S. Sohal and Elsevier-North Holland (Fig. 8.2); Dr C. Verdone-Smith, Dr H. E. Enesco and Pergamon Press (Fig. 8.3); Professor D. Armstrong (Figs. 8.4 and 8.6); Professor D. Armstrong and Elsevier-North Holland (Fig. 8.5).

J.M.C.G. is indebted to his wife Pushpa, and children Samantha and Mark, for their encouragement and support during the preparation of this book. B.H. is grateful to Pat Allen, Norma Belcher, and Liz Haigh for their invaluable help with typing. Both authors thank Dr Graham Burton for his advice on chemical structures and vitamin E reactions.

Contents

1 Oxygen is poisonous—an introduction to oxygen toxicity and free radicals

The oxygen in the Earth's atmosphere is itself a 'free radical' and a major promoter of radical reactions in living cells. It is therefore appropriate to begin by making some general comments about oxygen, after which the nature and definition of radicals shall be considered.

1.1. Oxygen and the Earth

Except for those organisms that are especially adapted to live under anaerobic conditions, all animals and plants require oxygen for the efficient production of energy. Free oxygen appeared in the Earth's atmosphere in significant amounts about 2×10^9 years ago, probably due to the evolution of oxygen-evolving photosynthetic organisms. The appearance of oxygen must have been accompanied by the appearance of a layer of ozone (O_3) in the high atmosphere, and the absorption of damaging solar ultraviolet radiation by oxygen and ozone probably permitted the evolution of more complex terrestrial organisms. Oxygen is now the most prevalent element in the Earth's crust (atomic abundance 53.8 per cent) and the percentage of oxygen in the atmosphere has risen to 21 per cent in dry air. If the barometric pressure of dry air at sea level is 760 mm Hg (1 mm Hg = 1 torr), the partial pressure of oxygen would thus be about 159 mm Hg.

Oxygen is also found dissolved in seas, lakes, rivers, and other bodies of water; the oxygen content of surface water is generally in equilibrium with the atmosphere. The solubility of oxygen in sea-water at $10\,°C$ corresponds to a concentration of $0.284\ \mathrm{mmol\,l^{-1}}$, and decreases at higher temperatures (e.g. $0.212\ \mathrm{mmol\,l^{-1}}$ at $25\,°C$). Oxygen is more soluble in fresh water, e.g. for distilled water: $0.258\ \mathrm{mmol\,l^{-1}}$ at $25\,°C$, $0.355\ \mathrm{mmol\,l^{-1}}$ at $10\,°C$. Of course, the oxygen concentration within living cells will depend on how far the oxygen has to move in order to get to them as well as on how quickly they consume it. For example, the O_2 tension in human venous blood is only 40 mm Hg (about $53\ \mathrm{\mu mol\,l^{-1}}\ O_2$), about 25 per cent of ambient. Within some eukaryotic cells, e.g. heart or liver, there is an oxygen gradient, decreasing in concentration from the cell membrane to the oxygen-consuming mitochondria. Oxygen is seven-to-eight times more soluble in organic solvents than in water, a point worth bearing in mind when considering oxidative damage to the hydrophobic interior of biological membranes (Chapter 4).

As the oxygen content of the atmosphere rose, it also exposed living matter to oxygen toxicity: oxidations in the cell harmful to the organism and in some cases lethal. There was considerable pressure upon organisms to evolve protective mechanisms against oxygen toxicity, or to retreat to environments that the oxygen did not penetrate. Studies of present-day anaerobes show us what must have happened to the numerous primitive species that failed to adapt and were lost during evolution.

1.2. Oxygen and anaerobic organisms

The term 'anaerobic organism' covers a wide range of biological variation. There are 'strict' anaerobes such as the bacteria *Treponema denticola* and several *Clostridia* that will grow in the laboratory only if oxygen is virtually absent. 'Moderate' anaerobes can grow in atmospheres up to about 10 per cent O_2 (e.g. *Bacteroides fragilis* or *Clostridium novyi* Type A), whereas microaerophiles require a low concentration of oxygen for growth but cannot tolerate 21 per cent O_2. Even 'strict anaerobes' display a wide spectrum of oxygen tolerance. Some are killed by even a brief exposure to oxygen whereas in others oxygen inhibits growth but does not kill the cells, e.g. *Methanobacterium AZ* ceases growth at 0.01 ppm O_2 but survives exposure for several days to 7 ppm dissolved O_2, equivalent to an atmospheric concentration of 20 per cent.

Any environment that can develop a low enough oxygen concentration can harbour anaerobes. For example, in the human mouth, strict anaerobes can be cultured from pockets in the gums, from decaying teeth, and from the deeper layers of dental plaque; whereas less strict anaerobes and microaerophiles can be found in the more superficial layers of plaque on the teeth. The human colon (over 90 per cent of faecal bacteria are anaerobes), rotting material, polluted waters and gangrenous wounds all provide places for anaerobic bacteria to thrive. Indeed, the treatment of gas gangrene due to *Clostridial* infections by exposure of the patient to pure oxygen at high pressure is based on the known sensitivity of these anaerobes to oxygen. As discussed below, however, such treatment is not without problems!

The damaging effects of oxygen on strict anaerobes seem to be due to the oxidation of essential cellular components. Anaerobes thrive in reducing environments and, by oxidizing such constituents as NAD(P)H, thiols, iron–sulphur proteins, and pteridines, the oxygen can 'drain away' the reducing equivalents that are needed for biosynthetic reactions within the cell. Some enzymes in anaerobes are inactivated by oxygen, e.g. the nitrogenase enzyme of *Clostridium pasteurianum* is inactivated due to the oxidation of essential components at its active site. This enzyme, which catalyses reduction of nitrogen to ammonia, is essential for the survival of the organism in environments poor in nitrogen compounds. Indeed, all nitrogenase enzymes are inactivated by oxygen to some extent, but not all nitrogen-fixing species are strict anaerobes. Indeed, a study of nitrogen-fixing organisms has shown a variety of ways around this problem. *Clostridium pasteurianum,* as we have seen, adopts a simple solution and keeps away from oxygen! Several aerobic, nitrogen-fixing (and other) bacteria surround themselves with a thick slime capsule to restrict the entry of oxygen; some cyanobacteria locate their nitrogenase in specialized, thick-walled, oxygen-resistant cells known as 'heterocysts'. In the root nodules of leguminous plants an oxygen-binding protein, leghaemoglobin, is present to control the free oxygen concentration and to prevent the nitrogen-fixing bacteroids of the nodule from being damaged. The nitrogen-fixing aerobe *Azotobacter* has one of the highest respiration rates of any micro-organism, which may serve to consume all the oxygen entering the cell and prevent it reaching the nitrogenase. The photosynthetic cyanobacterium *Gloeocapsa* contains both nitrogenase and an oxygen-evolving photosynthetic apparatus within the same cell, but its life cycle is such that nitrogenase is only highly active when the rate of photosynthesis is low.

Anaerobes can teach us a great deal about the evolution of protective mechanisms against oxygen toxicity and we shall consider them again when reviewing the various protective mechanisms thought to exist.

1.3. Oxygen and aerobes

Oxygen supplied at concentrations greater than those in normal air has long been known to be toxic to plants, animals, and to aerobic bacteria such as *Escherichia coli*. Plots of the logarithm of survival time against logarithm of the oxygen pressure have shown inverse, approximately linear, relationships, for protozoa, mice, fish, rats, rabbits, and insects. Indeed, there is considerable evidence that even 21 per cent O_2 has slowly-manifested damaging effects. Figure 1.1 shows one example of oxygen effects: the dry matter accumulation in the leaves of soybean plants is actually increased if they are placed in sub-normal oxygen concentrations. All plant tissues are damaged at oxygen concentrations above normal; there is inhibition of chloroplast development, decrease in seed viability and root growth, membrane damage, and eventual shrivelling and dropping-off of leaves. Green plants produce oxygen during photosynthesis and can expose themselves and their surroundings

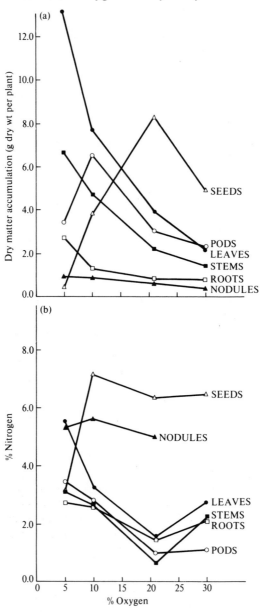

Fig. 1.1. Dry matter accumulation and nitrogen content of various parts of soyabean plants grown in chambers containing different percentages of atmospheric oxygen. Seed production is less sensitive to oxygen than other parts of the plant. From B. Quebedeaux, U. D. Havelka, K. L. Livak and R. W. F. Hardy (1975) *Plant Physiology* **56,** 761–4, with permission.

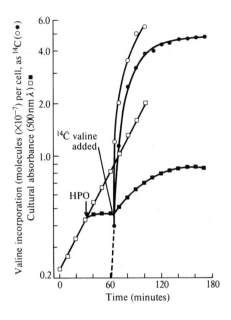

Fig. 1.2. Inhibition of the growth of *E. coli* cells by exposure to high-pressure oxygen. The growth medium was mineral salts, glucose, and amino acids (no valine) at 37 °C. At the point marked HPO the atmosphere was changed from air to one of 80 per cent O_2 at 5 atm total pressure. At the point indicated valine was added and growth was restored. Closed symbols: HPO experiment; open symbols: normal air control. From O. R. Brown and F. Yein (1978) *Biochemical and biophysical Research Communications* **85**, 1219–21, with permission.

to it, e.g. oxygen bubbles from some aquatic plants have been reported to interfere with the breeding of mosquitoes.

The growth of *E. coli* and other aerobic bacteria is slowed by exposure to pure oxygen at 1 atmosphere pressure; Fig. 1.2 shows that exposure of this organism to high-pressure oxygen causes immediate growth inhibition. Oxygen enhances the damaging effects of ionizing radiation both to bacteria and to animal cells in culture; Fig. 1.3 shows this effect for cultured Chinese hamster ovary cells exposed to X-rays. As will be discussed in Chapter 2, the effects of oxygen and those of ionizing radiation on organisms have some similarities.

The toxicity of O_2 to animals, including Man, has been of interest in relation to diving, underwater swimming and escape from submarines, and, more recently, in the use of oxygen in the treatment of cancer, gas gangrene, multiple sclerosis and lung diseases, and in the design of the gas supply in spacecraft. Rises in the oxygen partial pressure to which an organism is subjected can be due not only to an increase in the percentage

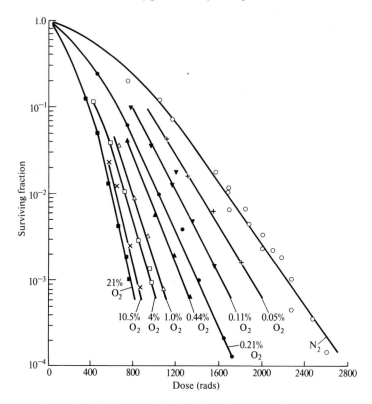

Fig. 1.3. The 'oxygen effect' in exposing cultured Chinese hamster ovary cells to ionizing radiation. Figure by courtesy of Dr H. B. Michaels.

of oxygen in the air but also, as in diving, to an increase in the total pressure. High-pressure oxygen frequently causes acute central nervous system toxicity, producing convulsions. Figure 1.4 shows a rat in this sorry state. Oxygen at 1 atmosphere pressure does not usually produce such convulsions, but oxygen concentrations of 50 per cent or above, corresponding to an inspired partial pressure of 360 mm Hg, gradually damage the lungs. Exposure of humans to pure oxygen at 1 atmosphere pressure for as little as six hours causes chest soreness, cough, and sore throat in a few people; further exposure leads in all cases to damage to the alveoli of the lungs. This is manifested at first as an increased thickness of the air–blood barrier caused by oedema. Further oxygen exposure causes death of alveolar epithelial cells and an eventual laying down of inelastic fibrous material in the lungs. Such damage can never be repaired. Figure 1.5 shows the gradual development of pulmonary oxygen toxicity as seen on chest X-rays. Recent clinical and experimental

Fig. 1.4. An adult female Sprague–Dawley rat with convulsive paralysis, especially of the forelimbs, induced by exposure to pure oxygen at 5 atm pressure. Photograph by courtesy of Dr J. D. Balentine.

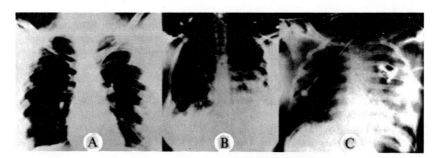

Fig. 1.5. A: Initial chest X-ray of a patient with mild respiratory discomfort after administration of oxygen for a nonpulmonary condition. There are no visible significant abnormalities. B: Further exposure causing X-ray-visible damage with diffuse, irregular pulmonary densities of various sizes in both lungs. C: Late radiological manifestations of pulmonary oxygen toxicity with extension and joining-up of the lesions. Radiological manifestations are due to fluid accumulation (oedema), atelectasis, and accumulation of cellular debris in alveolar spaces and in the terminal airways. There is laboured, gasping breathing often accompanied by frothy, bloody sputum. The damaged lungs cannot absorb sufficient oxygen for the body, resulting in cyanosis. From G. L. Huber and D. B. Drath, Chapter 14 in *Oxygen and living processes, an interdisciplinary approach* (ed. D. L. Gilbert), Springer-Verlag, New York, 1981, with permission.

observations suggest that oxygen may worsen lung damage caused by other means even at concentrations thought to be 'safe'.

Other tissues do not escape damage when animals are exposed to high oxygen concentrations, however. The form of blindness known as retrolental fibroplasia (from the Latin for 'formation of fibrous tissue behind the lens') arose abruptly in the early 1940s among infants born prematurely, and quickly became widespread. Not until 1954 was it realised that this disease is associated with the use of high oxygen concentrations in incubators for premature babies, and more careful control of oxygen use has greatly decreased its incidence. Elevated O_2 appears to inhibit the growth of retinal blood vessels. On return to a normal atmosphere there is an excessive regrowth of the vessels, which sometimes occurs to an extent that causes detachment of the retina and subsequent blindness. The new vessels lack structural integrity and often bleed.

Table 1.1 lists some typical effects of oxygen on other animal tissues. High oxygen concentrations also cause a general 'stress reaction' in animals, which stimulates the action of some endocrine glands. Removal of, for example, the thyroid gland decreases the toxic effects of O_2 in some animals whereas administration of thyroxine, cortisone or adrenalin often makes them worse. Exposure of pregnant animals to elevated oxygen concentrations has been reported to increase the incidence of foetal abnormalities.

The damaging effects of oxygen on aerobic organisms vary considerably with the type of organism used, its age, physiological state, and diet. Different tissues of an animal are affected in different ways. For example, the effective oxygen concentration in the swim-bladder of the rat-tail fish at a depth of 3000 m is 2500 times greater than ambient, yet the bladder remains undamaged. The fish as a whole cannot tolerate anywhere approaching this oxygen concentration, and so its swim-bladder must be specially protected. Cold-blooded animals, such as turtles and crocodiles, are relatively resistant to oxygen toxicity at low environmental temperatures, but become more sensitive at higher temperatures. Young rats are more resistant to oxygen than are adult rats: adult humans are less sensitive than are adult rats. Oxygen toxicity is also influenced by the presence in the diet of varying amounts of vitamins A, E, and C, heavy metals, anti-oxidants (now added to many human foods) and polyunsaturated fatty acids. For example, rats fed on a fat-free test diet supplemented with cod-liver oil could tolerate pure O_2 much better than if the supplement consisted of coconut oil. In rats, an elevated blood glucose concentration has been reported to delay the onset of convulsions caused by hyperbaric oxygen.

Table 1.1. Some typical effects on animal tissues of exposure to high oxygen concentrations

Species used	Nature of exposure	Organ examined	Results found
Adult, male rats	Pure O_2 at 5 atm for 75 min	Heart	Mitochondrial swelling followed by damage to myofibrils
Cats	Pure O_2 at 8 atm for 50 min	Kidney	Swelling of tubules, glomerular abnormalities
Rats	Pure O_2 at 0.33 atm for 3 days	Liver	Mitochondrial damage
Monkeys	Pure O_2 at 0.5 atm for up to 22 days	Liver	Proliferation and abnormality of smooth endoplasmic reticulum, decrease in glycogen content
Male hamsters	70% O_2 for 3–4 weeks	Testes	Degeneration of seminiferous epithelium, cessation of sperm production
Humans	'Hyperbaric oxygen therapy'	Ear	Haemorrhages of inner ear, deafness
Guinea pigs	70% O_2 at 1 atm for 6–36 days	Bone-marrow	Inhibition of erythroid cell development

1.4. What causes the toxic effects of oxygen?

Perhaps the earliest suggestion made to explain oxygen toxicity was that oxygen inhibits cellular enzymes. Indeed, direct inhibition by oxygen accounts for the loss of nitrogenase activity in oxygen-exposed *Clostridium pasteurianum* (see above) and in a few other cases. Figure 1.2 shows that the inhibition of growth observed on exposing *E. coli* to high-pressure oxygen can be relieved by adding the amino acid valine to the culture medium, apparently because its synthesis is impaired due to a rapid inhibition of the enzyme dihydroxyacid dehydratase in the metabolic pathway for its synthesis. Even when valine is supplied, however, growth soon ceases because of a slower inhibition of other cellular enzymes, and supplementation of the culture medium with niacin and thiamin can then permit further growth. The onset of oxygen-induced convulsions in animals is correlated with a decrease in the cerebral content of the neurotransmitter GABA (γ-aminobutyric acid), perhaps because of an inhibition of the enzyme glutamate decarboxylase (glutamate $\rightarrow$ GABA + CO_2) by oxygen. In neither of these cases, however, has it been shown that the enzyme inhibition *in vivo* is due to oxygen itself rather than, say, to an increased production of oxygen radicals (see below).

Perhaps the best example of a direct effect of oxygen itself on aerobes comes from green plants. During photosynthesis, illuminated green plants fix carbon dioxide (CO_2) into sugars by a complex metabolic pathway known as the Calvin cycle. The first enzyme in this pathway, ribulose bisphosphate carboxylase, combines carbon dioxide with a five-carbon sugar (ribulose-1,5-bisphosphate) to produce two molecules of phospho-glyceric acid (Chapter 5). Oxygen is an inhibitor of this reaction competitive with carbon dioxide, and so at elevated oxygen concentrations there is less carbon dioxide fixation and less plant growth. This simple mechanism accounts for part, but not all, of the decreased leaf-growth at elevated oxygen concentrations (Fig. 1.1).

In general, however, the rates of enzyme inactivation by oxygen in aerobic cells are too slow and too limited in extent to account for the rate at which toxic effects develop and many enzymes are totally unaffected by O_2 at all. This led Rebecca Gershman and Daniel L. Gilbert in the USA to propose in 1954 that most of the damaging effects of oxygen could be attributed to the formation of free oxygen radicals. Let us now consider exactly what 'free radicals' are. (In order to understand the discussion in the next section, it is essential to appreciate clearly what is meant by such chemical terms as 'covalent bond', 'Pauli principle', 'atomic orbital', 'antibonding molecular orbital', 'spin quantum number', 'Hund's rule', and 'transition metal'. Readers requiring explanation of such terms are advised to consult the Appendix before reading further in this chapter.)

1.5. What is a free radical?

The term 'radical' is often used loosely in chemistry to refer to various groups of atoms that behave as a unit, such as the carbonate radical (CO_3^{2-}), nitrate radical (NO_3^-), and the methyl radical (CH_3—). We shall avoid this use and define a 'free radical' as follows: *a free radical is any species capable of independent existence that contains one or more unpaired electrons.* (An unpaired electron is one that occupies an atomic or molecular orbital by itself.)

The presence of one or more unpaired electrons causes the species to be attracted slightly to a magnetic field (i.e. to be *paramagnetic*), and sometimes makes the species highly reactive. Consideration of the above broad definition shows that there are many free radicals in chemistry and biology (e.g. the hydrogen atom; see Appendix). Radicals can easily be formed when a covalent bond is broken if one electron from each of the pair shared remains with each atom, a process known as *homolytic fission*. The energy required to dissociate the covalent bond can be provided by heat, electromagnetic radiation or other means as will be discussed further in subsequent chapters. Many covalent bonds only dissociate at high temperatures, e.g. 450–600 °C is often required to rupture C–C, C–H, or C–O bonds. Many studies of radical reactions have been carried out in the gas phase at high temperatures; combustion is well known to chemists as a free-radical process.

If A and B are two atoms covalently bonded (A_X^X representing the electron pair), homolytic fission can be written as

$$A\,{}_X^X\,B \rightarrow A^X + B_X$$

A^X is an A-radical, often written as A·, and B_X is a B-radical (B·) Homolytic fission of one covalent bond in the water molecule will yield a hydrogen radical (H·) and a hydroxyl radical (OH·). The opposite of homolysis (homolytic fission) is *heterolytic fission* in which one atom receives both electrons when a covalent bond breaks i.e.

$$A\,{}_X^X\,B \rightarrow A_X^{X-} + B^+$$

The extra electron gives A a negative charge and B is left with a positive charge. Heterolytic fission of water gives the hydrogen ion H^+ and the hydroxide ion OH^-. In fact, pure water is very slightly ionized in this way and contains 10^{-7} moles per litre each of H^+ and OH^- ions at 25 °C.

Let us now look at some atoms and molecules of biological importance to see how far they fit into our definition of radicals.

1.5.1. Oxygen and its derivatives

Inspection of Fig. 1.6 shows that the oxygen molecule, as it occurs naturally, certainly qualifies as a radical: it has two unpaired electrons each located in a different π^* antibonding orbital. These two electrons have the same spin quantum number (or, as is often written, they have *parallel spins*). This is the most stable state, or *ground state*, of oxygen. Oxygen is a good oxidizing agent, the basic definitions being:

> *Oxidation:* loss of electrons by an atom or molecule (e.g. the conversion of a sodium atom to the ion Na^+).
> *Reduction:* gain of electrons by an atom or molecule (e.g. the conversion of a chlorine atom to the ion Cl^-).

An oxidizing agent therefore is good at absorbing electrons from the molecule it oxidizes (as is chlorine) whereas a reducing agent (such as sodium) is an electron donor. (These definitions are simplified, but sufficient for our purpose.) If oxygen attempts to oxidize another atom or molecule by accepting a pair of electrons from it, both of these electrons must be of antiparallel spin so as to fit in to the vacant spaces in the π^* orbitals (Fig. 1.6). A pair of electrons in an atomic or molecular orbital would not meet this criterion however, since they would have opposite spins in accordance with Pauli's principle. This imposes a restriction on electron transfer which tends to make oxygen accept its electrons one at a time, and also means that oxygen reacts only sluggishly with many non-radicals. Theoretically, the complex organic compounds of the human body should immediately combust in the oxygen of the air (as

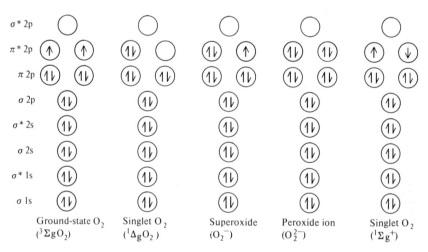

Fig. 1.6. Bonding in the diatomic oxygen molecule.

occult magazines, such as *The Unexplained,* occasionally claim that people have done) but the spin restriction slows this down, fortunately!

More reactive forms of oxygen, known as *singlet oxygens,* can be generated by an input of energy (see Chapter 2 for details of how this is done). The $^1\Delta gO_2$ state (Fig. 1.6) has an energy 22.4 kcal above the ground state. The $^1\Sigma g^+$ state is even more reactive, 37.5 kcal above the ground state. By our definition, $^1\Delta gO_2$ is not a radical; there are no unpaired electrons. In both forms of singlet oxygen the spin restriction is removed and so the oxidizing ability is greatly increased.

If a single electron is added to the ground-state O_2 molecule, it must enter one of the π^* antibonding orbitals (Fig. 1.6). The product is the *superoxide radical* O_2^-. With only one unpaired electron, superoxide is actually less of a radical than is O_2 itself, despite its name. Its properties are considered in the next chapter. Addition of one more electron will give O_2^{2-}, the *peroxide ion* which, as may be seen from Fig. 1.6, is not a radical. Since the extra electrons in O_2^- and O_2^{2-} are entering antibonding orbitals, the strength of the oxygen–oxygen bond is decreasing (see Appendix for an explanation of this). In ground-state O_2 the atoms are effectively bonded by two covalent bonds, but in O_2^- only by one-and-a-half (there is an extra electron in an antibonding orbital), and in O_2^{2-} by one bond only. Hence the oxygen–oxygen bond in O_2^{2-} is quite weak. Addition of another two electrons to O_2^{2-} would eliminate the bond entirely since they would go into the σ^*2p orbitals, so giving $2O^{2-}$ species. Usually in biological systems the two-electron reduction product of oxygen is hydrogen peroxide (H_2O_2), and the four-electron product, water. To summarize:

$$O_2 \xrightarrow[\text{reduction}]{\text{one-electron}} O_2^-$$

$$O_2 \xrightarrow[\substack{\text{reduction} \\ \text{(plus 2H}^+)}]{\text{two-electron}} H_2O_2 \text{ (protonated form of } O_2^{2-})$$

$$O_2 \xrightarrow[\substack{\text{reduction} \\ \text{(plus 4H}^+)}]{\text{four-electron}} 2H_2O \text{ (protonated form of } O^{2-}).$$

Hydrogen peroxide is a pale-blue covalent viscous liquid, boiling point 150 °C. It mixes readily with water and acts as an oxidizing agent, which gives it some antibacterial properties. In fact, the weak antiseptic activity of honey, which has been used in wound treatment since ancient times, is probably due to the formation of hydrogen peroxide by enzymes contained within it.

Since the O–O bond is relatively weak (see above), hydrogen peroxide

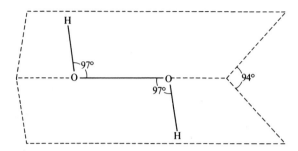

Fig. 1.7. Structure of hydrogen peroxide.

decomposes easily: homolytic fission would give the hydroxyl radical

$$H_2O_2 \xrightarrow{\text{energy}} 2OH^\cdot.$$

The structure of hydrogen peroxide is shown in Fig. 1.7. Although a covalent molecule, it does ionize at highly alkaline pH values (way above the biological range, however)

$$H_2O_2 \rightarrow 2H^+ + O_2^{2-}.$$

1.5.2. Ozone and oxides of nitrogen

Ozone (O_3), a pale-blue gas, is an important shield for solar radiation in the higher reaches of the atmosphere. It is produced by the photodissocia-tion (i.e. splitting-up caused by light energy) of molecular O_2 into oxygen atoms, which then react with oxygen molecules

$$O_2 \xrightarrow{\text{solar energy}} 2O$$
$$O_2 + O \rightarrow O_3$$

The two oxygen–oxygen bonds in the ozone molecule are of equal length and intermediate in nature between those of an oxygen–oxygen single bond and a double bond. Ozone has an unpleasant smell and severely damages the lungs. It is a much more powerful oxidizing agent than is ground-state oxygen, i.e. it is only helpful to us in its proper place. Significant amounts of ozone can form in the lower atmosphere in urban air as a result of a series of complex photochemical events resulting from pollution. The ozone molecule is not a radical and is diamagnetic (weakly repelled by a magnetic field), although the damage that it causes is often mediated by free-radical production (Chapter 6).

Recently there has been a great fuss over the use of fluorinated hydrocarbons in, for example, aerosol sprays because they may help to deplete the ozone layer in the upper atmosphere. Photodissociation of gases such as CF_2Cl_2 and $CFCl_3$ produces chlorine atoms in the atmosphere, which cause a breakdown of ozone. Two of the oxides of nitrogen, nitric oxide (NO) and nitrogen dioxide (NO_2), can also deplete ozone. Nitric oxide is a colourless gas, a weak reducing agent (i.e. it tends to donate electrons to other molecules) and it reacts with oxygen to give nitrogen dioxide. Nitrogen dioxide is a dense, brown poisonous gas and a powerful oxidizing agent. Sources of nitrogen oxides in the higher atmosphere include solar flares and supersonic aircraft exhausts. They are also both found in cigarette smoke and since both NO and NO_2 molecules possess odd numbers of electrons, they fall into our definition of free radicals. Further consideration of their noxious effects can be found in Chapter 6.

1.5.3. Transition metals

All the metals in the first row of the d-block in the Periodic Table contain unpaired electrons and can thus qualify as radicals, with the sole exception of zinc (see Appendix for further explanation if necessary). Copper does not really fit the definition of a transition element since its 3d-orbitals are full, but it readily forms the Cu^{2+} ion by loss of two electrons, one from the 4s- and one from a 3d-orbital. This leaves an unpaired electron. Many transition elements are of great biological importance (Table 1.2) and so it is worthwhile examining their properties.

The transition elements are all metals. Their most important feature from a radical point of view is their variable valency, which allows them to undergo changes in oxidation state involving one electron. For example, iron has two common valencies in which the electronic configurations are as follows (see Appendix for an explanation of this notation, if necessary):

Table 1.2. Biological importance of some d-block elements

Metal	Biochemical significance
Copper (Cu)	Essential in human diet. Required for enzymes such as superoxide dismutase, cytochrome oxidase, lysine oxidase, dopamine-β-hydroxylase, and caeruloplasmin. About 80 mg Cu in adult human body. Highest concentrations in liver and brain. Overall blood content in males 0.106 mg/100 ml. Toxic in excess
Zinc (Zn)	Non-transition element, fixed valency of 2. Suggested that it sometimes inhibits iron-dependent radical reactions by displacing iron from its binding site. Essential in human diet; found in RNA polymerase, carbonic anhydrase, superoxide dismutase. Plasma zinc approx 0.112 mg/100 ml. Toxic in excess
Vanadium (V)	Essential in animals but requirement in man not yet established. Accumulated in large amounts in some tunicates. Involved in cholesterol metabolism. Inhibits strongly the ATPase enzyme which exchanges Na^+ and K^+ ions across cell membranes: this might be a physiological regulatory mechanism
Chromium (Cr)	Probably essential in diet, involved in regulation of glucose metabolism. Normal serum Cr is 1–5 ng/ml (1 nanogram (ng) $= 10^{-9}$ g)
Manganese (Mn)	Essential in animals, probably in man as well (normal blood level 9 µg/ml). Needed for mitochondrial superoxide dismutases, also activates a number of hydrolase and carboxylase enzymes. Free and total Mn contents in liver cells of fed rats estimated as about 0.71 and 34 nmoles ml^{-1} of cell water respectively. (1 nanomole $= 10^{-9}$ moles)
Iron (Fe)	Essential in human diet: deficiency causes simple anaemia. Most abundant transition metal in humans. Normal serum iron in males approx. 0.127 mg/100 ml mostly bound to the protein transferrin. Regulation of iron content of the body is done by regulation of iron uptake in the gut. Needed for haemoglobin, myoglobin, cytochromes, several enzymes and non-haem-iron proteins (see Chapter 2)
Cobalt (Co)	Essential as a component of vitamin B_{12} but little else known.
Nickel (Ni)	Probably essential in animals, requirement in man not yet established. Found in urease in plant cells and in several bacterial enzymes, such as hydrogenases and carbon monoxide dehydrogenase
Molybdenum (Mo)	Essential in trace amounts for some flavin metalloenzymes, e.g. xanthine oxidase, nitrogenase, sulphite oxidase, nitrate reductase

The existence of a ferryl (iron IV) species has also been suggested. In solution in the presence of air, the iron(III) state is the most stable, whereas iron(II) salts are weakly reducing and ferryl compounds are powerful oxidizing agents. If a solution of an iron(II) salt, e.g. 'ferrous sulphate' ($FeSO_4$), is left exposed to the air it slowly oxidizes to the iron(III) state. This is a one-electron oxidation, and oxygen dissolved in the solution is reduced to the superoxide radical O_2^- (Fig. 1.6).

$$Fe^{2+} + O_2 \rightleftharpoons Fe^{2+}\!\!-\!\!O_2 \rightleftharpoons Fe^{3+}\!\!-\!\!O_2^- \rightleftharpoons Fe^{3+} + O_2^-.$$
$$\text{intermediate complexes}$$

Copper has two common valencies, copper(I) and copper(II), formerly known as 'cuprous' and 'cupric'

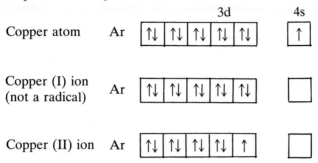

Again, the one-electron difference between these valency states allows copper to take part in radical reactions. Under appropriate conditions, for example, copper salts can both accept electrons from, and donate electrons to, the superoxide radical O_2^-,

$$Cu^{2+} + O_2^- \rightarrow Cu^+ + O_2$$
$$Cu^+ + O_2^- \rightarrow Cu^{2+} + O_2^{2-}$$
$$O_2^{2-} + 2H^+ \rightarrow H_2O_2$$

Net reaction:

$$O_2^- + O_2^- + 2H^+ \rightarrow H_2O_2 + O_2.$$

The copper salt, by changing its valency, is causing the net combination of two O_2^- radicals and two H^+ ions to form H_2O_2 and O_2. The copper salt is acting as a *catalyst:* it remains unchanged in amount and chemical nature at the end of the reaction whilst speeding it up enormously. The variable valency of transition metals helps them to be effective catalysts of many reactions involving oxidation and reduction and they are used for this purpose at the active sites of many enzymes catalysing such reactions. The radical reactions they promote can overcome the spin restriction on direct reaction of oxygen with non-radical species.

Manganese has a most stable valency state in aqueous solution of Mn^{2+}; more oxidized species such as Mn(III), Mn(IV) and Mn(VII) also exist. Again radical reactions are possible, e.g. Mn^{2+} and O_2^-:

$$Mn^{2+} + O_2^- + 2H^+ \rightarrow Mn^{3+} + H_2O_2.$$

Zinc, by contrast, has only one valency (Zn^{2+}) and does not promote radical reactions. It has been suggested that zinc may inhibit some radical reactions *in vivo* by displacing other transition metal ions such as iron from the binding sites at which they are promoting such reactions.

Fenton reaction

A mixture of hydrogen peroxide and an iron(II) salt reacts with many organic molecules, as was first observed by Fenton in 1894. The reactivity is most likely due to formation of the hydroxyl radical:

$$Fe^{2+} + H_2O_2 \rightarrow Fe^{3+} + OH^{\cdot} + OH^-$$

Traces of Fe^{3+} can react further with H_2O_2:

$$Fe^{3+} + H_2O_2 \rightarrow Fe^{2+} + O_2^- + 2H^+.$$

Even more reactions are possible:

$$OH^{\cdot} + H_2O_2 \rightarrow H_2O + H^+ + O_2^-$$
$$O_2^- + Fe^{3+} \rightarrow Fe^{2+} + O_2$$
$$OH^{\cdot} + Fe^{2+} \rightarrow Fe^{3+} + HO^-.$$

Thus this simple mixture of an iron salt and hydrogen peroxide, which can almost certainly form in biological systems under certain circumstances (Chapter 2), can provoke a whole series of radical reactions. The overall sum of these, unless some other reagent is added, is an iron-catalysed decomposition of hydrogen peroxide,

$$2H_2O_2 \xrightarrow[\text{catalyst}]{\text{Fe-salt}} O_2 + 2H_2O.$$

Other reagents present can react with the various radicals and change the mechanism of the reaction.

Copper(I) salts also react with H_2O_2 to make hydroxyl radicals:

$$Cu^+ + H_2O_2 \rightarrow Cu^{2+} + OH^{\cdot} + OH^-.$$

In the next chapters, we shall examine the biological significance of some inorganic and organic radicals in detail.

1.6. Further reading

Balentine, J. D. (1978). Experimental pathology of oxygen toxicity. In *Oxygen and physiological function* (ed., F. F. Jobsis). Professional Information Library, Dallas, Texas.

Balentine, J. D. (1982). *Pathology of oxygen toxicity.* Academic Press, New York.

Deneke, S. M., and Fanburg, B. L. (1980). Normobaric oxygen toxicity of the lung. *New Engl. J. Med.* **303,** 76.

Dulka, J. J., and Risby, T. H. (1976). Ultratrace metals in some environmental and biological systems. *Analyt. Chem.* **48,** 640A.

Gallon, J. R. (1981). The oxygen-sensitivity of nitrogenase: a problem for biochemists and micro-organisms. *Trends in biochem. Sci.* **January 1981** 19.

Gilbert, D. L. (ed.) (1981). *Oxygen and living processes: an interdisciplinary approach.* Springer-Verlag, New York.

Halliwell, B. (1981). Free radicals, oxygen toxicity and ageing. In *Age Pigments* (ed. R. S. Sohal) p. 1. Elsevier/North Holland Biomedical Press, Amsterdam.

Haugaard, N. (1968). Cellular mechanisms of oxygen toxicity. *Physiol. Rev.* **48**, 229.

Morris, J. G. (1976). Oxygen and the obligate anaerobe. *J. appl. Bact.* **40**, 229.

Morris, J. G. (1979). Oxygen and growth of the oral bacteria. In *Saliva and dental caries* (eds. I. Kleinberg *et al.*). Information Retrieval Ltd., New York.

Prasad, A. S. (ed.) (1978). *Trace Elements and Iron in Human Metabolism.* Plenum Press, USA.

Simons, T. J. B. (1979). Vanadate. A new tool for biologists. *Nature (Lond.)* **281**, 337.

Thauer, R. K., Diekert, G., and Schönheit, P. (1980). Biological role of nickel. *Trends in Biochem. Sci.* **November 1980** 304.

Walling, C. (1975). Fenton's reagent revisited. *J. Am. Chem. Soc.* **8**, 125.

Willson, R. L. (1978). Iron, zinc, free radicals and oxygen in tissue disorders and cancer control. In *Iron metabolism,* p. 331. CIBA Foundation Symposium, Elsevier, Amsterdam.

2 The chemistry of oxygen radicals and other oxygen-derived species

Before we can really understand what oxygen radicals and related species can and cannot do in biological systems, we must first look at their chemical properties.

2.1. Reaction rates and rate constants

In describing the properties and reactions of radicals, frequent reference will be made to the rates at which reactions proceed, so it is worthwhile beginning by establishing clearly how rates are usually expressed. The rate of a reaction can be measured either by following the loss of the starting materials (*reactants*), or by following the formation of the products. Reaction rate is then simply defined as the amount of product formed in unit time, or as the amount of reactant used up in unit time. Time is usually quoted in seconds and the amounts in *moles*, one mole of a substance being its molecular weight (relative molecular mass is a more precise term) expressed in grams. One mole of any substance contains the same number of molecules, that number being *Avogadro's number*, numerically equal to the enormous figure of 6.023×10^{23}.

The rate of a reaction will obviously depend on the concentration of reactants present. To take a simple case, suppose one mole of a substance A in solution in a volume of one litre is reacting to form another substance B

$$A \rightarrow B$$

Suppose further that after 1 second, 0.01 mole of A has been converted into B. The reaction rate (R) can then be expressed either as 0.01 moles of B formed in 1 litre in 1 second ($R = 0.01 \, mol \, l^{-1} \, s^{-1}$ for B), or as 0.01 moles of A used up in 1 litre in 1 second ($R = 0.01 \, mol \, l^{-1} \, s^{-1}$ for A).

If the concentration of A is doubled, it is likely that the rate of reaction will double. The exact mathematical relationship between the rate of a reaction and the concentration of the reactants is known as the *rate law*. In this case R is proportional to the concentration of A, expressed as moles of A per litre of solution ($mol \, l^{-1}$). This is mathematically equivalent to saying that R is equal to the concentration of A multiplied by a constant, the *rate constant* for the reaction, i.e. the rate law is:

$$R = k_1 \, [A]$$

where k_1 is the rate constant at the temperature of the experiment, and [A] means the concentration of A in moles per litre. Once the reaction has started, A is used up, [A] falls and so R will fall. Hence rate measurements are always made in the first few seconds of a reaction so that the concentration of reactants has not changed significantly from that originally present (so-called *initial rate measurements*). Rate constants, and hence rates of reactions, increase as temperature is raised.

In the rate law

$$R = k_1 [A]$$

the rate of the reaction depends only on the first power of the concentration of A; another way of saying this is that the reaction is *first order with respect to A*. If the rate law had been found by experiment to be

$$R = k_2 [A]^2 \quad \text{Second order}$$

(i.e. the rate is proportional to the square of the concentration of A), the reaction would be *second order with respect to* A, and if

$$R = k_3 [A]^3 \quad \text{third order}$$

then it would be *third order with respect to* A. k_1 would be called a *first-order rate constant* with units of s^{-1}, k_2, a *second-order rate constant* with units of $l\,s^{-1}\,mol^{-1}$ (sometimes written as $M^{-1}\,s^{-1}$ where M (molar) is another way of writing moles per litre), and k_3 is a *third-order rate constant*. It is these rate constants that are usually published in the scientific literature.

Now consider another simple reaction in which there are two reactants, e.g.

$$A + B \rightarrow \text{products}$$

This type of equation often represents the reaction of a radical A with some other material B and it usually follows the rate law

$$R = k_2 [A] [B]$$

where k_2 is a second-order rate constant. The reaction is first order with respect to A, first order with respect to B and *second order overall*. The constant k_2 has the units $l\,s^{-1}\,mol^{-1}$ $(M^{-1}\,s^{-1})$.

As an example of the information that can be gleaned from published rate constants, let us look at the formation of hydroxyl radicals from H_2O_2 in the presence of either Fe^{2+} or Cu^+ ions, as described in Chapter 1. The published approximate second-order rate constants are:

$$H_2O_2 + Fe^{2+} \rightarrow Fe^{3+} + OH^- + OH^{\cdot} \qquad k_2 = 76 \; l\,mol^{-1}\,s^{-1}$$
$$H_2O_2 + Cu^+ \rightarrow Cu^{2+} + OH^- + OH^{\cdot} \qquad k_2 = 10^9 \; l\,mol^{-1}\,s^{-1}.$$

If equal concentrations of H_2O_2 are mixed with equal concentrations of Fe^{2+} or Cu^+, the initial rate of hydroxyl radical formation in the latter case will be greater by a factor of $10^9/76$, i.e. a factor of 1.32×10^7. Values of the rate constant can be applied to see how fast a reaction might occur under biological conditions. As discussed later in this chapter, the concentrations of hydrogen peroxide and Fe^{2+} in, say, liver cells are likely to be low, in the range of μmoles per litre (10^{-6} moles l^{-1}) or less. If 1 μmol l^{-1} of H_2O_2 comes into contact with 1 μmol l^{-1} of Fe^{2+}, how much OH$^\cdot$ radical will be formed? The rate law is:

$$R = k_2[H_2O_2][Fe^{2+}]$$
$$= 76(10^{-6})(10^{-6}) = 7.6 \times 10^{-11} \text{ moles } l^{-1} s^{-1}.$$

This seems a tiny figure, but remember that 1 mole of a substance contains 6.023×10^{23} molecules. Hence the *number* of hydroxyl radicals formed per litre per second is 4.58×10^{13}—much more impressive! If the cell volume is 10^{-12}–10^{-11} litre (average volumes for a liver cell) this still means 46 to 458 hydroxyl radicals formed per cell every second. Of course, as the reaction proceeds both Fe(II) and H_2O_2 will be used up and the rate of OH$^\cdot$ production will fall unless they are continuously replenished. Thus, even reactions with low rate-constants (such as the Fenton reaction) can be biologically important.

2.2. Measurement of reaction rates for radical reactions

Many radical reactions proceed extremely quickly and so special techniques are required to measure their rates. Two techniques have commonly been used, *stopped flow* and *pulse radiolysis*, and they are considered briefly below.

2.2.1. Pulse radiolysis

In this technique, the compound to be studied is placed in a reaction cell in solution and a radical is formed directly in the cell by a 'pulse' of ionizing radiation, e.g. from a linear accelerator. By the appropriate choice of experimental conditions, specific radicals can be generated and their reactions followed over a microsecond (10^{-6} s) time scale or even longer. Since many radicals absorb light at different wavelengths than their parent compound, the progress of the reaction is followed by changes in the light absorbance of the cell components, usually displayed on an oscilloscope. Figure 2.1(a) shows an outline of the apparatus. Radiation sources useful for pulse radiolysis generally provide pulses of electrons in the energy range 1–30 mega-electron-volts. Exposure of

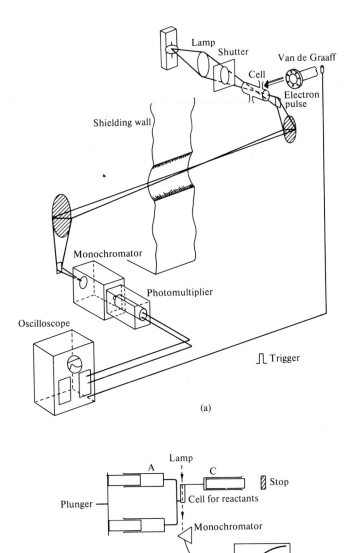

Fig. 2.1. Techniques for measuring fast reactions. Light is shone through a reaction cell and analysed for changes in light absorbance by the monochromator and photomultiplier. The results are displayed on an oscilloscope. In pulse radiolysis (part (a)) reaction is started by generating a radical in the reaction cell by means of a burst of ionizing radiation, e.g. from a Van de Graaff accelerator. In stopped flow (part (b)) the reactants are contained in separate syringes (A and B) and they enter the reaction vessel when the plunger is pressed, eventually passing into the collecting syringe C. Figure (a) is taken from *Progress in reaction kinetics 11*, (1982), pp. 73–135, with permission. Figure (b) by courtesy of Professor R. L. Willson.

water to these produces ionization and excitation within 10^{-16} second

$$H_2O \rightarrow H_2O^+ + e^- + H_2O^*$$

where e^- represents an electron and H_2O^* an excited water molecule. Such excited molecules undergo homolytic fission in 10^{-14}–10^{-13} second to give hydrogen atoms (which can equally well be called hydrogen radicals since they contain one unpaired electron) and hydroxyl radicals:

$$H_2O^* \rightarrow H^{\boldsymbol{\cdot}} + OH^{\boldsymbol{\cdot}}.$$

Within the same timescale H_2O^+ also reacts to give $OH^{\boldsymbol{\cdot}}$

$$H_2O^+ + H_2O \rightarrow H_3O^+ + OH^{\boldsymbol{\cdot}}.$$

The electrons become surrounded by clusters of water molecules within 10^{-12}–10^{-11} second. These hydrated electrons can be written as e^-_{aq} where 'aq' is an abbreviation for 'aqueous'. Hence three different radicals are produced on 'pulsing' an aqueous solution: $H^{\boldsymbol{\cdot}}$, $OH^{\boldsymbol{\cdot}}$, and e^-_{aq}. Alterations of pH, and addition of various compounds, can 'select out' a particular radical for further study. For example, if the aqueous solution is saturated with nitrous oxide (N_2O) gas before pulsing, e^-_{aq} are removed by the reactions

$$e^-_{aq} + N_2O \rightarrow N_2 + O^-$$
$$O^- + H^+ \rightarrow OH^{\boldsymbol{\cdot}} \quad \text{\textit{hydroxyl radicals are produced}}$$

and converted into hydroxyl radicals. By contrast, if the solution is saturated with oxygen gas and also contains 0.1 mole per litre of sodium formate (an ionic solid containing sodium ions and formate ions, $HCOO^-$) the following reactions occur to produce the superoxide radical

$$e^-_{aq} + O_2 \rightarrow O_2^- \quad \text{\textit{saturated w/ O_2 gas}}$$
$$H^{\boldsymbol{\cdot}} + HCOO^- \rightarrow H_2 + CO_2^{\boldsymbol{\cdot}-}$$
$$OH^{\boldsymbol{\cdot}} + HCOO^- \rightarrow H_2O + CO_2^{\boldsymbol{\cdot}-}$$
$$CO_2^{\boldsymbol{\cdot}-} + O_2 \rightarrow CO_2 + O_2^{\boldsymbol{\cdot}-} \quad \text{\textit{superoxide radical is produced}}$$

Thus relatively 'clean' sources of $OH^{\boldsymbol{\cdot}}$ or O_2^- can be produced and studies of their reactions with various compounds can be made. This technique has been especially useful in investigating reactions of $OH^{\boldsymbol{\cdot}}$ and O_2^- with biological molecules. Usually the reaction is observed directly by following the rise in light absorbance of a reaction product, or the loss of absorbance of a reactant. If this is not possible, a 'competition method' may be used. For example, $OH^{\boldsymbol{\cdot}}$ reacts with thiocyanate ion (CNS^-) to give a strongly absorbing product and the rate constant for this reaction is known. If another compound (X) that reacts with $OH^{\boldsymbol{\cdot}}$ is added, then it

competition method ↓

will intercept some of the OH˙, and the absorbance change due to the CNS⁻ reaction will be smaller. Knowing the concentrations of CNS⁻ and X, and the above rate constant, the rate constant for the reaction between X and OH˙ can easily be calculated. The dye ABTS is also well suited for such 'pulse competition' studies.

The pulse radiolysis method can be used to study other radicals as well. For example, removal of one electron from ascorbic acid (vitamin C) produces an ascorbate radical. The absorption spectrum and properties of this radical can be observed by generating it in a pulse radiolysis apparatus, either by reducing dehydroascorbic acid (ascorbate that has lost two electrons) with e^-_{aq} or by oxidizing ascorbate with the OH˙ radical

$$OH˙ + vit\ C \rightarrow vit\ C˙^- + OH^-$$

or with another oxidizing radical such as Br_2^-, formed by adding bromide (Br^-) ions.

$$OH˙ + 2Br^- \rightarrow OH^- + Br_2^-$$

2.2.2. Stopped-flow methods

Stopped-flow methods are often used where the rates of radical reactions are too slow to be measured conveniently by pulse radiolysis, yet too fast to be measured by standard biochemical techniques. Solutions of the compounds to be reacted are contained in syringes (A and B in Fig. 2.1(b)). To initiate the reaction the plungers are pushed so that the syringe contents are forced simultaneously into a quartz cell, where they mix, and then into a collecting syringe (C in Fig. 2.1(b)). Experimental conditions are chosen so that, while the solutions are flowing, any given volume of mixed solution does not stay in the cell long enough for significant reaction to occur. When the plunger of the collecting syringe strikes the 'stop', the flow halts abruptly and the reaction within the cell proceeds to completion. Its absorbance changes can be measured and recorded on an oscilloscope and so reaction rates can be calculated. For example, a solution of superoxide ion (as its potassium salt $K^+O_2^-$) in an organic solvent can be mixed with a compound in aqueous solution and the rate of reaction measured (see Section 2.9 for further details).

2.3. Reactions of the hydroxyl radical

Rate constants for OH˙ reactions have mainly been determined by pulse radiolysis methods. Inspection of Table 2.1 shows that this radical reacts with extremely high rate-constants with almost every type of molecule found in living cells: sugars, aminoacids, phospholipids (lecithin),

Table 2.1. Typical second-order rate constants for reactions of the hydroxyl radical

Compound tested	pH	Rate constant ($M^{-1} s^{-1}$)	Compound tested	pH	Rate constant ($M^{-1} s^{-1}$)
Carbonate ion, CO_3^{2-}	10.7	2.0×10^8	Glycylglycine	2	7.8×10^7
Bicarbonate ion, HCO_3^-	6.5	1.0×10^7	Glycyltyrosine	2	5.6×10^9
Fe^{2+}	2.1	2.5×10^8	Guanine	—	1.0×10^{10}
H_2O_2	7	4.5×10^7	Haemoglobin	—	3.6×10^{10}
Adenine	7.4	3.0×10^9	Histidine	6–7	3.0×10^9
Adenosine	7.7	2.5×10^9	Hydroxyproline	2	2.1×10^8
AMP	5.4	1.8×10^9	Lactate ion	9	4.8×10^9
Arginine	7	2.1×10^9	Lecithin	—	5.0×10^8
Ascorbic acid	1	7.2×10^9	Methanol	7	4.7×10^8
Benzene	7	3.2×10^9	Methionine	7	5.1×10^9
Benzoic acid	3	4.3×10^9	Nicotinic acid	—	6.3×10^8
Butan-1-ol (*n*-butanol)	7	2.2×10^9	Phenol	7	4.2×10^9
Catalase	—	2.6×10^{11}	Phenylalanine	6	3.5×10^9
Citric acid	1	3.0×10^7	Propan-1-ol	7	1.5×10^9
Cysteine	1	7.9×10^9	Pyridoxal phosphate	—	1.6×10^9
Cystine	2	3.2×10^9	Ribonuclease	—	1.9×10^{10}
Cytidine	2	2.0×10^9	Ribose	7	1.2×10^9
Cytosine	7	2.9×10^9	Serum albumin	—	2.3×10^{10}
Deoxyguanylic acid	7	4.1×10^9	Thiourea	7	4.7×10^9
Deoxyribose	—	1.9×10^9	Thymine	7	3.1×10^9
Ethanol	7	7.2×10^8	Tryptophan	6	8.5×10^9
Glucose	7	1.0×10^9	Uracil	7	3.1×10^9
Glutamic acid	2	7.9×10^7	Urea	9	$<7.0 \times 10^5$
Glutathione	1	8.8×10^9			

Values are mostly taken from the compilation by M. Anbar and P. Neta (1967) *Int. J. appl. Radiation and Isotopes* **18**, 493–523.

nucleotides, and organic acids. Indeed, it is one of the most reactive chemical species known. For example it is pointless to try to demonstrate hydroxyl radical reactions *in vitro* in solutions containing Tris buffer, since OH˙ attacks this buffer rapidly and a Tris-derived radical is produced. Reactions of OH˙ can be classified into three main types: *hydrogen abstraction, addition,* and *electron transfer.* These reactions illustrate an important principle of radical chemistry: *reaction of a free radical with a non-radical species produces a different free radical, which may be more or less reactive than the original radical.* Radicals produced by reactions with OH˙ are usually less reactive, however, since OH˙ is such an aggressive species.

As an example of hydrogen abstraction, consider the reaction of OH˙ with alcohols. The OH˙ 'pulls off' a hydrogen atom (H) and combines with it to form water, leaving behind an unpaired electron on the carbon atom, e.g. for the alcohol ethanol

$$\underset{\substack{|\\\text{H}}}{\overset{\substack{\text{H}\\|}}{\text{H—C}}}\text{—}\underset{\substack{|\\\text{H}}}{\overset{\substack{\text{H}\\|}}{\text{C}}}\text{—O—H} + \text{OH}^{\bullet} \longrightarrow \underset{\substack{|\\\text{H}}}{\overset{\substack{\text{H}\\|}}{\text{H—C}}}\text{—}\underset{\substack{|\\\text{H}}}{\overset{\substack{\bullet\\}}{\text{C}}}\text{—O—H} + \text{H}_2\text{O}$$

or for methanol

$$\text{CH}_3\text{OH} + \text{OH}^{\bullet} \rightarrow {}^{\bullet}\text{CH}_2\text{OH} + \text{H}_2\text{O}.$$

Further reactions of the carbon radical can then occur, e.g. reaction with oxygen

$$^{\bullet}\text{CH}_2\text{OH} + \text{O}_2 \rightarrow {}^{\bullet}\text{O}_2\text{CH}_2\text{OH}$$

or the joining-up of two radicals to form a non-radical product, the two unpaired electrons between them forming a covalent bond.

$$\text{CH}_3\dot{\text{C}}\text{HOH} + \text{CH}_3\dot{\text{C}}\text{HOH} \longrightarrow \underset{\substack{|\\\text{CH}_3\text{CHOH}}}{\text{CH}_3\text{CHOH}}$$

The reaction of OH˙ with lecithin (also known as phosphatidylcholine), an important phospholipid found in biological membranes, is of the hydrogen-abstraction type, and the carbon radicals left behind undergo a series of reactions leading to membrane damage (this is discussed further in Chapter 4). Attack of OH˙ on a sugar such as deoxyribose, found in DNA, produces a huge variety of different products, some of which have been shown to be mutagenic in bacterial test systems.

The reaction of OH˙ with aromatic ring structures proceeds by addition, and similar reactions occur with the purine and pyrimidine bases present in DNA and RNA. For example OH˙ can add on across a double bond in the pyrimidine base thymine

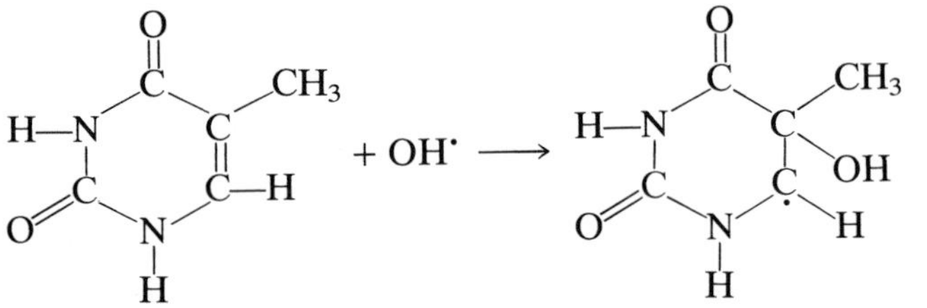

The thymine radical then undergoes a series of further reactions. Thus hydroxyl radical severely damages the bases and sugars of DNA, and also induces strand breakage. If damage is so extensive that it cannot be repaired, the cell may die. Even survivable damage can result in mutation.

Hydroxyl radicals take part in electron-transfer reactions with inorganic and organic compounds, e.g. with the chloride ion

$$Cl^- + OH^\cdot \rightarrow Cl^\cdot + OH^-$$
$$Cl^\cdot + Cl^- \rightarrow Cl_2^{\cdot -}$$

It is clear from Table 2.1 that the reactivity of OH˙ radicals is so great that, if they are formed in living systems, they will react immediately with whatever biological molecule is in their vicinity, producing secondary radicals of variable reactivity. For example, their reaction with carbonate ion (CO_3^{2-}) produces carbonate radicals ($CO_3^{\cdot -}$), which are powerful oxidizing agents. Let us see how OH˙ radicals could be produced.

2.4. Production of hydroxyl radicals in living systems

2.4.1. Ionizing radiation

Since the major constituent of living cells is water, exposure of them to ionizing radiation such as X-rays or γ-rays will result in hydroxyl radical production, as described previously. Hydroxyl radicals are responsible for a large part of the damage done to cellular DNA and to membranes by ionizing radiation. Single- and double-strand breaks in DNA are considered to be very important damaging events, especially as double-strand breaks cannot be repaired by the cell. Oxygen, normally present in most biological systems, aggravates the damage done by ionizing radiation. This in part occurs because oxygen allows formation of O_2^- from e^-_{aq}, and O_2^- in biological systems can produce more OH· radicals as discussed in Chapter 5. (Further reasons for this 'oxygen effect' are explored in Chapter 6.)

2.4.2. Reaction of metal ions with hydrogen peroxide

As discussed in Section 2.10, hydrogen peroxide is formed in many aerobic cells. Thus the reactions of Cu^+ and Fe^{2+} ions with H_2O_2 are feasible sources of OH· *in vivo* provided that these metal ions are present. Is this so?

Iron distribution: is iron available for the Fenton reaction?

An adult human contains approximately four grams of iron, about two-thirds of which circulates in the oxygen-carrying pigment haemoglobin. A further ten per cent is found in myoglobin, and a very small amount in various iron-containing enzymes and the transport-protein transferrin (see below). The rest is present in intracellular storage proteins, ferritin and haemosiderin. These are found mainly in liver, spleen, and bone-marrow, but also to some extent in most other tissues; and some ferritin is found in the blood plasma.

   Non-haem iron in the diet exists in the oxidized, Fe(III), form and it must be dissolved and reduced before it can be absorbed. The hydrochloric acid in the stomach achieves solubilization and dietary vitamin C (ascorbic acid, a reducing agent) reduces some of the iron to the Fe(II) state and facilitates its absorption, which occurs mainly in the upper jejunum. Iron taken up by the gut enters the plasma-protein *transferrin*, which functions as a carrier molecule. Transferrin is a glycoprotein and each molecule has two separate binding sites to which Fe(III) attaches extremely tightly. Under normal conditions the transferrin present in the bloodstream is only about thirty per cent loaded with iron on average, so that the amount of free iron salts available in the blood plasma would be expected to be virtually zero, a result confirmed by experiment. A protein similar to transferrin, known as *lactoferrin*, is found in several body fluids and in milk and is produced by phagocytic cells (Chapter 7). Lactoferrin also binds two moles of Fe(III) per mole of protein.

   Iron from transferrin must enter the various cells of the body for use in synthesizing iron enzymes and proteins. Transferrin binds to receptors on the cell surface and is internalised, entering the cytoplasm in a vacuole. The contents of the vacuole are then acidified to facilitate the release of iron, which probably becomes bound to various cellular constituents such as citrate, ATP, GTP, or other phosphate esters. The iron-free transferrin (*apotransferrin*) is then ejected from the cell whilst the small pool of non-protein-bound iron can be used in the synthesis of iron proteins. For example, mitochondria take up iron salts rapidly for incorporation into cytochromes and non-haem iron proteins and these organelles contain small 'pools' of iron in the matrix. Any iron not required by cells can be stored in the protein *ferritin*.   Ferritin is a LARGE Protein

Ferritin is a large protein (molecular weight 444 000) that can store up to 4500 moles of iron per mole of protein. It consists of a protein shell (*apoferritin*) enclosing the iron, which exists in the form of a hydrated iron(III)oxide–phosphate complex. The protein shell is made up of twenty-four subunits and has 'pores' that allow access to the interior. Ferritin appears to function in preventing an excessive intracellular accumulation of non-protein-bound iron, and synthesis of apoferritin can be stimulated by loading cells with iron salts. It is rarely saturated with iron *in vivo*. Iron enters ferritin as Fe(II), which becomes oxidized by the protein to Fe(III) and deposited in the interior. Similarly, iron can be removed from ferritin as Fe(II) by the action of a number of biological reducing agents, including cysteine, reduced flavins, and ascorbate. Ferritin can be degraded in lysosomes, the protein shell being attacked to leave an insoluble product known as *haemosiderin*.

The small pool of intracellular non-protein-bound iron, presumably attached to such chelating agents as ATP, GTP and citrate, could conceivably provide iron for the Fenton reaction, since such physiological iron chelates have been shown to react with H_2O_2 *in vitro*. No doubt this is why the pool is kept as small as possible. The authors have developed an assay, based on the use of the antibiotic bleomycin, to detect iron complexes capable of catalysing radical reactions in biological fluids. Table 2.2 summarizes the principles of this assay and the results of applying it to human extracellular fluids. No bleomycin-detectable iron is present in serum or plasma from normal humans, as would perhaps be expected from the low percentage saturation of transferrin. Bleomycin-detectable iron is, however, present in samples of human cerebrospinal fluid, some samples of human sweat and in knee-joint synovial fluid from patients with rheumatoid arthritis. Bleomycin-detectable iron presumably represents iron bound to low-molecular-weight chelating agents (e.g. ADP or citrate) or perhaps *loosely* bound to certain proteins (e.g. albumin). Iron bound to transferrin, lactoferrin, haemoglobin or iron-containing enzymes does not register in the bleomycin assay. Indeed, addition of commercial apotransferrin to human synovial or cerebrospinal fluids makes the iron unreactive on subsequent assay by the bleomycin method.

Bleomycin-detectable iron has also been found in extracts of several bacterial strains. Indeed, Repine *et al.* in the USA have shown that the killing of a *Staphylococcus aureus* strain by H_2O_2 is potentiated by

Table 2.2. Bleomycin-detectable iron in extracellular fluids

Fluid tested	Concentration of non-protein-bound iron salts (μmoles per litre)
Human blood serum or plasma	0
Human synovial fluid[1] (rheumatoid patients)	2.8 ± 1.2
Human cerebrospinal fluid (normal)	2.2 ± 1.3
Human sweat (from the trunk of ✻?✻ athletes)	4.6 ± 2.9

[1] The iron content of normal synovial fluid has not been established since there is no ethical reason for taking samples from normal joints.

The presence of iron is tested for by utilizing the fact that the antibiotic bleomycin requires the presence of iron salts in order to degrade DNA (see Chapter 8 for a more extensive discussion). Suitable calibration of the DNA degradation allows the iron content of body fluids to be determined. Results are abstracted from Gutteridge *et al.* (1982) *Lancet* **ii**, 459–60; (1982) *Biochem. J.* **206**, 605–9; (1981) *Biochem. J.* **199**, 263–5. Bleomycin-detectable iron has been shown to be capable of catalysing radical reactions.

growing the cells on a medium rich in iron salts. Similarly, iron chelating agents can decrease the DNA damage observed when mouse cells are treated with H_2O_2. Thus the Fenton reaction is a biological reality if H_2O_2 is available *in vivo*.

The amount of iron in the body is determined by the amount assimilated into the body from the gut—there is no obvious physiological mechanism for disposing of excess iron. The consequences of this are seen in *iron overload*. This can happen as a result of grossly excessive dietary iron intake (children eating all their mother's iron tablets, or the Bantu tribe of Africa who drink acidic beer out of iron pots) but is more often due to medical treatment of other diseases. For example, the *thalassaemias* are inborn conditions in which the rate of synthesis of one of the haemoglobin chains is diminished, the prefix α- or β-thalassaemia being used to identify the chain that is synthesized abnormally slowly. Thalassaemia major and minor refer to the homozygous and hetero-zygous states respectively. The disease was named from a Greek word meaning 'the sea' since many patients who have it are of Mediterranean ancestry. Untreated patients with thalassaemia major die of anaemia in infancy, but can be kept alive by regular blood transfusions. Since each unit of blood contains about 0.2 g iron, the patients become overloaded with iron, leading to saturation of ferritin, haemosiderin, and transferrin, and often the appearance of non-protein-bound iron salts in the blood. Iron accumulates especially in the liver, spleen, and heart, the latter organ being very sensitive to it so that many thalassaemics treated by

non-protein-bound iron salts in the blood

transfusion die in their twenties or thirties from heart failure. Similar problems arise in the treatment of other chronic anaemias by transfusion. A slower-developing iron overload of dietary origin is seen in patients suffering from an inherited disease (*idiopathic haemochromatosis*) in which much more dietary iron than usual is absorbed from the gut. The metabolic abnormality causing the increased iron uptake is unknown. The time taken for iron overload to develop in haemochromatosis depends on the patient's diet, but in the plasma of patients showing symptoms the transferrin is 80 per cent or more loaded with iron and iron complexes capable of reacting in the bleomycin assay are often present.

It may well be that the pathology of iron overload is related to increased hydroxyl radical formation *in vivo*, and there is evidence for oxygen radical damage in the spleen of iron-overloaded patients. Excessive accumulation of haemosiderin in the liver also seems to weaken the lysosomal membranes.

There has therefore been considerable clinical interest in <u>chelating agents</u>, <u>compounds which bind to iron salts and prevent them from undergoing radical reactions</u>. Table 2.3 shows several chelating agents in which interest has been expressed. The most widely used is desferrioxamine B, a powerful chelator of iron(III) extracted from *Streptomyces pilosus*. Children with β-thalassaemia major in Britain were first given desferrioxamine in 1962, and it appears to have been successful in prolonging their lifespan. Desferrioxamine and its iron(III) complex (*ferrioxamine*) are rapidly excreted, mainly in urine but also in bile, so removing iron from the body. Large doses of desferrioxamine are required, it cannot be given by mouth and its low cell-uptake limits its usefulness, so there is now interest in other chelators such as rhodotorulic acid and PIH (Table 2.3). It is interesting to note that the commonly used chelating agent EDTA (Table 2.3) does not prevent the reaction of iron salts with H_2O_2 or with oxygen radicals.

Copper availability

Since non-protein-bound iron salts exist intra-cellularly and in some extracellular fluids, an iron-dependent generation of $OH^\cdot$ from H_2O_2 is biologically feasible. Can the same be said for copper?

The adult human body contains about 80 mg copper. It is absorbed from the diet in the stomach or upper small intestine, probably in the form of complexes with aminoacids, such as histidine, or small peptides. These simple copper complexes enter the blood and most of the copper binds tightly to serum albumin in equilibrium with a small 'pool' of copper complexes. In the liver the copper is taken up and incorporated into the glycoprotein *caeruloplasmin*, which is then released into the circulation (normal plasma content is 200–400 mg l^{-1}). About ninety-five per cent of total plasma copper is found in this protein, the rest being attached to albumin or aminoacids as mentioned above.

Caeruloplasmin contains six or seven moles of copper per mole of protein. It does not exchange copper readily nor will it bind extra copper, so it does not have the characteristics of a copper transport protein (by contrast transferrin can easily bind Fe(III) and release it again within cells). It seems that cells must take up and degrade caeruloplasmin in order to obtain copper from it.

Caeruloplasmin can catalyse the oxidation of a variety of polyamines and polyphenols, the dioxygen taken up being reduced to water. It will also oxidize Fe(II) ions to Fe(III) ions (its *ferroxidase* activity).

Whereas the non-enzymic oxidation of Fe(II) produces superoxide radical (Section 1.5.3), that catalysed by caeruloplasmin releases only water. The ferroxidase action of caeruloplasmin is probably important *in vivo* in assisting the binding of iron onto transferrin and possibly onto serum ferritin. Wilson's disease :

Wilson's disease is an inherited metabolic defect characterized by low concentrations of caeruloplasmin in the blood. Copper is deposited in the liver, kidney, and brain, causing damage that leads to lack of co-ordination, tremors, and progressive mental retardation. Again, there is evidence that free-radical reactions are involved in the pathology of Wilson's disease. Treatment of Wilson's disease involves a copper re-stricted diet and the use of the chelating agent penicillamine (Table 2.3) to promote copper excretion.

So is free copper available normally to make $OH^{\cdot}$ from H_2O_2? Table 2.4 shows the results of an experiment designed to test this. Formation of free hydroxyl radicals from Cu^+ ions and H_2O_2 is inhibited by added histidine or albumin at physiological concentrations. Hence the copper ions in blood serum cannot give rise to 'free' hydroxyl radicals. Similarly, the protein concentration within the cells is very high and if any free copper ions are available they should rapidly become protein-bound, unable to catalyse formation of free hydroxyl radicals. By contrast, albumin does not prevent Fe(II)-dependent formation of free hydroxyl radicals. It may be, however, that the bound copper ions can still generate $OH^{\cdot}$ which reacts immediately with the binding molecule and thus cannot be detected in free solution. Consistent with this, mixtures of Cu^+ and H_2O_2 have been observed to damage many proteins. Such 'site-specific' damage by OH radicals may be very important *in vivo*.

2.4.3. Formation of hydroxyl radicals from ozone

Ozone (O_3) has several damaging effects on biological systems, including its ability to cross-link tyrosine residues in proteins by oxidizing their

Table 2.3. Chelating agents that can bind metal ions

Name	Formula	Comments
Penicillamine	$\mathrm{HS-C(CH_3)_2}$ $\mathrm{H_2N-CHCOOH}$	Useful in promoting urinary excretion of copper salts. Also binds iron (II).
EDTA (ethylenediaminetetraacetic acid), ion	$^{-}\mathrm{OOCCH_2}$ $\diagdown$ $^{\mathrm{CH_2COO^-}}$ $\mathrm{N(CH_2)_2N}$ $^{-}\mathrm{OOCCH_2}$ $\diagup$ $_{\mathrm{CH_2COO^-}}$	Chelates several metal ions. Copper–EDTA chelates usually less active than free copper ions in radical reactions, whereas chelates of EDTA with Fe(II) or Fe(III) still react with H_2O_2 or superoxide.
DETAPAC (diethylenetriaminepentaacetic acid), ion	$^{-}\mathrm{OOCCH_2}$ $\diagdown$ $_{\mathrm{N(CH_2)_2N(CH_2)_2N}}$ $^{\mathrm{CH_2COO^-}}$ $^{-}\mathrm{OOCCH_2}$ $\quad$ $_{\mathrm{CH_2COO^-}}$ $\quad$ $_{\mathrm{CH_2COO^-}}$	Chelates several metal ions other than iron and copper. Not much used clinically as has several toxic side effects including magnesium depletion.
Rhodotorulic acid		Both this and desferrioxamine (below) are examples of *siderophores*, i.e. compounds produced by some micro-organisms in order to chelate iron from the growth medium and bring it into the cell. Both bind iron as Fe(III).

2,3-Dihydroxybenzoic acid, ion

Has given promising results in treatment of thalassaemias. Can be given orally.

Desferrioxamine B

A linear molecule that 'bends round' to complex Fe^{3+} with six oxygen ligands, forming a bright-red ligand known as ferrioxamine. Commercially available (CIBA) as 'desferal'—desferrioxamine B methane-sulphonate (mol. wt. 657).

PIH (pyridoxal isonicotinoyl hydrazone)

Shown to be an effective iron chelator in animals. Can be given orally.

Table 2.4. Copper-dependent formation of hydroxyl radicals

Reagent added	Rate of OH˙ formation observed (as percentage of maximum)
None	100
Histidine (400 μmol l⁻¹)	10
(200 μmol l⁻¹)	22
Albumin (0.5 g l⁻¹)	7
(0.25 g l⁻¹)	22

Copper ions (Cu^{2+}) were mixed with hydrogen peroxide in the presence of a reducing agent to generate Cu^+ and the rate of OH˙ formation was measured. In the data shown above, the reducing agent was ascorbic acid although superoxide radical could also be used. Normal blood serum contains $28–45$ mg l⁻¹ of albumin and $50–240$ μmol l⁻¹ of histidine. It may be seen that concentrations at or below these suppress the formation of detectable hydroxyl radicals. Data from Rowley and Halliwell (1983) *Archs Biochem. Biophys.* **225**, 279–84.

—OH groups to give O,O'-dityrosine. There is also evidence that ozone can produce hydroxyl radicals in aqueous solution, which could contribute to toxicity. Hydroxyl radical formation is favoured at very high pH values, but might occur to some extent at physiological pH.

2.4.4. Ethanol metabolism

Most adult humans ingest ethanol to improve their lifestyle and even in teetotallers it is produced in small amounts by the gut flora. Ethanol absorbed from the gut is metabolized in the liver, mainly by the action of an alcohol dehydrogenase enzyme that oxidizes it to ethanal (acetaldehyde). A minor contribution to ethanol metabolism is made by a 'microsomal ethanol oxidizing system' (MEOS), so called because it is found in the microsomal pellet upon subcellular fractionation of liver homogenates. When liver cells are disrupted, the plasma membrane and endoplasmic reticulum are torn up and the pieces form membrane vesicles which can be sedimented by high-speed centrifugation. Thus the microsomal fraction so obtained is a heterogeneous collection of membrane vesicles from different parts of the cell; and this should be borne in mind when interpreting the results of experiments upon this fraction. We have seen more than one paper in which the authors spoke of microsomes as if they were a discrete subcellular organelle.

The MEOS system probably originates from the endoplasmic reticulum, which contains systems that generate hydrogen peroxide (Section 2.10). There is considerable evidence that some of the ethanol is oxidized by hydroxyl radicals generated from this hydrogen peroxide. Since the action of MEOS *in vitro* is partially inhibited by desferrioxamine, the

OH˙ may be produced by a Fenton reaction using iron bound to the microsomal membranes.

2.5. Detection of hydroxyl radicals in biological systems

2.5.1. Electron spin resonance and spin-trapping

Electron spin resonance (ESR) is a technique that can be applied to free radicals, since it detects the presence of unpaired electrons. An unpaired electron has a spin of either $+\frac{1}{2}$ or $-\frac{1}{2}$ and behaves as a small magnet. If it is exposed to an external magnetic field, it can align itself either parallel or anti-parallel (in opposition) to that field, and thus can have two possible energy levels. If electromagnetic radiation of the correct energy is applied, it will be absorbed and used to move the electron from the lower energy level to the upper one. Thus an absorption spectrum is obtained, usually in the microwave region of the electromagnetic spectrum. For reasons that need not concern us, ESR spectrometers are set up to display *first-derivative spectra*, which show not the absorbance but the *rate of change of absorbance*, i.e. a point on the derivative curve corresponds to the gradient (slope) at the equivalent point on the absorption plot.

The condition to obtain an absorbance is:

$$\Delta E = g\beta H$$

where ΔE is the energy gap between the two energy levels of the electron, H is the applied magnetic field, and β a constant known as the *Bohr magneton*. The value of g (the *'splitting factor'*) for a free electron is 2.00232 and nearly all biologically important radicals have values close to this. Thus, if this equation is obeyed, an absorption spectrum results. For a single electron this can be crudely represented as:

but, if presented as its first derivative (as ESR machines do) it will appear as:

A number of atomic nuclei, such as those of hydrogen and nitrogen, also behave like small magnets and will align either parallel or antiparallel to the applied magnetic field. Thus in a hydrogen atom the single unpaired electron will actually see two different magnetic fields: the one applied plus that from the nucleus, or the one applied minus that from the nucleus. Thus there will be two energy absorptions and the single line becomes a doublet, i.e.

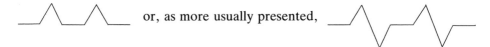 or, as more usually presented,

If the unpaired electron 'sees' two hydrogen nuclei, each can be aligned in the same way with the applied field, in opposite ways, or one in the same way and one opposite i.e.

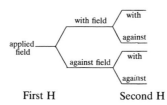

First H   Second H

giving a three-line spectrum in which the intensities are in the ratio 1:2:1, i.e.

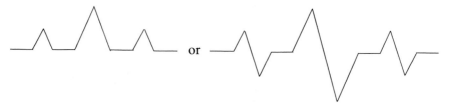 or

In the methyl radical, $CH_3^\bullet$, the unpaired electron on the carbon can 'see' three hydrogen nuclei and the ESR spectrum contains four lines. Remembering that the field of each hydrogen nucleus can align for or against the applied magnetic field, a 'tree' diagram like that above can be used to predict the spectrum i.e.

First   Second   Third
H   H   H

which therefore consists of four lines with intensity ratios $1:3:3:1$, or

The number of lines in the ESR spectrum of a radical is called the *hyperfine structure* and is often very large in complicated radicals containing many nuclei. A radical can be identified from its ESR spectrum by looking at the g value, hyperfine structure, and line shape.

ESR is a very sensitive method and can detect radicals at concentrations as low as 10^{-10} mol l^{-1}, provided that they stay around long enough to be measured. For very unstable radicals a number of techniques are available to detect their presence. One can use flow systems whereby the radicals are continuously generated in the spectrometer so as to maintain a steady-state concentration (see Fig. 2.1(b) for the principles of this). Another approach is to generate the radical in a frozen transparent solid matrix which prevents it from colliding with other species and undergoing reaction. Such a 'rapid-freezing' technique was used in 1969 by P. F. Knowles and others in England to identify the superoxide radical being produced in an enzyme-catalysed reaction (xanthine oxidase catalysing the oxidation of xanthine) by observing its ESR spectrum. By allowing the matrix to warm up, reactions of the radical can then be observed.

Another approach is *spin trapping*. A highly reactive radical, difficult to observe by normal ESR, is allowed to react with a compound to produce a long-lived radical. Reaction of nitroso compounds (R·NO) with radicals often produces nitroxide radicals that have a long lifetime

$$R\!-\!N\!=\!O \quad + \quad R'' \quad \longrightarrow \quad \begin{matrix} R \\ \diagdown \\ R' \end{matrix} \! N\!-\!O\cdot$$

(R symbolizes	very reactive	
'rest of	radical	nitroxide radical
molecule')		(fairly stable)

Nitrones $\left(\begin{matrix} R' \\ | \\ R\!=\!\overset{+}{N}\!-\!O^- \end{matrix} \right)$ also produce nitroxide radicals in a similar reaction.

Spin-trapping methods have often been used to detect the presence of superoxide and hydroxyl radicals in biological systems, and also the formation of organic radicals during lipid peroxidation (this is discussed further in Chapter 4). Table 2.5 shows some of the trapping molecules that have been used, DMPO being especially popular. The 'ideal' trap should react rapidly and specifically with the radical one wants to study, to produce a product that is stable and has a highly characteristic ESR spectrum. It is also worth noting that if a biological process is dependent on, say, hydroxyl radicals, then addition of a trap that reacts with such radicals will inhibit the process to an extent that depends on how much trap is added and on the rate constant for reaction. For example, the spin trap DMPO inhibits ethanol oxidation by the MEOS system as it removes the OH˙ required. None of the spin-traps at present in use is ideal, although better ones are being developed. For example, DMPO reacts with both

Table 2.5. A selection of the 'spin traps' that have been used in biological systems

Name	Abbreviation	Structure
tert-Nitrosobutane (nitroso-*tert* butane)	tNB (NtB)	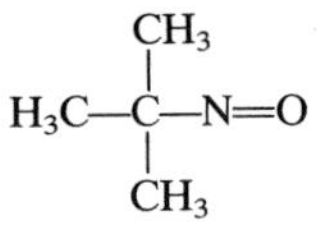 $H_3C-\underset{\underset{CH_3}{\mid}}{\overset{\overset{CH_3}{\mid}}{C}}-N{=}O$
α-Phenyl-*tert*-butylnitrone	PBN	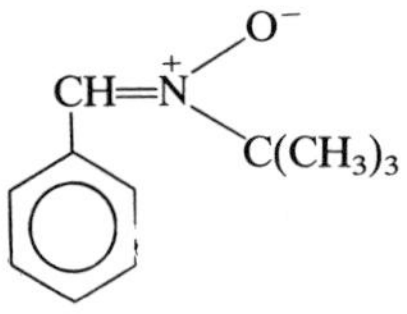
5,5-Dimethylpyrroline-*N*-oxide	DMPO	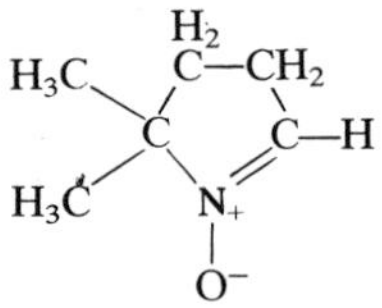
tert-Butylnitrosobenzene	BNB	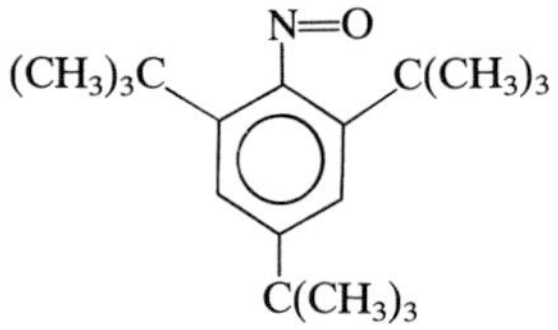
α-(4-Pyridyl-1-oxide)-*N*-tert-butylnitrone	4-POBN	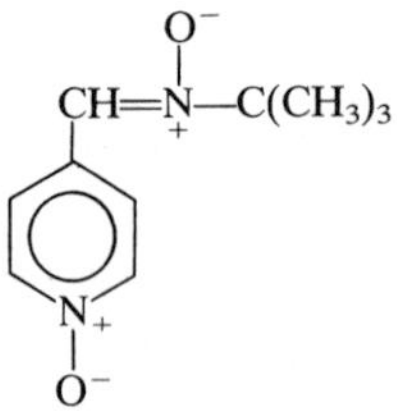

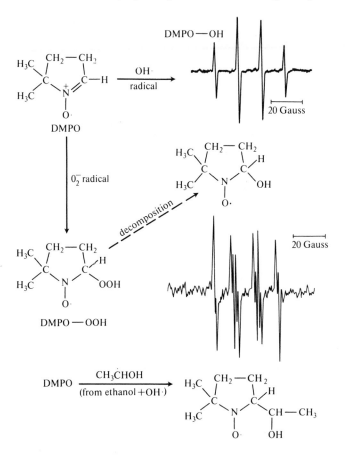

Fig. 2.2. Reactions of the spin-trap DMPO with superoxide, hydroxyl, and ethanol radicals. The ESR spectra of the DMP—OH and DMP—OOH adducts are shown diagrammatically to illustrate the difference between them.

OH˙ and O_2^- radicals to form products with different ESR spectra (Fig. 2.2). The second-order rate constants for the reactions are very different however, approximately $10 \, M^{-1} s^{-1}$ for O_2^- and 3.4×10^9 for OH˙. Unfortunately, the product of reaction of DMPO with O_2^- is unstable and decomposes to form the same product as is given by direct reaction with OH˙. One way around this problem is to add ethanol to the system. If the 'OH'' signal observed was due to a reaction of DMPO directly with OH˙, then ethanol scavenges OH˙ and should abolish the signal. Further, the reaction of OH˙ with ethanol produces a hydroxyethyl radical (Fig. 2.2) that reacts with DMPO to give another adduct with a different ESR spectrum. This spectrum should be observed as the OH˙ signal disappears. If, however, the signal arose from decomposition of a O_2^- adduct then ethanol should have no effect, since it does not react with O_2^-. This is by no means the only 'artefact' identified in the use of DMPO, however. Others include the possibility that decomposition of the DMPO—OOH adduct (formed by reaction with O_2^-) into the DMPO—OH adduct actually releases OH˙ radical, i.e. addition of DMPO to a system producing O_2^- might *cause* small amounts of OH˙ to be formed.

Since the trapping molecules commonly used (Table 2.5) are commercially available, any laboratory with access to ESR facilities can carry out spin-trapping experiments and 'identify' all sorts of oxygen and organic radicals in biological systems. Interpreting these results is not a job for the chemically naive, however.

2.5.2. Aromatic hydroxylation

Aromatic compounds (i.e. compounds containing benzene rings; Appendix, Section A.2.4) react extremely rapidly with hydroxyl radicals, the first product being a *hydroxycyclohexadienyl* radical, e.g. for benzene

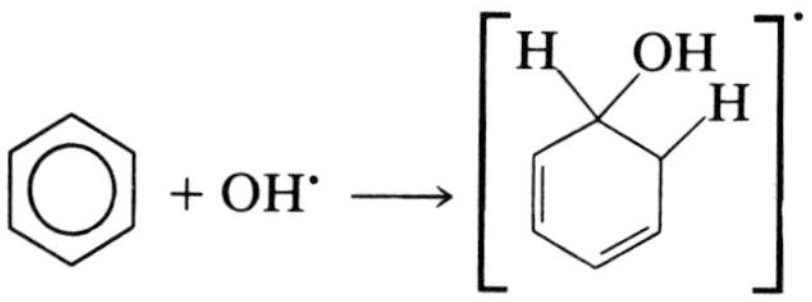

Such radicals can undergo several reactions, e.g. dimerization to give a product that can decompose under appropriate conditions to give biphenyl,

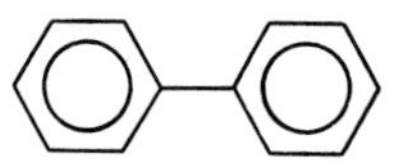

or undergoing a disproportionation reaction to give a mixture of phenol and benzene

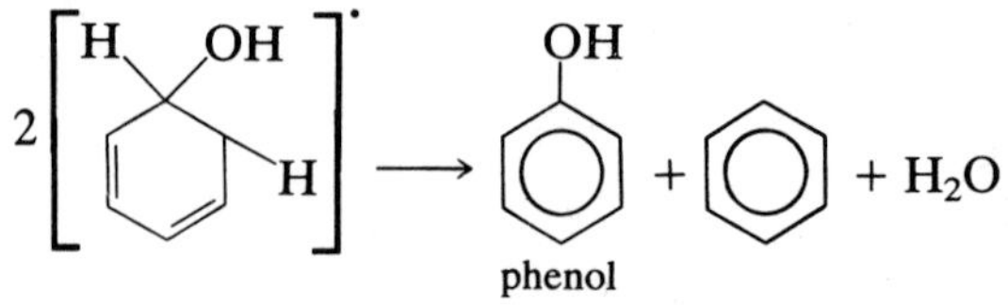

Disproportionation is any reaction in which one molecule is reduced and an identical molecule oxidized. In this case one radical molecule is being reduced to benzene and another oxidized to phenol. The attack of OH· upon phenol itself produces a mixture of hydroxylated products, the 1,3-product being found in much smaller yields than is the 1,2- or 1,4 product.

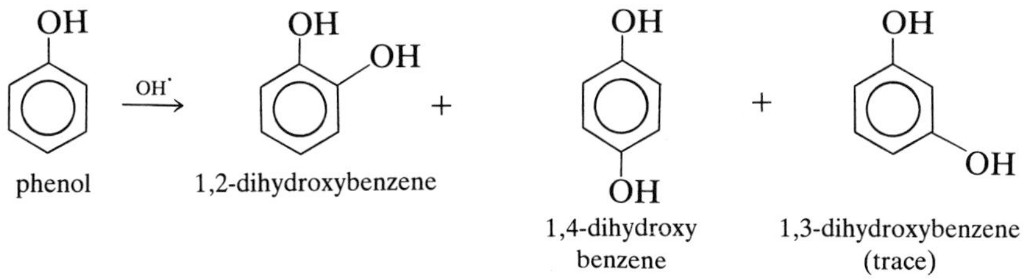

phenol   1,2-dihydroxybenzene   1,4-dihydroxy benzene   1,3-dihydroxybenzene (trace)

If benzoic acid 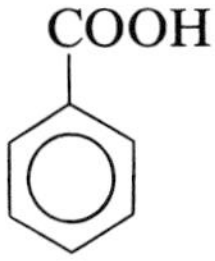 is used, one of the side reactions results in loss

of the carboxyl group as carbon dioxide with probable formation of phenol, although this is likely to be only a minor reaction pathway.

The formation of hydroxylated products from aromatic compounds has been used to detect the formation of hydroxyl radicals in several biochemical systems, the products being identified either by gas–liquid chromatography, by 'colour reactions', or by their fluorescence spectra. An alternative approach has been to add benzoic acid labelled with the isotope ^{14}C in the carboxyl group, and measure the production of radioactive $^{14}CO_2$ by radiochemical techniques. Table 2.6 shows the former method being used to detect hydroxyl radicals produced during the oxidation of xanthine catalysed by the enzyme xanthine oxidase.

2.5.3. Conversion of methional and related compounds into ethene gas (C_2H_4)

This assay method, developed in Professor Fridovich's laboratory in the USA, was the first used to show formation of hydroxyl radicals in a biological system. Methional, which has the structure CH_3—S—CH_2—CH_2—CHO, reacts with hydroxyl radicals in a very complex series of reactions to form ethene ($H_2C{=}CH_2$) gas, which may be measured by

Table 2.6. Formation of hydroxyl radicals during the oxidation of xanthine by xanthine oxidase

Reagent added	Rate constant for reaction with OH˙ ($M^{-1}s^{-1}$)	Amount of hydroxylated products formed (nmoles)	Percentage inhibition of hydroxylation
None	—	102	—
Mannitol (5 mM)	1.0×10^9	64	37
Sodium formate (5 mM)	2.7×10^9	49	52
Thiourea (5 mM)	4.7×10^9	24	76
Urea (5 mM)	$<7.0 \times 10^5$	102	0

Experiments were carried out at pH 7.4; for further details see B. Halliwell (1978) *FEBS Lett.* **92,** 321–6. Hydroxyl radicals were detected by measuring the formation of

hydroxylated products from 2-hydroxybenzoic acid, . The amount of OH˙

reacting with this compound can be decreased by adding other reagents that react with OH˙. The more quickly these other reagents react with OH˙, the more inhibition of the formation of hydroxylated products will they produce.

gas–liquid chromatography. The amino acid methionine (CH_3—S—

CH_2—CH_2—$\overset{\overset{+}{N}H_3}{\underset{|}{CH}}$—$COO^-$) and the compound KTBA (2-keto-

4-thiomethylbutanoic acid, $CH_3 \cdot S \cdot CH_2 \cdot CH_2 \cdot \overset{\overset{O}{||}}{C} \cdot COOH$) also release ethene upon exposure to hydroxyl radicals. Ethene formation from these compounds is not specific as a test for the hydroxyl radical since there are several other oxidizing radicals that can convert them into ethene. However, Fridovich's group showed that formation of ethene in the system that they studied was inhibited by compounds that react with the hydroxyl radical in the way expected from their known rate-constants (Table 2.1). If the ethene formation is due to OH˙, then addition of, say, sodium formate at a concentration that competes with methional for the OH˙ should decrease ethene formation. Addition of an equal concentration of ethanol should inhibit less, because it is less reactive with OH˙ than is formate. Thiourea should inhibit more, and urea, which reacts only slowly with OH˙ (Table 2.1) should have little inhibitory effect. If the ethene formation is inhibited by OH˙ scavengers to extents that can be correlated with known rate-constants for reactions of these compounds with OH˙ (Table 2.1) then it can probably be attributed to OH˙. (Table

2.6 shows the principle of the use of these compounds in a different method of detecting OH'.) If the scavengers do not inhibit, the oxidizing species is probably something else. It is essential to use a wide range of different OH' 'scavengers' in such experiments (Table 2.6).

2.5.4. Other methods

Hydroxyl radical reacts with many other compounds to give products that can be easily detected. It is essential that 'scavenger' experiments (Table 2.6) be performed to show whether or not the radical producing the product is indeed OH' or something else. Table 2.7 summarizes some of these other methods, several of which have been applied to biological systems with varying degrees of success.

2.6. Production of singlet oxygen

We saw in Chapter 1 that two singlet states of oxygen exist in which the spin restriction that slows reaction of this molecule with non-radicals is removed, resulting in increased reactivity. The $^1\Sigma g^+$ state of oxygen is extremely energetic and rapidly decays to the $^1\Delta g$ state, so only the latter need usually be considered in biological systems. Hence references to just 'singlet oxygen' below can be taken to refer to this state. Although $^1\Delta gO_2$ is not a free radical (Chapter 1) it can be formed in some radical reactions and can trigger off others, so its chemistry is worth discussing.

One system that can be used to generate singlet oxygen in the laboratory is a mixture of hydrogen peroxide and the hypochlorite ion, OCl^-, which is formed when chlorine gas is passed into cold alkaline solution

$$Cl_2 + OH^- \rightarrow Cl^- + OCl^- + H^+$$
$$OCl^- + H_2O_2 \rightarrow Cl^- + H_2O + O_2(\text{singlet}).$$

The singlet oxygen arises from the hydrogen peroxide molecules. This reaction might be biologically relevant, since OCl^- can be formed by the enzyme myeloperoxidase during phagocytosis (Chapter 7). The decomposition of the compound potassium peroxochromate, K_3CrO_8, in aqueous solution also produces singlet oxygen. Its usefulness as a source of this species is restricted, however, since it also generates O_2^- and OH' radicals as it decomposes.

Singlet oxygen is most often generated in the laboratory by *photosensitization reactions*. If certain molecules are illuminated with light of a given wavelength they absorb it and the energy raises the molecule into an 'excited state'. The excitation energy can then be transferred onto an

Table 2.7. Methods for detection of hydroxyl radicals

Method	Principle of method	Comments
Bleaching of *p*-nitroso dimethylaniline (PNDA)	PNDA reacts rapidly with OH• but not with O_2^- or singlet O_2. Reaction is accompanied by loss of the yellow colour ('bleaching') (see Bors *et al.* (1979) *Eur. J. Biochem.* **95**, 621–7).	Use in biological systems has found many bleaching reactions not inhibited by compounds that react with OH•, suggesting that it is bleached non-specifically by many biological molecules. We do not recommend it for use.
Deoxyribose method	Reaction of OH• with deoxyribose produces products that, when heated with thiobarbituric acid at low pH, give a colour (see Halliwell and Gutteridge (1981) *FEBS Lett.* **128**, 347–52).	Highly sensitive method, but inhibition studies (Table 2.6) must be done to prove that the reacting species is OH•.
Tryptophan method	Reaction of OH• with tryptophan produces a characteristic set of products (see Singh *et al.* (1981) *Bull. Eur. Physiopath. Resp.* **17**, 31–41).	Tryptophan also reacts with singlet O_2 but the products are different.
Dimethylsulphoxide (DMSO) method	OH• radicals react with DMSO, to give which decomposes to give, among other products, methane gas, measured by gas–liquid chromatography (see Klein *et al.* (1981) *Biochemistry* **20**, 6006–12).	Inhibition studies (Table 2.6) must be done to prove that the reacting species is OH•. Methanal (H·CHO) is also produced and can be measured instead.

adjacent oxygen molecule, converting it to the singlet state whilst the photosensitizer molecule returns to the ground state. Popular sensitizers of singlet oxygen formation in the laboratory include the dyes *acridine orange, methylene blue, rose bengal,* and *toluidine blue*; but many compounds found *in vivo* are also effective, such as the water-soluble vitamin riboflavin and its derivatives FMN (flavin mononucleotide) and FAD (flavin adenine dinucleotide), chlorophylls a and b, the bile pigment bilirubin, retinal, and various porphyrins, both free and bound to proteins (Fig. 2.3). The singlet oxygen produced on illumination of these substances with light of the correct wavelength can react with other molecules present or it can attack the photosensitizer molecule itself. The chemical changes thereby produced are known as *photodynamic effects.* Hence illuminated solutions of flavins lose their orange colour and chlorophylls their green colour as they are attacked; this is called *photobleaching.* Reactions of this type can cause the dyes in your clothes or curtains to fade when exposed to sunlight. Indeed, exposure of cells in culture to high-intensity visible light causes damage, especially to the mitochondria, which are rich in haem proteins and flavin-containing proteins. Haem-containing enzymes such as catalase are also inactivated. Some of the media used for growth of cell cultures contain added riboflavin, which makes these effects worse since the fluorescent lighting found in many laboratories is quite intense enough to cause flavin-sensitized singlet oxygen formation. Not all photosensitization damage need arise by means of singlet oxygen production, however, since the excited state of the photosensitizer can often cause damage itself. Such damage is said to proceed by a *type I mechanism* as opposed to that caused by singlet oxygen production, described as a *type II mechanism.* Both mechanisms may operate simultaneously and the relative import-ance of each type of damage will depend on the nature of the molecule present, the efficiency of energy transfer to oxygen, and on the oxygen concentration (remember that oxygen concentrates in membranes, where many sensitizers are located). In addition, illumination of several porphyrins, and at least one acridine dye, in aqueous solution has been reported to produce hydroxyl radical.

Photosensitization reactions, probably involving singlet oxygen, are important in many biological situations. The most obvious is the chloroplasts of higher plants, which contain chlorophylls and a high oxygen concentration—the problems they face are discussed in detail in Chapter 5. The retina of the eye is obviously exposed to light and its rod cells contain the pigment retinal (Fig. 2.3) bound to a protein known as 'opsin', the whole being called *rhodopsin* or 'visual purple'. Retinal can sensitize singlet oxygen formation and thus damage itself. Indeed, visual cells are known to be injured by prolonged exposure to bright light. The

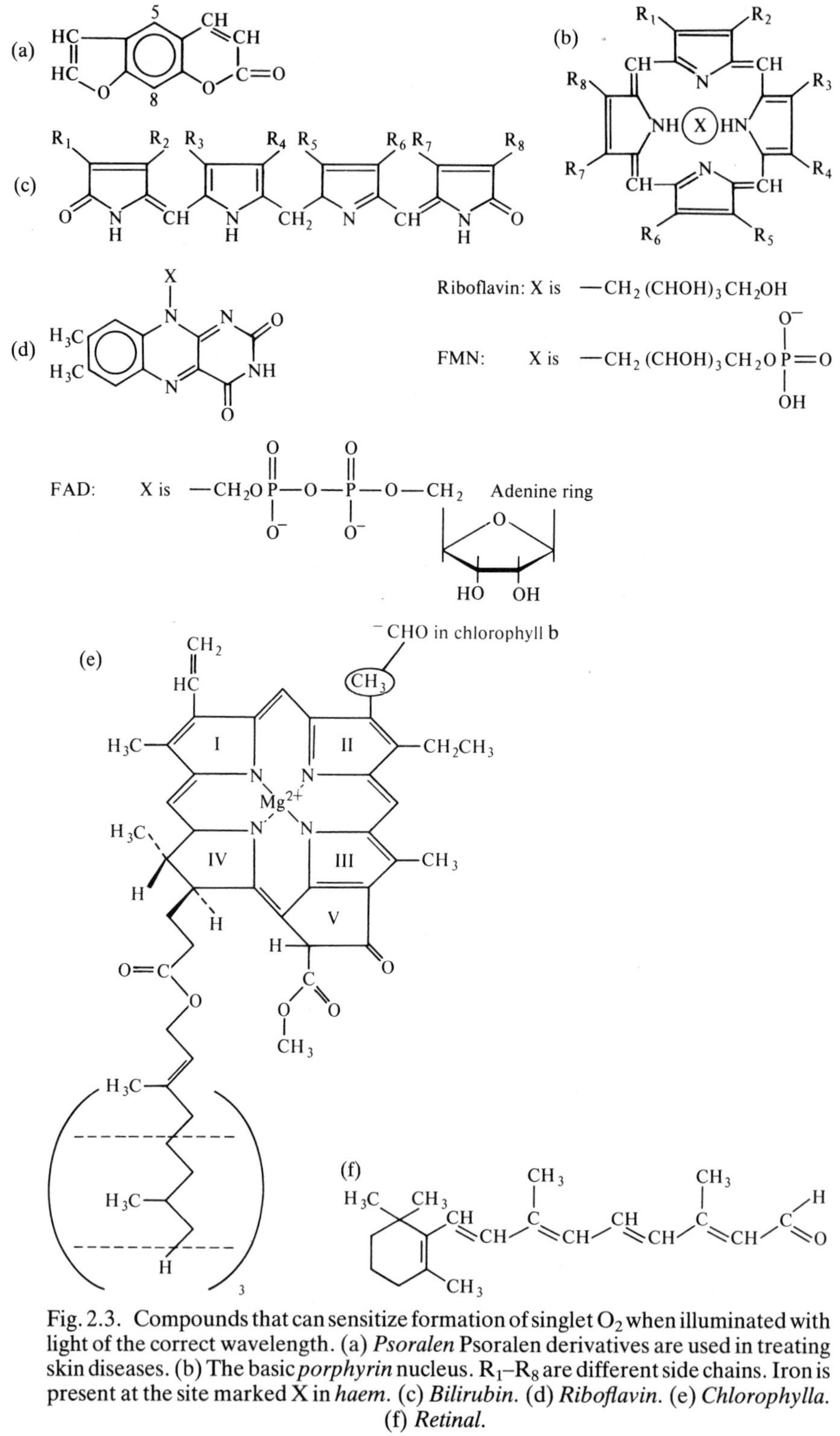

Fig. 2.3. Compounds that can sensitize formation of singlet O_2 when illuminated with light of the correct wavelength. (a) *Psoralen* Psoralen derivatives are used in treating skin diseases. (b) The basic *porphyrin* nucleus. R_1–R_8 are different side chains. Iron is present at the site marked X in *haem*. (c) *Bilirubin*. (d) *Riboflavin*. (e) *Chlorophylla*. (f) *Retinal*.

outer segment membranes of the retinal rods contain a large number of highly unsaturated lipids (i.e. containing a lot of carbon–carbon double bonds $\diagup C = C \diagdown$) which could readily be attacked by singlet oxygen. It has also been suggested that the lens of the human eye contains sensitizers of singlet oxygen formation. Indeed, singlet-oxygen-mediated oxidation of lens proteins *in vitro* gives products similar to those that accumulate in the lens during aging (Chapter 5). Illumination of milk or milk products causes development of 'off-flavours' as the riboflavin present photosensitizes the degradation of milk proteins and lipids.

Several diseases can lead to excessive singlet oxygen formation. For example, the *porphyrias* are diseases caused by defects in the metabolism of haem and other porphyrins (Fig. 2.3). In these diseases, porphyrins are excreted in the urine and often accumulate in the skin, exposure of which to light causes damage leading to unpleasant eruptions, scarring, and thickening (Fig. 2.4). The severity of the damage depends on the exact structure of the porphyrin accumulated and thus will differ in different diseases. Illumination of some porphyrins has also been reported to produce the hydroxyl radical.

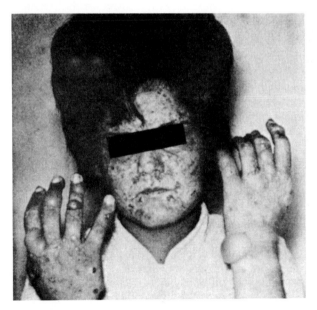

Fig. 2.4. Skin and tissue damage in a porphyria patient. The porphyrins accumulated in porphyrias can be found in all body organs. Among other symptoms this is recognizable by deep-red staining of the tooth roots and the teeth themselves. The photosensitizing actions of porphyrins causes skin blistering and swelling. From B. Franck (1982) *Angew-Chem. Int. Ed.* **21** 343, with permission.

Sometimes photochemical effects are made use of in medicine, as in *photodynamic therapy* in the treatment of 'cold sores' and genital sores caused by the virus *Herpes simplex*. The lesions are painted with a dye such as neutral red or proflavin, which enters and binds to the DNA of the virus. Subsequent illumination destroys the DNA, prevents viral replication and allows the lesion to heal. This therapy is not without problems, however. One unfortunate male patient with genital herpes had the whole of his penis painted with a red dye and exposed to incandescent light. This helped deter the herpes virus, but soon afterwards skin cancer developed elsewhere on his penis! A better-established use of photodynamic therapy is in treatment of the jaundice often developed by premature infants soon after birth. The yellow colour is due to the accumulation in the skin of the pigment *bilirubin* (Fig. 2.3). The bulk of bilirubin is derived from the breakdown of the haemoglobin in worn-out red blood cells, and from the destruction of other haem proteins. It travels in the blood tightly bound to serum albumin, which can bind two moles of bilirubin per mole of protein. The liver takes up the bilirubin and converts it into a water-soluble product by the action of the enzyme *glucuronyl transferase*, which catalyses the reaction

bilirubin + 2UDP-glucuronic acid → bilirubin diglucuronide + 2UDP

The diglucuronide is disposed of by excretion into the bile. In premature babies sufficient glucuronyl transferase is often not present in the liver so that the lipid-soluble bilirubin accumulates in the blood, and deposits in tissues with a high lipid content, such as the brain, where it can cause irreversible damage. The jaundice can be reduced by careful exposure of the babies to blue light from a sunlamp, whereupon the pigment deposited in the skin sensitizes its own destruction in a reaction that involves singlet oxygen. In addition to this, and probably more important in quantitative terms, is a light-induced rearrangement of the structure of bilirubin to give water-soluble products that can be excreted. This rearrangement is called *photoisomerization*.

A third example of the application of photosensitization reactions in medicine is the use of psoralens (Fig. 2.3) in the treatment of skin diseases such as psoriasis. The treatment consists of the combined application of ultraviolet light in the wavelength range 320–400 nm (sometimes known as UVA) and a psoralen, and is often referred to as PUVA therapy (Psoralen Ultra Violet A). Psoralens are a class of compounds produced by plants and they are powerful photosensitizers. There is considerable debate about the safety of PUVA therapy, e.g. does it increase the risk of skin cancer?

It has been observed that certain porphyrins are taken up by cancerous tumours. After injection of a porphyrin derivative known as 'HPD' (which stands for haematoporphyrin derivative), fluorescent products are strongly retained by tumour tissues and this can be used to detect the presence of the tumour by observing the fluorescence. Irradiation with light of a wavelength absorbed by HPD can damage the tumour, and such reactions are of potential use in cancer chemotherapy, especially for skin cancer. Both hydroxyl radicals and singlet oxygen have been suggested to cause the damage.

Photosensitization reactions are also of importance in veterinary medicine. Chlorophyll digestion in ruminants (which eat a lot of green plants) forms the pigment *phylloerythrin* which is absorbed from the gut and excreted in the bile. Liver damage or malfunction can allow it to enter the bloodstream and deposit in the tissues. When the animals are exposed to sunlight the pigment in the skin causes damage that results in reddening and swelling. The fungal product *sporidesmin* achieves a similar effect by inhibiting bile production. The *buckwheat* plant synthesizes a compound that sensitizes singlet oxygen formation and its consumption by animals must be avoided. The same is true of the *St. Johns-wort* plant which produces the sensitizer *hypericin*. Japanese scientists have reported a light-induced dermatitis in some people who have ingested large quantities of tablets made from the alga *Chlorella* and have attributed the damage to a product derived from chlorophyll.

2.7. Reactions of singlet oxygen

Singlet oxygen can interact with other molecules in essentially two ways: it can either combine chemically with them, or else it can transfer its excitation energy to them, returning to the ground state while the molecule enters an excited state. The latter phenomenon is known as *quenching*. Sometimes both can happen (Fig. 2.5). The best studied chemical reactions of singlet oxygen are those involving compounds that contain carbon–carbon double covalent bonds, $>C=C<$. Such bonds are present in many biological molecules, such as carotenes, chlorophyll, and the fatty-acid side-chains present in membrane lipids (Chapter 4). Compounds containing two double bonds separated by a single bond (known as *conjugated double bonds*) often react to give *endoperoxides*

endoperoxide

Diphenylisobenzofuran reacts in this way (Fig. 2.5). If one double bond is present, the ene-reaction can occur—the singlet oxygen adds on and the double bond shifts to a different position:

hydroperoxide

This is of importance in lipid peroxidation (Chapter 4).

If there is an electron-donating atom, such as N or S, adjacent to the double bond, singlet oxygen may react by *dioxetane* formation. Dioxetanes are unstable and decompose to give compounds containing the carbonyl group, $>C{=}O$ e.g.

dioxetane   carbonyl compounds

Tryptophan can react in this way (Table 2.7).

Figure 2.5 shows the reactions of singlet oxygen with several biologically important molecules. Damage to proteins by singlet oxygen is often due to oxidation of essential methionine, tryptophan, histidine, or cysteine residues. The reaction mechanisms are extremely complicated. Histidine has the highest second-order rate constant for reaction with singlet oxygen at $10^8 \, M^{-1} s^{-1}$, tryptophan 3×10^7, and methionine $1.7 \times 10^7 \, M^{-1} s^{-1}$. Photosensitization by dyes bound to the active site of enzymes has often been used to identify amino acid residues essential for the catalytic activity, it being especially useful in investigating the role of histidine residues. The histidine is destroyed at a controlled rate by careful illumination, and the effect on enzyme activity is measured.

2.8. Detection of singlet oxygen in biological systems

In distinguishing between damage caused by type I or by type II mechanisms in photosensitization reactions, or in investigating the role of singlet oxygen in other biological systems, scientists have often relied on the use of 'singlet-oxygen scavengers' and quenchers. Popular compounds have included DABCO, diphenylisobenzofuran, histidine, and azide (Fig.

Compound	Structure	Type of reaction
DABCO (1,4-diazabicyclooctane)	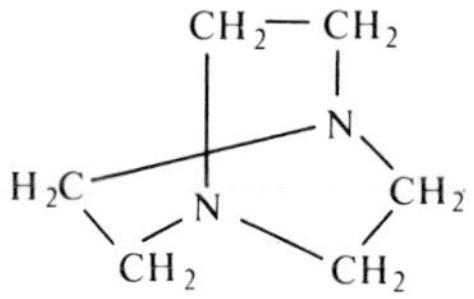	Quenching only. Fairly unreactive – about 50 mM needed in aqueous solution to quench half the $^1O_2$ formed
Azide ion	N_3^-	Mostly quenching. Much more effective than DABCO but inhibits many enzymes and scavenges OH·, thus not generally useful in biological systems
α-Tocopherol (vitamin E)	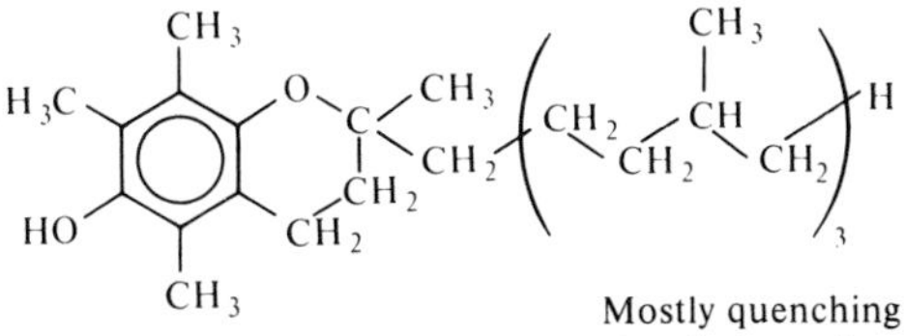	Mostly quenching but some chemical reaction to give 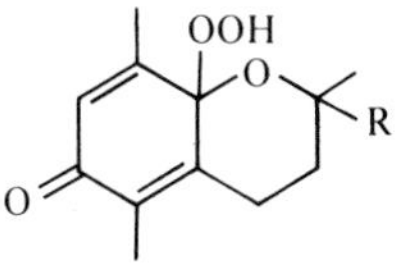 which decomposes to various products including α-tocopherylquinone (R is the side chain) 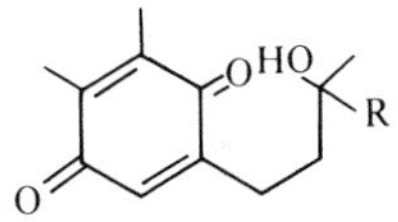
Phenols	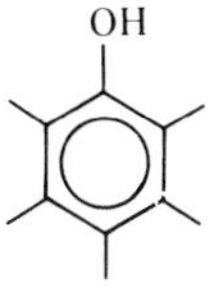	Both quenching and chemical reaction often of the type 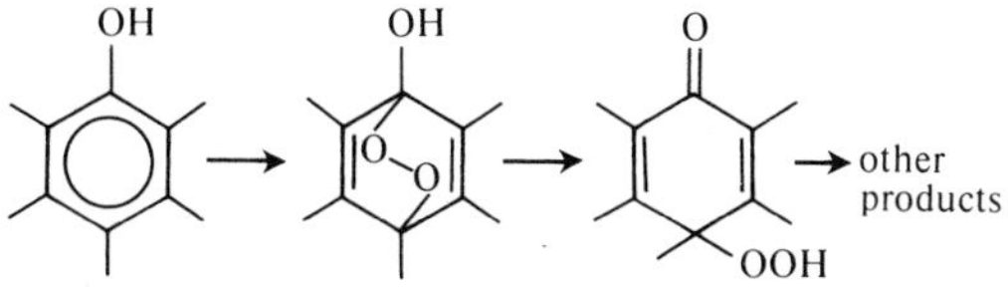
Bilirubin	(see Fig. 2.3)	Mostly quenching but some chemical reaction

Fig. 2.5. Compounds that react with singlet oxygen. In a structure such as that of cholesterol, each 'angle' represents a carbon atom, e.g.  is an abbreviation

for 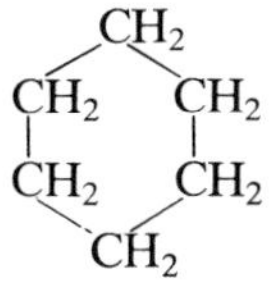 and  an abbreviation for 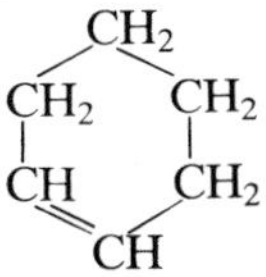 .

Compound	Structure	Type of Reaction
DNA		Complex mixture of products from all types of reaction with purine and pyrimidine bases, e.g. uracil hydroperoxide formation and subsequent decomposition
Cholesterol	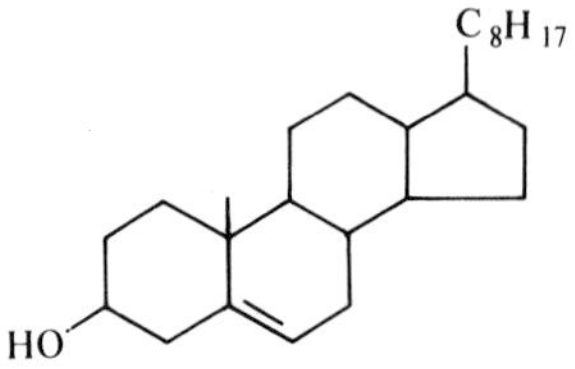	Major product is the $5-\alpha$ hydroperoxide
β-Carotene	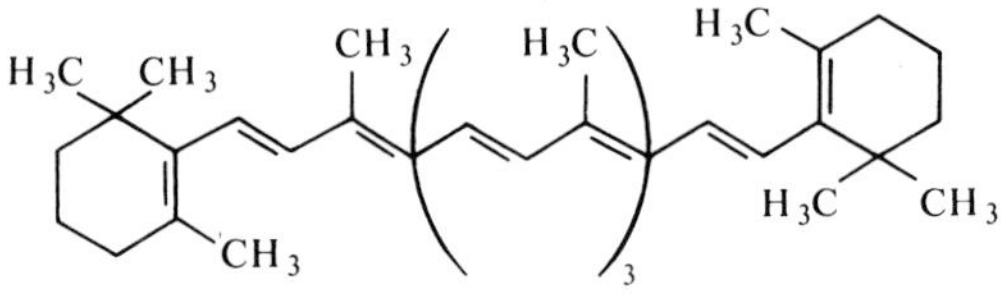	Mostly quenching, some (very complex!) chemical reactions
Diphenylisobenzofuran (other furans react similarly)	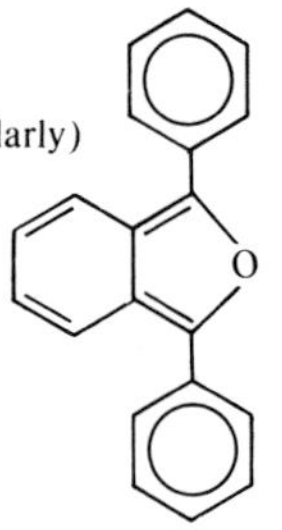	*Endo*-peroxide formation, accompanied by loss of the light absorption at 415 nm
Tryptophan	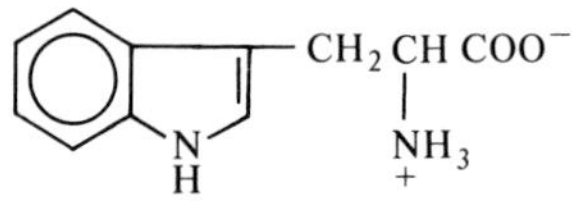	Reacts by several mechanisms, including an ene-reaction to give initially (R is the side-chain) and also formation of a dioxetane which decomposes to *N*-formylkynurenine

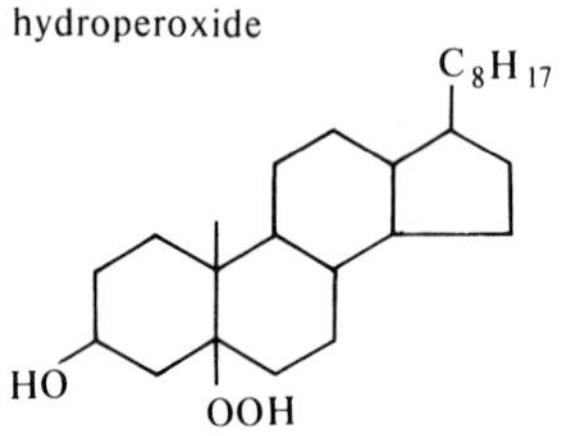

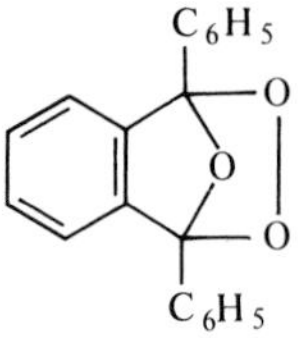

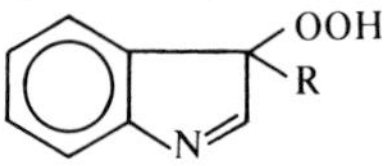

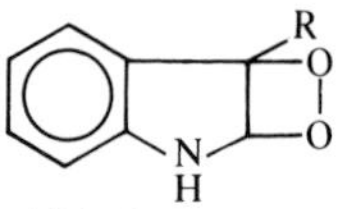

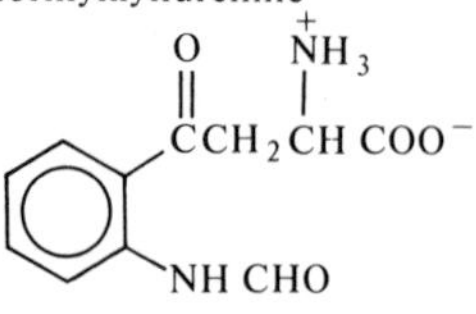

Fig. 2.5. (*Continued*)

Compound	Structure	Type of Reaction

Methionine — $CH_3-S-CH_2-CH_2-\overset{+NH_3}{\underset{COO^-}{CH}}$

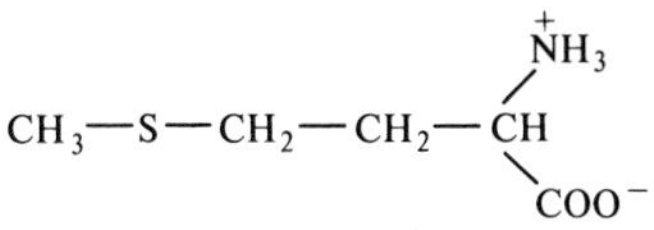

Forms methionine sulphoxide

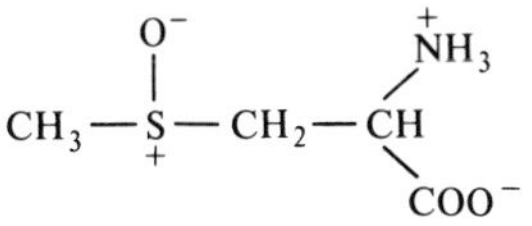

$CH_3-\overset{O^-}{\underset{+}{S}}-CH_2-\overset{+NH_3}{\underset{COO^-}{CH}}$

Cysteine — $HS-CH_2-\overset{+NH_3}{\underset{COO^-}{CH}}$ (R—SH)

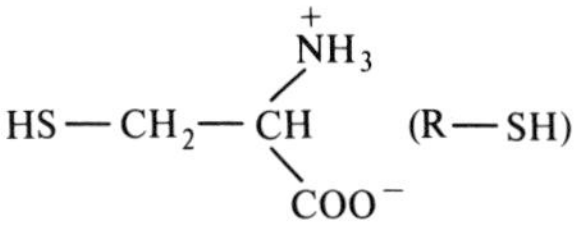

Reaction not well characterized: both disulphides (R—S—S—R) and sulphonic acids (R — SO$_3$H) are produced

Histidine — $N\!\!-\!\!\overset{+NH_3}{\underset{|}{CH_2CH\,COO^-}}$ (imidazole ring)

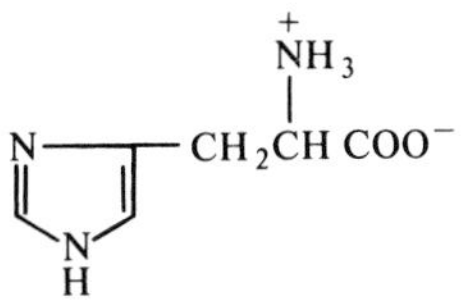

Probably reacts first to give an endoperoxide

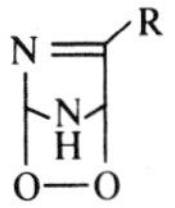

which decomposes to a complex mixture of products

NADPH — (nicotinamide structure with ribose, phosphates and adenine)

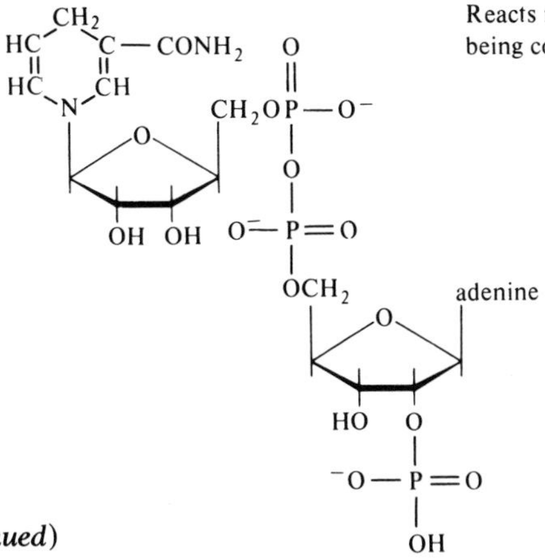

Reacts rapidly with singlet oxygen, being converted into NADP$^+$,

Fig. 2.5. (*Continued*)

2.5). Addition of these should inhibit a reaction dependent on singlet oxygen. If, when added at high concentrations, none of them inhibits the reaction under study, one may conclude that singlet oxygen is not required for it to proceed. If they do inhibit, this does *not* prove a role for singlet oxygen since all these compounds react with hydroxyl radical, often with a greater rate constant than the reaction with singlet oxygen.

For example, azide reacts to give an azide radical:

$$N_3^- + OH^{\cdot} \rightarrow N_3^{\cdot} + OH^-$$

Fortunately, the products of reaction of cholesterol and tryptophan with singlet oxygen are different from those obtained on reaction with $OH^{\cdot}$, so isolation and characterization of them allows distinction. This is hard work, however, and the use of singlet-oxygen scavengers, like that of spin traps, cannot be recommended to the chemically naive.

Another approach, particularly employed in studying photodynamic effects, is the use of deuterium oxide. The lifetime of singlet oxygen is longer in D_2O than in H_2O by a factor of ten or fifteen. Thus if a reaction in aqueous solution is dependent on singlet oxygen, carrying it out in D_2O instead should greatly potentiate the reaction. Theoretically a type I photodynamic reaction should be unaffected. Light emission (*luminescence*) has also been investigated. As singlet oxygen decays back to ground-state, some of the energy is emitted as light. The light from individual singlet-oxygen molecules appears in the infra-red, but a so-called *dimol luminescence* in which two singlet-oxygen molecules co-operate also occurs and produces light at 634 and 704 nm. Many other chemical reactions produce light, however (e.g. the treatment of several haem proteins with hydrogen peroxide) and so the mere production of light cannot be taken to imply singlet-oxygen formation unless the exact spectrum has been measured. Light emission has been detected from activated phagocytes (Chapter 7), Fenton's reaction, and peroxidizing lipids (Chapter 4); but in none of these cases has it been rigorously proven to originate from singlet oxygen. If, however, singlet oxygen is formed it might well excite adjacent molecules, which could emit light of different wavelengths as they return to the ground state. The decomposition of dioxetanes can produce carbonyl compounds in the excited state, which then emit light.

Light emission is measured by an electronic device known as a photomultiplier, either in a liquid scintillation counter, or in a device specifically designed to measure luminescence (a *luminometer*). Professor Britton Chance and Dr Alberto Boveris in the USA are engaged in detailed studies of the light emitted by intact animal organs *in situ* both normally and after exposure to radical stress, since this provides a continuously monitorable parameter that does not require damage to the tissue. This should become a valuable technique once the detailed reactions generating the light have been elucidated. Light emission is increased markedly on exposure of tissues to elevated oxygen concentrations.

2.9. Reactions of the superoxide radical

The 'free-radical' hypothesis of Gershman and Gilbert (Chapter 1) was extended by J. M. McCord and I. Fridovich in the USA into a 'superoxide theory of oxygen toxicity' following their discovery in 1968 of an enzyme, superoxide dismutase, which specifically catalyses removal of the superoxide radical O_2^- (Chapter 3). This theory proposes that O_2^- formation is a major factor in oxygen toxicity and that superoxide dismutase enzymes constitute an essential defence against it. The current status of this theory is evaluated in detail in the next chapter, but here we shall lay the background to it by discussing the chemistry of the superoxide radical.

Superoxide chemistry differs greatly according to whether reactions are carried out in aqueous solution or in organic solvents, in many of which it is very stable. Various methods are available for producing superoxide for chemical studies. Firstly, oxygen may be reduced electrochemically in an appropriate electrolytic cell in the presence of an organic solvent such as dimethylsulphoxide or acetonitrile. Secondly, tetramethylammonium superoxide, an ionic salt of formula $(CH_3)_4N^+O_2^-$, can be dissolved in a number of organic solvents. Thirdly, if potassium metal is burned in oxygen the pale-yellow ionic compound potassium superoxide, $K^+O_2^-$, is obtained. This is slightly soluble in organic solvents and its solubility can be increased by the addition of compounds called *crown ethers*. Essentially these are cyclic compounds, a hole in the centre of which can bind K^+. They are very soluble in organic solvents and so 'drag into solution' the central K^+ ion together with its associated O_2^-. A popular crown ether in KO_2 experiments is dicyclohexyl-18-crown-6 (Fig. 2.6). The reaction with O_2^- of another compound added to the organic solvent can be observed, or the O_2^--containing organic solvents can be mixed with an aqueous solution of the compound, often in stopped-flow

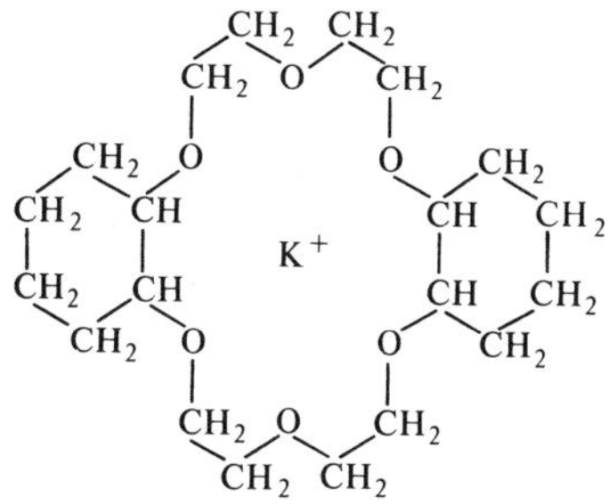

Fig. 2.6. Structure of dicyclohexyl-18-crown-6. The K^+ ion fits into the central hole.

experiments (Fig. 2.1). Lastly, O_2^- may be generated by pulse radiolysis of aqueous solutions as already described. Its presence may be detected by observing its UV-absorption spectrum (maximal absorption at 250 nm) or its ESR spectrum at low temperatures. Superoxide dissolved in the above organic solvents is very stable if water is kept away, but in aqueous solution it disappears rapidly. It has been suggested that the surface of the planet Mars holds oxygen as a superoxide complex which is stable in the absence of water.

One of the reactions of superoxide in aqueous solution is to act as a *base*; a base being defined as an acceptor of protons (H^+ ions). When O_2^- accepts a proton it forms the *hydroperoxyl radical* ($HO_2^{\cdot}$). $HO_2^{\cdot}$ can dissociate to release H^+ ions again, i.e. it can act as a supplier of protons (an *acid*). Hence when O_2^- and H^+ ions are mixed together, an equilibrium is set up

$$HO_2^{\cdot} \rightleftharpoons H^+ + O_2^-$$

The pH of a solution is defined as $-\log_{10}[H^+]$, the square brackets being used to denote concentration of H^+ in moles per litre. The pH at which the acid ($HO_2^{\cdot}$) and base (O_2^-) forms are present in equal concentrations is known as the pK_a, approximately equal to 4.8 in this system. Since pH is $-\log_{10}[H^+]$, a pH of 4.8 corresponds to $[H^+]$ equal to 1.58×10^{-5} mol l^{-1}. pH and pK_a are related by the *Henderson–Hasselbalch equation*

$$pH = pK_a + \log_{10}\frac{[\text{base}]}{[\text{acid}]}$$

or, in this case

$$pH = 4.8 + \log_{10}\frac{[O_2^-]}{[HO_2^{\cdot}]}.$$

At a pH of 3.8, $[O_2^-]/[HO_2^{\cdot}]$ is therefore 1/10, whereas at pH 5.8 it is 10/1. As the pH of most body tissues and fluids is in the range 6.4–7.5, then obviously the ratio of $[O_2^-]/[HO_2^{\cdot}]$ will be very large, e.g. 100/1 at pH 6.8. Hence any O_2^- generated will remain almost entirely in this form rather than becoming protonated.

The reason for the disappearance of O_2^- in aqueous solution is the so-called *dismutation reaction*. The overall reaction may be represented as

$$O_2^- + O_2^- + 2H^+ \rightarrow H_2O_2 + O_2$$

but in fact the rate constant for this reaction as written is virtually zero ($<0.3\,\text{M}^{-1}\,\text{s}^{-1}$). By contrast, the reaction:

$$HO_2^{\cdot} + O_2^- + H^+ \rightarrow H_2O_2 + O_2$$

has $k_2 = 8 \times 10^7 \, \text{M}^{-1} \text{s}^{-1}$, and the reaction

$$HO_2^{\bullet} + HO_2^{\bullet} \rightarrow H_2O_2 + O_2$$

has $k_2 = 8 \times 10^5 \, \text{M}^{-1} \text{s}^{-1}$.

Dismutation is thus most rapid at the acidic pH values needed to protonate O_2^- and will become slower as the pH rises (i.e. becomes more alkaline) and the concentration of $HO_2^{\bullet}$ in equilibrium with a given concentration of O_2^- decreases. For example, it may be calculated that in aqueous solution the dismutation reaction will have an overall rate constant of about $10^2 \, \text{M}^{-1} \text{s}^{-1}$ at pH 11 and about $5 \times 10^5 \, \text{M}^{-1} \text{s}^{-1}$ at pH 7.0. Any reaction undergone by O_2^- in aqueous solution will be in competition with this dismutation reaction, and it also follows that a system generating O_2^- must be producing H_2O_2.

As well as acting as a weak base, O_2^- in aqueous solution is a reducing agent, i.e. a donor of electrons. For example, it reduces cytochrome c, a haem protein. The iron at the centre of the haem ring is reduced from the Fe^{3+} to the Fe^{2+} state

$$\text{cyt c } (Fe^{3+}) + O_2^- \rightarrow O_2 + \text{cyt c } (Fe^{2+})$$

The copper protein plastocyanin (Chapter 5) is also reduced

$$\text{plastocyanin } (Cu^{2+}) + O_2^- \rightarrow O_2 + \text{plastocyanin } (Cu^+)$$

as is the yellow dye nitro-blue tetrazolium to produce a blue product known as formazan, although this reaction is much more complicated (Fig. 2.7). Superoxide reduces free Fe(III) salts to Fe(II) salts, although an accurate rate constant for this reaction has not been determined. Complexes of Fe(III) with the chelating agent EDTA (Table 2.3) are still reduced by O_2^-

$$Fe^{3+}\text{—EDTA} + O_2^- \rightarrow Fe^{2+}\text{—EDTA} + O_2;$$
$$k_2 = 1.3 \times 10^6 \, \text{M}^{-1} \text{s}^{-1} \text{ at pH 7}$$

whereas complexes of Fe(III) with DETAPAC or desferrioxamine (Table 2.3) are reduced much more slowly, if at all.

Superoxide in aqueous solution is also a very weak oxidizing agent (electron acceptor) e.g. it oxidizes ascorbic acid ($k_2 = 2.7 \times 10^5 \, \text{M}^{-1} \text{s}^{-1}$ at pH 7.4)

$$\text{ascorbate} + O_2^- + H^+ \rightarrow H_2O_2 + \text{ascorbate radical.}$$

It does not significantly oxidize NADH or NADPH, but it will interact with NADH bound to the active site of the enzyme lactate dehydrogenase (but not other dehydrogenases) to form an $NAD^{\bullet}$ radical (Table 2.8)

$$\text{enzyme—NADH} + O_2^- + H^+ \rightarrow \text{enzyme—NAD}^{\bullet} + H_2O_2$$

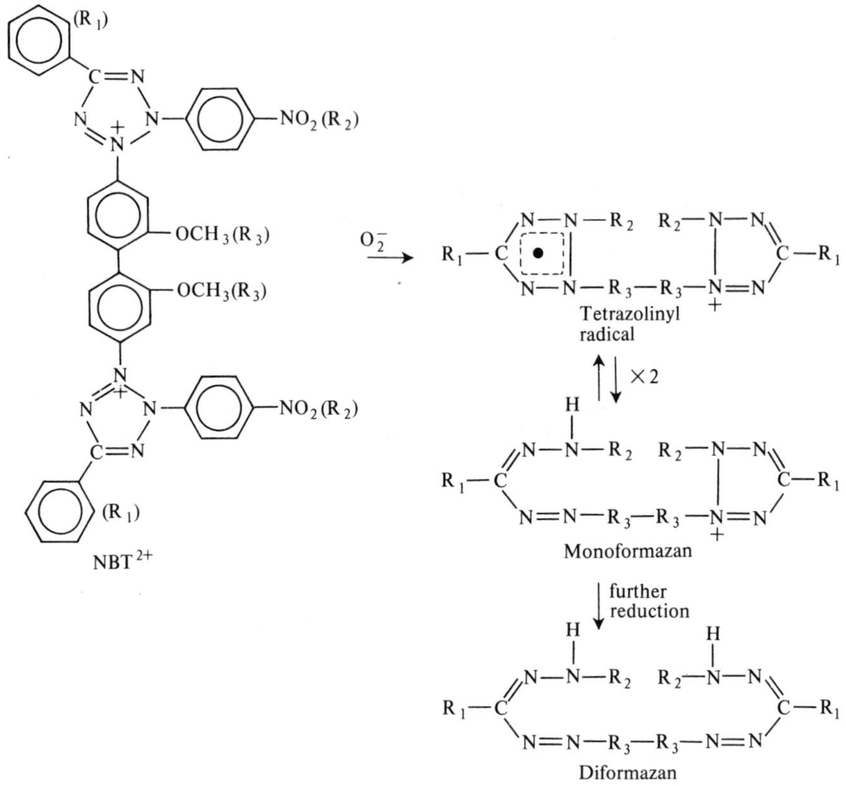

Fig. 2.7. Reduction of NBT^{2+} by superoxide radical. The detailed reaction mechanism is shown. It has been claimed that the tetrazolinyl radical can react with O_2 to form O_2^-, i.e. that the initial reaction is reversible, but this has been disputed.

O_2^- will oxidize compounds containing the thiol (—SH) group, but very slowly.

The protonated form of $O_2^-(HO_2^{\cdot})$ is a more powerful reducing agent and oxidant than is O_2^- itself, although not much $HO_2^{\cdot}$ will be present at physiological pH. For example, unlike O_2^-, $HO_2^{\cdot}$ will directly oxidize NADH ($k_2 = 1.8 \times 10^5 \, \text{M}^{-1} \text{s}^{-1}$). It reduces cytochrome c with a rate constant of 2×10^6 as compared with O_2^- at 2.6×10^5 (Table 2.8).

When superoxide is dissolved in organic solvents its ability to act as a base and as a reducing agent are increased. For example it can reduce dissolved sulphur dioxide (SO_2) gas in organic solvents but not in aqueous solution

$$SO_2 + O_2^- \rightarrow O_2 + SO_2^-.$$

Table 2.8. Rates of reaction of superoxide radical with various compounds in aqueous solution

Compound tested	pH	Second-order rate constant ($M^{-1} s^{-1}$)	Reaction
Cholesterol, membrane lipids, pyruvate, aspartate, histidine, methionine, tryptophan, tyrosine, H_2O_2	8–10	<1	Essentially none.
Cytochrome c (Fe(III))	8.5	2.6×10^5 (latest estimate)	Reduction to Fe(II) form.
Tetranitromethane	—	2×10^9	Reduction $C(NO_2)_4 + O_2^- \rightarrow C(NO_2)_3^- + NO_2 + O_2$.
Benzoquinone (and other quinones)	7	1×10^9	Reduction; an equilibrium is established (position depends on the quinone) 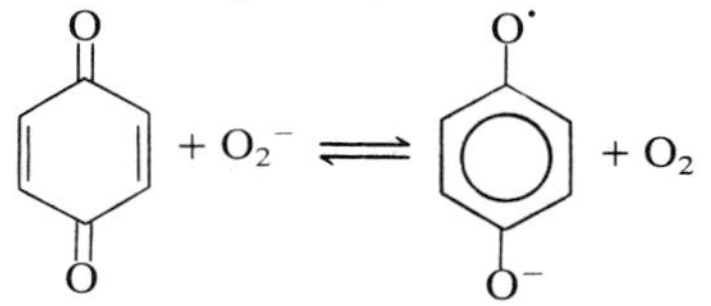
Catechol (and other diphenols)	7	1×10^9	Oxidation, to establish an equilibrium 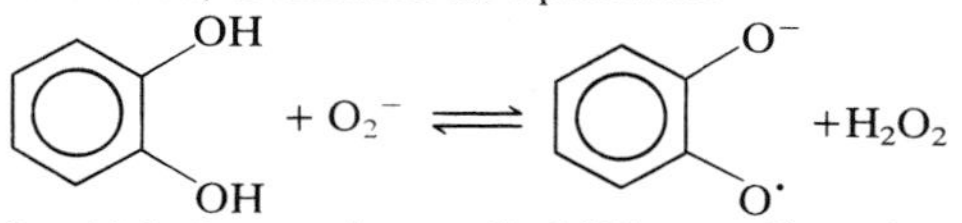
Ascorbic acid	7.4	2.7×10^5	Semidehydroascorbate radical (Chapter 3) produced.
NAD(P)H	7.4	<1	Essentially none.
NADH at active site of lactate dehydrogenase (other dehydrogenases not affected)	7.5	1×10^5	NAD$^{\bullet}$ radical formed which reduces O_2 to O_2^- (forming NAD$^+$) and starts a chain reaction.
Plastocyanin	7.7	1×10^6	Reduction of Cu(II) at active site to Cu(I).
Bilirubin	8.3	2.3×10^4	Bleaching; mechanism not characterized.
Nitrobluetetrazolium	7–11	6×10^4	NBT^{2+} reduced to a tetrazolium radical, two of which combine together to give a blue dye (monoformazan) and NBT^{2+}. For a detailed mechanism see Fig. 2.7.

Also, if protons are not readily available, then dismutation is prevented and the O_2^- stays around much longer. Further, it gains the ability to act as a *nucleophile*, a reagent that is attracted to centres of positive charge in a molecule. Consider, for example, an ester molecule of general formula

$$R—\overset{\overset{\displaystyle O}{\|}}{C}—O—R'$$

where R and R' are hydrocarbon groups. Since oxygen is more electronegative than carbon, the carbonyl group is slightly polarized (Appendix). O_2^- will be attracted to the $\delta+$ charge and will attack the molecule, the initial reactions being:

$$R—\underset{\underset{\displaystyle \uparrow}{\overset{\displaystyle \delta+}{C}}}{\overset{\overset{\displaystyle O^{\delta-}}{\|}}{C}}—O—R' \longrightarrow R—\underset{\underset{\displaystyle O_2^{\cdot}}{C}}{\overset{\overset{\displaystyle O^-}{\|}}{C}}—O—R' \longrightarrow R—\underset{\underset{\displaystyle O_2^{\cdot}}{C}}{\overset{\overset{\displaystyle O}{\|}}{C}} + R'—O^-$$

$$O_2^-$$

$$R—\overset{\overset{\displaystyle O}{\|}}{C}—OO^{\cdot} + O_2^- \longrightarrow R—\overset{\overset{\displaystyle O}{\|}}{C}—OO^- + O_2$$

Superoxide can displace chloride ion from chlorinated hydrocarbons such as chloroform (trichloromethane, $CHCl_3$) or tetrachloromethane (carbon tetrachloride, CCl_4), e.g.

$$Cl—\underset{\underset{\displaystyle Cl\ O_2^-}{\overset{\displaystyle |}{C}}}{\overset{\overset{\displaystyle Cl}{|}}{C}}—Cl \longrightarrow Cl—\underset{\underset{\displaystyle Cl}{\overset{\displaystyle |}{C}}}{\overset{\overset{\displaystyle Cl}{|}}{C}}—O_2^{\cdot} + Cl^-$$

followed by further displacements.

The nucleophilicity of O_2^- in aqueous solution is, by contrast, very low, in part because of competition by the dismutation reaction.

The oxidizing capacity of O_2^- in organic solvents is only seen with compounds that can donate H^+ ions, such as ascorbate, catechol, and α-tocopherol (Table 2.8). Tocopherol (Ht) is oxidized to a radical species (t·) possibly by the series of reactions

$$O_2^- + Ht \rightarrow t^- + HO_2^{\cdot}$$
$$O_2^- + HO_2^{\cdot} \rightarrow HO_2^- + O_2$$
$$O_2 + t^- \rightarrow O_2^- + t^{\cdot}$$
$$2t^{\cdot} \rightarrow dimer \rightarrow other\ products$$

Similar proton transfers occur in the oxidation of other substances by O_2^- dissolved in organic solvents.

In the next chapter we shall examine the superoxide theory of oxygen toxicity in relation to O_2^- chemistry.

2.10. Hydrogen peroxide in biological systems

Any biological system generating O_2^- will produce hydrogen peroxide by the dismutation reaction unless, of course, all the O_2^- is intercepted by some other molecule (e.g. a high concentration of cytochrome c). Hydrogen peroxide production, probably mainly via O_2^-, has been frequently observed from mitochondria and microsomes *in vitro*; and the amount produced increases as the surrounding oxygen concentration is raised. There are also several enzymes that produce hydrogen peroxide without the intermediacy of free O_2^- radical. These include glycollate oxidase, D-aminoacid oxidase, and urate oxidase. Professor Britton Chance's group in the USA studied the rate of hydrogen peroxide production in the isolated perfused rat liver by observing the spectral intermediates of the enzyme catalase (Chapter 3) and arrived at a figure of 82 nmol (82×10^{-9} moles) of H_2O_2 produced per minute per gram of liver in normally-fed animals. Inclusion of glycollate or urate in the perfusion medium increased this rate, as the above oxidases become active. Much less of this hydrogen peroxide arose from the endoplasmic reticulum than would be expected from the rate at which microsomes produce hydrogen peroxide *in vitro*: remember that microsomes are artefacts of subcellular fractionation, and the membrane rearrangements undergone by them during homogenization and centrifugation may increase their production of O_2^- and hence hydrogen peroxide (this is discussed further in Chapter 3). Because the liver has effective mechanisms for disposing of hydrogen peroxide, they estimated the steady-state H_2O_2 concentration as being in the range 10^{-7}–10^{-9} moles per litre, hydrogen peroxide being continuously generated and destroyed. In other animal cells, with less effective H_2O_2-removal mechanisms, more hydrogen peroxide is found, e.g. the lens of the human eye contains $24\ \mu mol\ l^{-1}\ H_2O_2$, and rabbit spermatozoa actually release hydrogen peroxide into the surrounding medium, again probably from O_2^-. Many bacteria and mycoplasmas (these are the smallest known organisms that exist independently and differ from bacteria in not having a cell wall) also release hydrogen peroxide into their surroundings, as do some blue-green algae upon illumination. Hydrogen peroxide is known to be produced during photosynthesis (Chapter 5) and during phagocytosis (Chapter 7), and hydrogen peroxide vapour has been detected in human expired air. The breathing of pure oxygen by experimental subjects increases the amount of hydrogen peroxide exhaled.

Hydrogen peroxide is a weak oxidizing agent and can inactivate a few enzymes directly, usually by oxidation of essential thiol (—SH) groups. Spinach chloroplast fructose bisphosphatase (Chapter 5) is one example. Hydrogen peroxide is often used at high concentrations as a disinfectant. Some bacterial strains are very sensitive to it, and some animal cells grown in culture (e.g. human fibroblasts) can be damaged by hydrogen peroxide added to the culture medium at concentrations in the micromolar range. Hydrogen peroxide produced by, for example, *Mycoplasma pneumoniae*, attacks epithelial cells in the trachea.

Hydrogen peroxide can penetrate cell membranes rapidly whereas O_2^- usually cannot. Once inside the cell it can probably react with Fe(II) or Cu(I) ions to form the hydroxyl radical, and this may be the origin of most of its toxic effects. Evidence for this comes from several observations. For example, the killing of bacterial spores by hydrogen peroxide can be related to their content of transition-metal ions. The killing of *Staphylococcus aureus* cells by hydrogen peroxide was much more efficient if they had been grown in medium with a high iron content, so increasing the intracellular amount of iron. Supplying the cells with dimethylsulphoxide, which penetrates into the cell and reacts with OH˙, decreased the toxicity, and the production of methane (Table 2.7) could be detected. Hydroxyl-radical scavengers that could not enter the cells, such as mannitol, had little effect. Hydrogen peroxide also enhances the damaging effect of near-ultraviolet radiation on bacteria and viruses— usually the damage seen in the presence of both agents is much greater than that done by each agent alone (a *synergistic effect*). It is possible that the UV light can cause homolytic fission of the hydrogen peroxide and hence increase the production of OH˙

$$H_2O_2 \xrightarrow[\text{energy}]{} 2OH^\cdot.$$

The damage appears to affect DNA especially, causing single-strand breaks and DNA–protein cross-links.

Let us now go on and look at the biological importance of some of the reactions discussed above.

2.11. Further reading

Aisen, P. and Liskowsky, I. (1980). Iron transport and storage proteins. *Ann. Rev. Biochem.* **49**, 357.

Anon (1977). Photodye herpes therapy—Cassandra confirmed? *J. Am. med. Ass.* **238**, 133.

Bors, W., Saran, M., and Michel, C. (1982). Assays of oxygen radicals. Methods and mechanisms. In *Superoxide dismutase* (ed. L. W. Oberley) Vol II, p. 31. CRC Press, Florida, USA.

Boveris, A. *et al.* (1980). Organ chemiluminescence: noninvasive assay for oxidative radical reactions. *Proc. nat. Acad. Sci. USA* **77**, 347.

Butler, J. and Halliwell, B. (1982). Reaction of iron–EDTA chelates with the superoxide radical. *Arch. Biochem. Biophys.* **218**, 174.

Cederbaum, A. J. and Dicker, E. (1983). Inhibition of microsomal oxidation of alcohols and of hydroxyl radical scavenging agents by the iron-chelating agent desferrioxamine. *Biochem. J.* **210**, 107.

Chang, L. Y. L. and Packer, L. (1979). Damage to hepatocytes by visible light. *FEBS Lett.* **97**, 124.

Clare, N. J. (1955). Photosensitisation in animals. *Adv. vet. Sci. comp. Med.* **2**, 182.

Duran, N. (1982). Singlet oxygen in biological processes. In *Chemical and biochemical generation of excited states* p. 345. Academic Press, New York.

Ebert, M., Keene, J. P., Swallow, A. J., and Baxendale, J. H. (eds) (1965). *Pulse Radiolysis*. Academic Press, London and New York.

Finkelstein, E., Rosen, G. M., and Rauckman, E. J. (1979) and (1982). Spin trapping of superoxide. *Mol. Pharmac.* **16**, 676 and **21**, 262.

Foote, C. S. (1979). Detection of singlet oxygen in complex systems: a critique. In *Biochemical and clinical aspects of oxygen* (ed. W. S. Caughey) p. 603. Academic Press, New York.

Foote, C. S. (1981). Photo-oxidation of biological model compounds. In *Oxygen and oxy-radicals in chemistry and biology* (eds M. A. J. Rodgers and E. L. Powers) p. 425. Academic Press, New York.

Gutteridge, J. M. C. and Stocks, J. (1981). Caeruloplasmin: physiological and pathological perspectives. *CRC crit. Rev. Clin. Lab. Sci.* **14**, 257.

Harrison, P. M. and Hoare, R. J. (1980). *Metals in biochemistry*. Chapman and Hall, London.

Hoigne, J. and Bader, H. (1975). Ozonation of water: role of hydroxyl radicals as oxidising intermediates. *Science* **190**, 782.

Kalyanaraman, B. (1982). Detection of toxic free radicals in biology and medicine. In *Reviews in biochemical toxicology* (eds. C. E. Hodgson, J. R. Bend, and R. M. Philpot) Vol IV, p. 73. Elsevier, New York.

Knowles, P. F., Gibson, J. F., Pick, F. M., and Bray, R. C. (1969). ESR evidence for enzymic reduction of oxygen to a free radical: the superoxide ion. *Biochem. J.* **111**, 53.

Korycka-Dahl, M. and Richardson, T. (1977). Photogeneration of superoxide anion in serum of bovine milk and in model systems containing riboflavin and amino acids. *J. Dairy Sci.* **61**, 400.

McCay, P. B. *et al.* (1980). Production of radicals from enzyme systems and the use of spin traps. In *Free radicals in biology* (ed. W. A. Pryor) Vol. IV, Chapter 5. Academic Press.

Mello Filho, A. C., Hoffmann, M. E. and Meneghini, R. (1984). Cell killing and DNA damage by hydrogen peroxide are mediated by intracellular iron. *Biochem. J.* **218**, 273.

Modell, B. *et al.* (1982). Survival and desferrioxamine in thalassaemia major. *Br. med. J.* **284**, 1081.

Oshino, N., Jamieson, D., and Chance, B. (1975). The properties of hydrogen peroxide production under hyperoxic and hypoxic conditions of perfused rat liver. *Biochem. J.* **145**, 53.

Pathak, M. A. and Joshi, P. C. (1984). Production of active oxygen species ($^1O_2$ and O_2^-) by psoralens and ultraviolet radiation. *Biochim. Biophys. Acta* **798,** 115.

Richmond, R., Halliwell, B., Chauhan, J., and Darbre, A. (1981). Superoxide-dependent formation of hydroxyl radicals. Detection of hydroxyl radicals by the hydroxylation of aromatic compounds. *Analyt. Biochem.* **118,** 328.

Sawyer, D. T. and Valentine, J. S. (1981). How super is superoxide? *Accts chem. Res.* **14,** 393.

Schafer, A. I. *et al.* (1981). Clinical consequences of acquired transfusional iron overload in adults. *New Engl. J. Med.* **304,** 319.

Scholes, G. (1983). Radiation effects on DNA. *Br. J. Radiol.* **56,** 221.

Turrens, J. F., Freeman, B. A., and Crapo, J. D. (1982). Hyperoxia increases H_2O_2 release by lung mitochondria and microsomes. *Archs Biochem. Biophys.* **217,** 411.

Wardman, P. (1978). Application of pulse radiolysis methods to study the reactions and structure of biomolecules. *Rep. Prog. Phys.* **41,** 259.

Zigler, J. S. and Goosey, J. D. (1981). Photosensitised oxidation in the ocular lens: evidence for photosensitisers endogenous to the human lens. *Photochem. Photobiol.* **33, 869.**

3 Protection against oxygen radicals in biological systems: the superoxide theory of oxygen toxicity

3.1. Protection by enzymes

3.1.1. Protection against hydrogen peroxide by catalase and peroxidases

We saw in Chapter 2 that hydrogen peroxide is damaging in living systems principally because it can give rise to the formation of OH$^{\cdot}$ radicals. It is therefore biologically advantageous for cells to control the amount of hydrogen peroxide that is allowed to accumulate.

Two types of enzyme exist to remove hydrogen peroxide within cells. They are *the catalases*, which catalyse the reaction:

$$2H_2O_2 \rightarrow 2H_2O + O_2$$

and *the peroxidases*, which bring about the general reaction:

$$SH_2 + H_2O_2 \rightarrow S + 2H_2O$$

in which SH_2 is a substrate that becomes oxidized. The oxygen produced by catalase is ground-state: no singlet oxygen can be detected.

Catalase

Most aerobic cells contain catalase activity, although a few do not, such as the bacterium *Bacillus popilliae, Mycoplasma pneumoniae,* the green alga *Euglena,* several parasitic helminths (eg. the liver fluke), and the blue-green alga *Gloeocapsa.* A few anaerobic bacteria, such as *Propionibacterium shermanii* also contain catalase, but most do not. In animals catalase is present in all major body organs, being especially concentrated in liver and erythrocytes. The brain, heart, and skeletal muscle contain only low amounts however, although the activity does vary between muscles and even in different regions of the same muscle (Table 3.1).

Most purified catalases have been shown to consist of four protein subunits, each of which contains a haem (Fe(III)—protoporphyin) group bound to its active site. Dissociation of the molecule into its subunits, which easily occurs on storage, freeze-drying, or exposure of the enzyme to acid or alkali, causes loss of catalase activity. The three-dimensional structures of catalase from beef liver and from the fungus *Penicillium vitale* have been determined by X-ray crystallography.

Table 3.1. Catalase and glutathione peroxidase activities in normal human tissues

Tissue		Catalase activity $(mg^{-1}$ protein)	Glutathione peroxidase activity $(mg^{-1}$ protein)
Liver	A	1300	190
	B	1500	120
Erythrocytes	A	990	19
	B	1300	19
Kidney cortex	A	430	140
	B	110	87
Adrenal gland	B	300	120
Kidney medulla	A	700	90
	B	220	73
Spleen	A	56	50
Lymph node	A	120	160
Pancreas	A	100	43
	B	120	110
Lung	A	210	53
	B	180	54
Heart	A	54	69
Skeletal muscle	A	36	38
	B	25	22
Brain grey-matter	A	11	71
	B	3	66
Brain white-matter	A	20	76
Adipose tissue	A	270	77
	B	560	89

Data were abstracted from Marklund *et al.* (1982) *Cancer Res.* **42**, 1955–61. Glutathione peroxidase was assayed with a hydroperoxide substrate (see text and Chapter 4). Results are expressed as enzyme activity per milligram (10^{-3} g) of protein. Two individuals were used, denoted A and B, as sources of tissue samples.

The catalase reaction mechanism may be written as follows

$$\text{catalase—Fe(III)} + H_2O_2 \xrightarrow{k_1} \text{compound I}$$

$$\text{compound I} + H_2O_2 \xrightarrow{k_2} \text{catalase—Fe(III)} + 2H_2O + O_2.$$

For rat liver catalase, the two second-order rate constants, k_1 and k_2, have values of $1.7 \times 10^7 \, M^{-1} s^{-1}$ and $2.6 \times 10^7 \, M^{-1} s^{-1}$ respectively. Formation of compound I leads to characteristic changes in the absorption spectrum of the molecule (Fig. 3.1). The exact structure of compound I is uncertain—the iron is oxidized to a nominal valency of Fe(V) but the extensive charge-delocalization in haem rings (Appendix) makes description of the exact structure difficult. It is probably intermediate in structure between a ferric peroxide (Fe(III)—HOOH) and Fe(V)=O. It is very difficult to saturate catalase with hydrogen peroxide—its maximal velocity (V_{max}) for the destruction of hydrogen

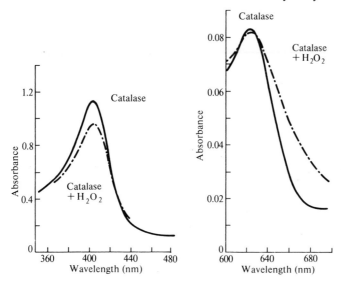

Fig. 3.1. Absorption spectra of purified rat liver catalase and catalase compound I. The large absorbance of catalase around 400 nm is known as the *Soret band*, and is seen with all haem proteins. An absorption spectrum is simply a plot of the amount of light absorbed by the protein as a function of the wavelength of that light.

peroxide is enormous. However, the above equations show that complete removal of hydrogen peroxide requires the impact of two molecules of hydrogen peroxide upon a single active site, which becomes less likely as hydrogen peroxide concentrations fall. The amount of compound I present in a mixture of catalase and hydrogen peroxide depends on the concentrations of catalase and hydrogen peroxide and on the rate constants k_1 and k_2. It may be calculated that at all reasonable concentrations, the rate of removal of hydrogen peroxide given by the equation:

$$\text{moles } H_2O_2 \text{ used } l^{-1}\,s^{-1} = 2k_2[H_2O_2][\text{compound I}]$$
$$= 2k_1[H_2O_2][\text{free catalase}]$$

If the concentration of hydrogen peroxide is fixed, the initial rate of removal of it will be proportional to the concentration of catalase present and hence will be higher in liver than in, say, brain or heart. Similarly, for a given concentration of catalase, the initial rate of hydrogen peroxide removal will be proportional to the hydrogen peroxide concentration. As a result, the specific activities (µmoles H_2O_2 decomposed per min per mg protein) quoted by manufacturers for their catalase preparations are meaningless unless they describe exactly how the assay was done. Hydrogen peroxide decomposition can be followed by the loss of its light absorbance at 240 nm, or by measuring the release of oxygen by using an oxygen-electrode.

Catalase activity can be inhibited by azide or cyanide, but these inhibit many other enzymes. A more useful inhibitor is *aminotriazole* (Fig. 3.2) which inhibits catalase activity when fed to animals or plant tissues. Its inhibitory action is exerted on compound I, so it will only inhibit catalase if hydrogen peroxide is present to allow generation of this intermediate. This H_2O_2-dependent inhibition of catalase by aminotriazole has been used *in vitro* to measure low rates of hydrogen peroxide production by various biological systems (e.g. wood-rotting fungi and *Mycoplasmas*) and the fact that aminotriazole inhibits catalase when it is fed to whole animals or plants indicates that hydrogen peroxide must be being produced *in vivo*.

Catalase is also capable of bringing about certain peroxidase-type reactions in the presence of a steady supply of hydrogen peroxide, which allows formation of compound I. Compound I will oxidize the alcohols methanol (CH_3OH) and ethanol (CH_3CH_2OH) to their corresponding aldehydes HCHO (formaldehyde or methanal) and CH_3CHO (acetaldehyde or ethanal); but propanols or butanols are much poorer substrates. In spinach leaves, formic acid (HCOOH, methanoic acid) is oxidized to carbon dioxide by the peroxidase action of compound I. It can also oxidize nitrite ion (NO_2^-) into nitrate (NO_3^-) *in vitro* and it has been suggested that it can oxidize elemental mercury (Hg) absorbed into the human body to form Hg^{2+} ions. The presence of peroxidatic substrates for catalase *in vivo* will decrease the concentration of compound I, causing more free catalase to be formed, and so this is yet another variable that must be considered in assessing how quickly hydrogen peroxide is removed. The separated catalase subunits show little catalase activity, but have peroxidase activity on a wider range of substrates, including NADH. This has no physiological significance, but is of interest in considering the active site chemistry. Another interesting suggestion is that in the presence of Mn(II) ions catalase shows *cinnabarinic acid synthetase* activity. 3-Hydroxyanthranilic acid is a compound which, when

Fig. 3.2. Structure of aminotriazole. The full name of this compound is 3-amino-1,2,4-triazole.

fed to animals, is converted into a carcinogenic molecule called cinnabarinic acid. The synthetase enzyme seems to be identical with catalase, and mice deficient in catalase activity show little cinnabarinate formation. Its formation in normal mice is also inhibited by the catalase inhibitor aminotriazole.

The catalase activity of animal and plant tissues is largely located in subcellular organelles bounded by a single membrane and known as *peroxisomes*. Although a significant proportion of the catalase activity detected in homogenates of animal and plant tissues is found not to be bound to organelles, this could be in part or in whole due to rupture of fragile peroxisomes during the homogenization (as subcellular fractionation techniques have improved, the amount of soluble catalase observed has decreased). Peroxisomes also contain some of the cellular H_2O_2-generating enzymes, such as glycollate oxidase (see legend to Fig. 3.3), urate oxidase, and the flavoprotein dehydrogenases involved in the β-oxidation of fatty acids (a metabolic pathway that operates in both mitochondria and peroxisomes in animal tissues). Mitochondria (at least in liver), chloroplasts, and the endoplasmic reticulum contain little, if any, catalase activity, so any hydrogen peroxide they generate *in vivo* cannot be disposed of in this way. It has been reported, however, that rat-heart mitochondria contain some catalase activity in the matrix. One must realize, however, that isolated subcellular fractions, especially microsomes, may be heavily *contaminated* with catalase activity and this can confuse experimental results. For example, the MEOS system (Chapter 2) was argued by some workers to be merely due to the peroxidase action of contaminating catalase on ethanol using hydrogen peroxide produced by the microsomal fraction, although further work has shown this not to be the case. Some ethanol can be oxidized by catalase *in vivo*, however (see below), although this would occur in the peroxisomes.

Professor Britton Chance in the USA has pioneered the direct observation of the absorption spectrum of catalase compound I in perfused animal organs or organs *in situ* as a means of assessing the rate of intracellular H_2O_2-production. Light is shone through a portion of, say, the liver or kidney, and the transmitted light analysed. The intracellular concentrations of hydrogen peroxide in liver quoted in Chapter 2 were obtained by this method. Figure 3.3 shows a typical experimental result. An alternative approach has been to measure the rate at which radioactively labelled (^{14}C) methanol and formate are oxidized to $^{14}CO_2$ by organs or tissues, based on the assumption that these oxidations are entirely due to the peroxidase action of catalase. It must be checked that this is true before these methods can be used, e.g. spinach-leaves contain an NAD^+-dependent formate dehydrogenase activity which catalyses oxidation of formate to carbon dioxide in addition to the catalase activity.

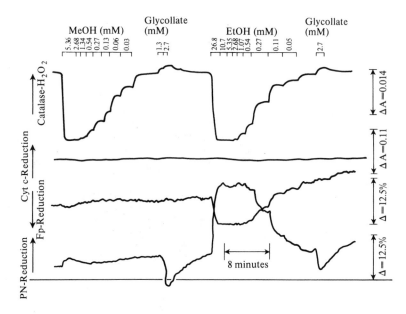

Fig. 3.3. Production of hydrogen peroxide in perfused rat liver. Rat liver was perfused with a bicarbonate–saline solution containing 2 mM L-lactate and 0.3 mM pyruvate at 30 °C. Light was shone through a liver lobe and the concentration of catalase compound I measured by dual-wavelength spectrophotometry. There is a steady concentration of compound I from which the endogenous rate of H_2O_2-production can be calculated. Inclusion of methanol in the perfusion medium reduces compound I concentration because it is a substrate for the peroxidase action of catalase. Ethanol has a similar effect, but it also causes an increased reduction of pyridine nucleotides (PN) since it is a substrate for the alcohol dehydrogenases which convert NAD^+ to NADH (Chapter 2). There is also some reduction of flavoproteins (Fp). Infusion of glycollate raises the steady-state concentration of compound I because it is oxidized in the peroxisomes to form hydrogen peroxide by glycollate oxidase

$$\text{glycollate} + O_2 \rightarrow \text{glyoxylate} + H_2O_2$$

Key: PN, pyridine nucleotides; Fp, flavoproteins; cyt-c, cytochrome c. Data from Oshino and Chance (1973) *Arch. Biochem. Biophys.* **154,** 117–31, with permission.

Glutathione peroxidase

The enzyme glutathione peroxidase was discovered in animal tissues in 1957 by G. C. Mills in the USA. It is not present in higher plants or bacteria, although it has been reported in some algae and fungi. Its substrate is the low-molecular-weight thiol compound glutathione, which is found in animals, plants, and some bacteria (e.g. in *E. coli* but not in most anaerobic bacteria) at concentrations that are often in the millimolar

Table 3.2. Presence of glutathione and enzymes using it in different organisms

System studied	[GSH]	Ratio GSH/GSSG	Glutathione peroxidase activity	Glutathione reductase activity
Spinach chloroplasts	3.0 mM	>10/1	Absent	High
Rat tissues				
liver	7–8 mM	>10/1	High	High
erythrocyte	2 mM	>10/1	Moderate	Moderate
heart	2 mM	>10/1	Moderate	Moderate
lung	2 mM	>10/1	Moderate	Moderate
lens	6–10 mM	>10/1	Moderate	Moderate
spleen	4–5 mM	>10/1	—	—
kidney	4 mM	>10/1	Moderate	Moderate
brain	2 mM	>10/1	Moderate	Moderate
skeletal muscle	1 mM	>10/1	Low	Low
blood plasma	0.02–0.03 mM	~5/1	Absent	Low
adipose tissue	3.2 µg per 10^6 cells	>100/1	Low	Low
Human tissue				
liver	4 µmol g^{-1} wet weight	>10/1	High	High
lens	6–10 mM	>10/1	Moderate	Moderate
erythrocytes	240 µg ml^{-1} blood	>10/1	Moderate	Moderate
Trypanosoma cruzi	5 nmol per 10^8 cells	—	Absent	Low
N. crassa	20 µmol g^{-1} dry weight	150/1	Absent	Moderate
E. coli				
aerobically-grown	27 µmol g^{-1}	>10/1	Absent	—
anaerobically-grown	7 µmol g^{-1}			

Whenever possible, concentrations are expressed as millimoles per litre (mM) but these cannot always be calculated from published data. These GSH values are not to be taken too literally, since they (i) decrease with age in animals, (ii) are different at different times of day in animals and at different points of the growth cycle in bacteria and fungi, (iii), in liver, fall on starvation, and (iv) will vary in the different cell types present in animal organs. Consumption of ethanol also decreases liver GSH concentrations.

(10^{-3} moles per litre) range (Table 3.2). The structure of glutathione is shown in Fig. 3.4. Most free glutathione *in vivo* is present as GSH rather than GSSG, but up to one-third of the total cellular glutathione may be present as 'mixed' disulphides with other compounds that contain —SH groups, such as cysteine, coenzyme A, and the —SH of the cysteine residues of several proteins. If R—SH is used to represent these other molecules, then the mixed disulphides have the general formula:

$$\text{glu—cys—gly}$$
$$|$$
$$\text{S—S—R}$$

Glutathione peroxidase catalyses the oxidation of GSH to GSSG at the

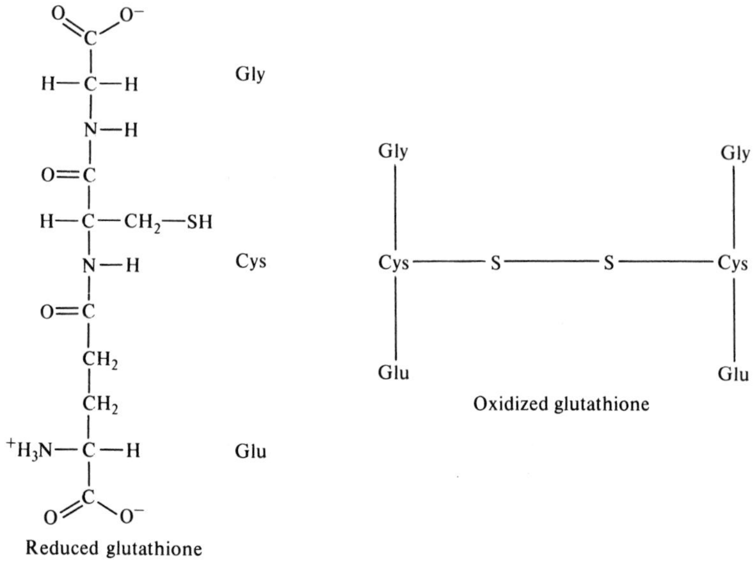

Fig. 3.4. Structure of glutathione. Glutathione is a simple tripeptide (glutamic acid–cysteine–glycine) in its reduced form, usually abbreviated to GSH. In the oxidized form, GSSG, two GSH molecules join together as the —SH groups of cysteine are oxidized to form a disulphide bridge, —S—S—.

expense of hydrogen peroxide,

$$H_2O_2 + 2GSH \rightarrow GSSG + H_2O.$$

It is found at high activity in liver, moderate activity in heart, lung, and brain, and low activity in muscle (Tables 3.1 and 3.2). The enzyme is specific for GSH as a hydrogen donor but will accept other peroxides as well as hydrogen peroxide (Chapter 4). It is made up of four protein sub-units, each of which contains one atom of the element selenium at its active site. Selenium is in group VI of the Periodic Table (see Appendix) and has properties intermediate between those of a metal and a non-metal. It is probably present at the active site as selenocysteine, the amino acid cysteine in which the normal sulphur atom has been replaced by a selenium atom (R—SeH instead of R—SH, where R is

$$\overset{+}{N}H_3$$
$$—CH_2—CH—COO^-).$$ The GSH apparently reduces the selenium and the reduced form of the enzyme then reacts with hydrogen peroxide with an approximate rate constant of $5 \times 10^7 \, M^{-1} s^{-1}$. Trace amounts of selenium are required in animal diets, although it is toxic in excess, and

several of the symptoms of selenium deficiency can be explained by the resulting lack of glutathione peroxidase. It must not be assumed that this is the only biochemical role performed by selenium, however (see Chapter 4 for a further discussion).

The ratios of GSH/GSSG in normal cells are kept high (Table 3.2) so there must be a mechanism for reducing GSSG back to GSH. This is achieved by *glutathione reductase* enzymes, which catalyse the reaction

$$GSSG + NADPH + H^+ \rightarrow 2GSH + NADP^+$$

Glutathione reductases can also catalyse reduction of certain 'mixed disulphides', such as that between GSH and coenzyme A. The NADPH required is mainly provided in animal tissues by a complex metabolic pathway known as the *oxidative pentose phosphate pathway*. The first enzyme in this pathway is *glucose-6-phosphate dehydrogenase,*

glucose-6-phosphate + $NADP^+ \rightarrow$ 6-phosphogluconate

$$+ NADPH + H^+$$

followed by *6-phosphogluconate dehydrogenase,*

6-phosphogluconate + $NADP^+ \rightarrow CO_2 + NADPH$

$$+ H^+ + \text{ribulose-5-phosphate.}$$

The rate at which the pentose phosphate pathway operates is controlled by the supply of $NADP^+$ to glucose-6-phosphate dehydrogenase. As glutathione reductase operates and reduces the $NADPH/NADP^+$ ratio the pentose phosphate pathway speeds up to replace the NADPH.

Glutathione reductases contain two protein subunits, each with the flavin FAD (Chapter 2) at its active site. Apparently the NADPH reduces the FAD, which then passes its electrons onto a disulphide bridge (—S—S—) between two cysteine residues in the protein. The two —SH groups so formed then interact with GSSG and reduce it to 2GSH, re-forming the protein disulphide.

Cooperation between catalase and glutathione peroxidase in animal tissues in the removal of hydrogen peroxide

Brain and spermatozoa contains little catalase activity but more glutathione peroxidase, so the question as to which enzyme is more important in removing hydrogen peroxide *in vivo* is easily answered. A number of animal tissues contain both enzymes, however, so how do they cooperate with each other? Mammalian erythrocytes (red blood cells) contain no subcellular organelles, and both catalase and glutathione peroxidase enzymes float around in the cell sap, although it is possible that some catalase might be attached to the inside of the erythrocyte membrane. The normal low rate of production of hydrogen peroxide in these cells (via superoxide dismutase—see Fig. 3.15) seems to be mainly dealt with by glutathione peroxidase. Indeed, humans suffering from an inborn defect in the catalase gene which produces an unstable mutant enzyme and so decreases erythrocyte catalase activities, show no significant harmful effects. If the concentration of hydrogen peroxide is raised, e.g. by supplying these cells with a drug that increases intracellular H_2O_2 generation, then catalase becomes more important.

Mammalian erythrocytes operate the pentose phosphate pathway in order to provide NADPH for glutathione reduction. However, over a hundred million people, principally in tropical and Mediterranean areas, have an inborn defect in the gene coding for glucose-6-phosphate dehydrogenase, so that its activity in erythrocytes is reduced below normal. This deficiency does cause some damage to the erythrocyte membranes, but it is not usually severe enough to cause clinical symptoms unless the rate of H_2O_2-production in erythrocytes is increased, e.g. by certain drugs. If the rate of H_2O_2-production exceeds the capacity of the enzyme to generate NADPH, then GSH/GSSG ratios fall, and glutathione peroxidase stops working, leading to destruction of the red blood cells (*haemolysis*), anaemia, and jaundice (Chapter 2) due to the excessive degradation of released haemoglobin. The survival of the defective gene in such large numbers in tropical areas has often been suggested to occur because erythrocytes with lowered dehydrogenase activity are resistant to the presence of malarial parasites within them. This could possibly be because the malarial parasite uses NADPH from its host erythrocyte to maintain its own internal GSH concentration, and so cannot live within cells deficient in NADPH generation. Indeed, as a result of suggestions made by I. A. Clark in Australia, there is now considerable interest in the role of free-radical generators in the eradication of malarial parasites (see Chapter 6 for further discussion). An inborn deficiency of erythrocyte glutathione peroxidase or in the synthesis of GSH itself is much less common, but produces severe haemolysis.

Liver contains high concentrations of both catalase and glutathione peroxidase. Whereas catalase is in the peroxisomes, the latter enzyme is found mainly in the cytosol but also in the matrix of mitochondria (about one-third of the total in rats but rather less in humans). The distribution of GSH is similar. Thus hydrogen peroxide produced by, say, glycollate oxidase and urate oxidase in the peroxisomes is largely disposed of by catalase, whereas hydrogen peroxide arising from mitochondria, the endoplasmic reticulum, or soluble (cytosolic) enzymes such as superoxide dismutase (Section 3.1.3) is acted upon by the peroxidase. The capacity of the glutathione system to cope in other tissues depends on the activity of peroxidase, glutathione reductase, and the pentose phosphate pathway enzymes. The glutathione content of tissues also varies at different times of day (Table 3.2). In lung, eye, and muscle the capacity of the system is restricted. For example, inhibition of catalase present in the eyes of rabbits by feeding aminotriazole to the animals caused the concentration of hydrogen peroxide in the aqueous humour of the eye to rise from about 0.06 to 0.15 mM even though glutathione reductase or peroxidase activities were unaffected. The glutathione system here cannot cope with the extra load caused by loss of catalase activity. Feeding young rabbits with aminotriazole can cause cataracts to develop (cataract is defined as loss of lens transparency), perhaps due to oxidation of lens proteins (see Chapter 5).

The rate of operation of the glutathione peroxidase system *in vivo* can be assessed in a number of ways. One approach has been to measure the pentose phosphate pathway activity by supplying $[1\text{-}^{14}\text{C}]$-labelled glucose to the tissue and measuring the release of radioactive $^{14}\text{CO}_2$ in the 6-phosphogluconate dehydrogenase reaction (see above). An increased pathway activity has been observed upon exposing isolated perfused rat lung, ox retina, or erythrocytes to elevated oxygen concentrations, presumably as more NADPH is consumed by glutathione reductase as it deals with increased GSSG production from glutathione peroxidase. An alternative approach has been to measure GSSG release: if cells are treated with chemical reagents that oxidize internal GSH to GSSG (such as *diamide*) they rapidly eject the GSSG into the surrounding medium. In the whole liver, GSSG is released into the bile. This rate of release of GSSG in perfused organs can be taken as a measure of glutathione peroxidase activity if glucose is omitted from the perfusing medium, so that NADPH cannot be produced by the pentose phosphate pathway for glutathione reductase activity. Exposure of isolated perfused liver and lung to elevated oxygen concentrations causes a rapid increase in GSSG release. It must be noted that glutathione peroxidase acts on lipid peroxides in addition to hydrogen peroxide (Chapter 4) and so the increased GSSG release cannot be entirely attributed to the latter molecule. Figure 3.5 shows the rate of GSSG release when hydrogen peroxide is infused into a perfused rat liver—the saturation of the effect is probably related to the increased action of catalase at the higher H_2O_2-concentrations. Inclusion of glycollate in the perfusing medium at physiological concentrations causes only a small increase in GSSG release, indicating that hydrogen peroxide generated by glycollate oxidase in the peroxisomes is largely disposed of by catalase, as expected. If animotriazole is used to inhibit catalase, glycollate infusion then does cause a marked increase in GSSG release.

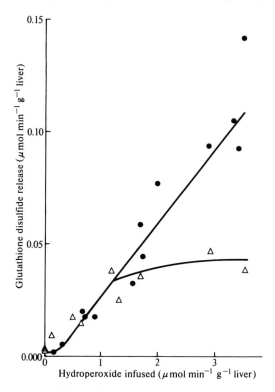

Fig. 3.5. Release of GSSG from the isolated perfused liver during infusion of hydrogen peroxide. Hydrogen peroxide was included in the perfusion medium at the concentrations stated (open triangles). The closed circles show the results when an organic hydroperoxide was used (see Chapter 4 for discussion).

Cytochrome c peroxidase

The haem-containing enzyme cytochrome c peroxidase is found between the inner and outer membrane of yeast mitochondria, which contain no catalase or glutathione peroxidases. It is also found in some bacteria. Cytochrome c peroxidase reacts rapidly with hydrogen peroxide to form a stable enzyme–substrate complex that has an absorption maximum at 419 nm, whereas that of the free enzyme is at 407 nm. On addition of reduced cytochrome c [Fe(II)] it is oxidized to the Fe(III) form and the enzyme returns to its resting state:

$$\text{enzyme} + H_2O_2 \rightarrow \text{complex} \ (\lambda_{max} \ 419 \ \text{nm})$$

$$\text{complex} + 2 \ \text{cyt-c}(Fe^{2+}) \rightarrow \text{enzyme} + 2 \ \text{cyt-c}(Fe^{3+}) + 2OH^-$$

Spectrophotometric measurement of the intermediate complex has been used as a highly accurate and sensitive method for measuring rates of

H_2O_2-formation. This method has been applied to protozoa and to isolated mitochondrial and peroxisomal fractions in the laboratory.

NADH peroxidase and oxidase

Several bacteria, such as *Lactobacillus casei* or *Streptococcus faecalis* contain a peroxidase which uses hydrogen peroxide to oxidize NADH into NAD^+. An aerotolerant mutant of the anaerobe *Clostridium perfringens* was reported to have gained such an NADH peroxidase. This enzyme should not be confused with the *NADH oxidase* enzyme found in some bacteria, in which NADH is oxidized to NAD^+, and oxygen simultaneously reduced to water. Several aerotolerant anaerobic bacteria synthesize NADH oxidase on exposure to oxygen; by its action it reduces oxygen to water and so removes it from the environment of the bacteria. Not until this enzyme has been overwhelmed by too much oxygen can irreversible damage occur. The NADH oxidase is useful because it allows the cell to survive exposure to limited amounts of oxygen, but the NADH it requires must be provided by metabolic pathways such as glycolysis, as is also true for NADH peroxidase. Drainage of the cell's reducing equivalents away from biosynthetic reactions into the NADH peroxidase or oxidase reactions could con- tribute to the growth-inhibitory effects of oxygen. This is better than cell death however!

'Nonspecific' peroxidases

Plants and bacteria often harbour haem-containing peroxidases (Fig. 3.6) that are capable of acting on a very wide range of substrates. They are usually assayed in cell extracts by using artificial substrates, which they oxidize in the presence of hydrogen peroxide to give coloured products.

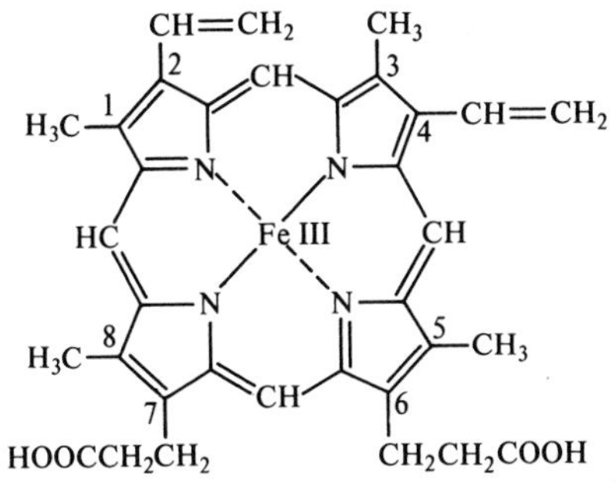

Fig. 3.6. Haem peroxidases. Most peroxidases contain the structure shown, ferriprotoporphyrin IX, at their active site. Of those that do not, e.g. myeloperoxidase, most contain some other haem-derivative.

Such artificial substrates include guaiacol (produced by certain millipedes), benzidine, and *o*-dianisidine. Often the substrates of these peroxidases *in vivo* have not been identified. 'Nonspecific' peroxidases have also been found in a few animal systems. For example, *lactoperoxidase* is found in milk and saliva. It may function to regulate the growth of some strains of bacteria since, among many other substrates, it can oxidize thiocyanate (SCN^-) ions, which are found in both milk and saliva, into hypothiocyanite ($OSCN^-$), which is very toxic to some bacterial strains, including *E. coli, Streptococci,* and *Salmonella typhimurium.* Lactoperoxidase may be one of the factors in milk that protects babies against infections of the gastrointestinal tract. The hydrogen peroxide that it needs to oxidize SCN^- seems to be derived from some of the other strains of bacteria present, which excrete hydrogen peroxide into their surroundings. *Myeloperoxidase,* another 'nonspecific' peroxidase, is found in phagocytic cells and will be further considered in Chapter 7. *Thyroid peroxidase* is found in the thyroid gland and probably serves to oxidize iodide ion (I^-) into iodine and to attach it to the thyroid hormones. A *uterine peroxidase* has also been described. Its function is unknown but the amount present can be increased by administration of oestrogens. In general, however, 'nonspecific' peroxidases are not widely distributed in animal tissues.

Perhaps the most-studied nonspecific peroxidase in biology is *horseradish peroxidase,* obtained from the roots of the horseradish plant (*Armoracia lapathifolia*). Several different forms of the enzyme exist, each containing bound carbohydrates, but they all have very broad substrate specificity. For example, they will oxidize guaiacol, pyrogallol, NADH, thiol compounds, phenols, and the plant hormone, indoleacetic acid (auxin). Oxidations by horseradish peroxidase, and probably by most other plant peroxidases, can in almost all cases be represented by the following series of reactions, in which SH_2 is the substrate

$$peroxidase + H_2O_2 \rightarrow compound\ I$$

$$compound\ I + SH_2 \rightarrow SH^{\cdot} + compound\ II$$

$$compound\ II + SH_2 \rightarrow SH^{\cdot} + peroxidase$$

$$SH^{\cdot} + SH^{\cdot} \rightarrow S + SH_2$$

The iron in the haem ring of 'resting' peroxidase is in the oxidation state Fe(III) (Fig. 3.6). Hydrogen peroxide removes two electrons to give compound I; the exact structure of this is unknown but it probably contains iron in the Fe(IV) oxidation state, the extra oxidizing capacity being located elsewhere on the protein molecule. The two electrons are replaced in two one-electron steps, in each of which a substrate molecule forms a radical, $SH^{\cdot}$. Compound II is the intermediate state of the enzyme.

The substrate-derived radicals then usually undergo a disproportiona-tion reaction, one reducing the other to SH_2 and simultaneously oxidizing itself to S. This is not always the case, however, since the SH˙ radicals can sometimes reduce O_2 to O_2^-. By the dismutation reaction of O_2^- (Chapter 2) hydrogen peroxide will be formed, and so only catalytic quantities of hydrogen peroxide need be added to start the reaction. Oxidation of NADH by horseradish peroxidase can probably be represented (in an oversimplified way) by the equations below. No hydrogen peroxide need be added to start the reaction since traces of hydrogen peroxide are always present in NADH solutions:

$$\text{peroxidase} + H_2O_2 \rightarrow \text{compound I}$$

$$\text{compound I} + \text{NADH} \rightarrow \text{compound II} + \text{NAD}^\cdot + H^+$$

$$\text{compound II} + \text{NADH} \rightarrow \text{peroxidase} + \text{NAD}^\cdot + H^+$$

$$2\text{NAD}^\cdot + 2O_2 \rightarrow 2\text{NAD}^+ + 2O_2^- \ (k_2 = 1.9 \times 10^9 \, \text{M}^{-1}\text{s}^{-1})$$

$$2O_2^- + 2H^+ \rightarrow H_2O_2 + O_2 \ \text{(dismutation)}$$

overall reaction: $2\text{NADH} + O_2 \rightarrow 2\text{NAD}^+$

This is one example of the so-called *oxidase reactions of peroxidase* as compared to its normal reactions in which equal amounts of H_2O_2 and SH_2 must be provided and no oxygen is taken up. Oxidase reactions occur when the SH˙ radicals (e.g. NAD˙ above) can reduce oxygen directly.

Lignification of plant cell walls involves the polymerization of a number of phenols derived from the aromatic amino acid phenylalanine,

$$\overset{\displaystyle COO^-}{\underset{\displaystyle CH_2-CH-NH_3^+}{|}}$$

Peroxidase bound to the cell-walls functions to oxidize these phenols into *phenoxy radicals* which polymerize to form the lignin. One source of the hydrogen peroxide required for this oxidation may be the simultaneous oxidation by peroxidase of NADH generated by a malate dehydrogenase enzyme, also bound to the cell walls. Plant peroxidases are involved in the degradation of the hormone, indoleacetic acid (auxin), and thus in the regulation of plant growth. This reaction is much more complicated than the usual peroxidase reactions however.

Apart from these cases, the identity of the *in vivo* substrates of plant and bacterial 'nonspecific' peroxidases is unknown, making it very difficult to assess their contribution to H_2O_2-removal *in vivo*. A low-level luminescence (Chapter 2) has been observed from root and stem tissues of a wide variety of plants, and has been suggested to arise in part from reactions carried out by peroxidase. For example, oxidation of auxin can lead to formation of a dioxetane-like species (Chapter 2) which decomposes to a carbonyl compound in an excited state. As this decays to the ground state, light is emitted. It can also sensitize the formation of singlet oxygen as it decays (see Chapter 2)

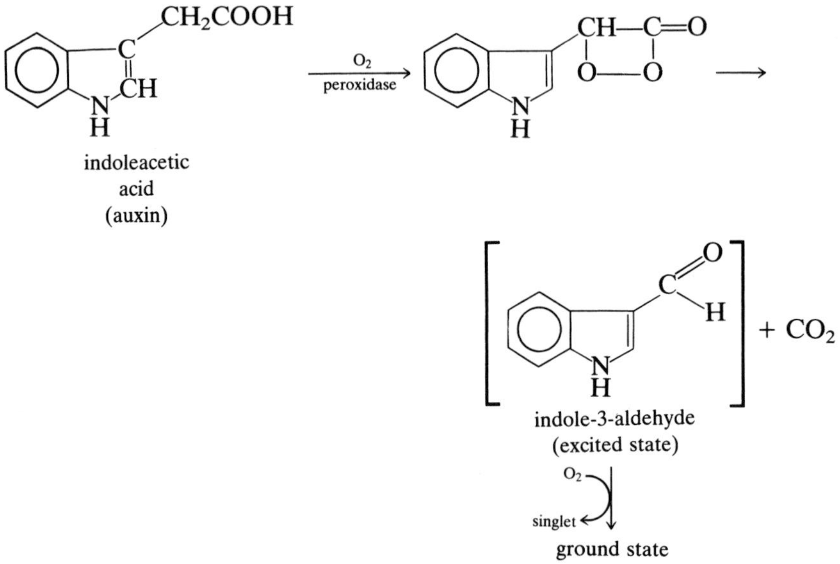

indoleacetic
acid
(auxin)

indole-3-aldehyde
(excited state)

ground state

Indole-3-aldehyde is not the only product of the action of peroxidase on auxin, however.

Horseradish peroxidase is much used in the laboratory as a method of measuring H_2O_2-production, often employing as its substrate the compound *scopoletin*. Scopoletin emits light (fluoresces) at 450 nm when illuminated with light at 360 nm, but the product of its oxidation by peroxidase does not fluoresce. After suitable calibration, this loss of fluorescence can be used to measure the rate of H_2O_2-production in a system. Of course, if that system contains other substrates for the peroxidase that can compete with scopoletin, an underestimate of H_2O_2-production will result. For example, this method cannot be applied to measure H_2O_2-production in chloroplasts since they contain large amounts of ascorbic acid, which is oxidized by the peroxidase.

Table 3.3 summarizes the various methods that have been used to measure H_2O_2-production by biological systems.

Chloroperoxidase and bromoperoxidase

Chloroperoxidase is a non-specific peroxidase first isolated from the fungus *Caldariomyces fumago*. It catalyses the usual peroxidase reactions

Table 3.3. Methods for detecting H_2O_2-production in biological systems: a summary

Method	Principle of the method	Systems to which it has been applied
Observation of intracellular catalase compound I, or oxidation of $[^{14}C]$-methanol or $[^{14}C]$-formate.	See Section 3.1.1	Bacteria, liver (perfused and *in situ*), organ slices or homogenates.
Cytochrome c peroxidase	See Section 3.1.1	Animal and plant mitochondria, protozoa, peroxisomes, microsomes.
Horeseradish peroxidase + scopoletin	Section 3.1.1	Animal and plant mitochondria, sub-mitochondrial particles, phagocytes, protozoa, microsomes.
O_2-electrode method	Add large excess of catalase and measure release of oxygen: $$2H_2O_2 \rightarrow 2H_2O + O_2$$	Only useful if little catalase present to start with. Used to study H_2O_2-removal in chloroplasts.
Catalase inhibition	If a reaction requires H_2O_2, then it should be inhibited by catalase	Catalase very slow at destroying low concentrations of H_2O_2, so a large amount must be added. Often used to investigate the role of H_2O_2 in radical reactions (see Section 3.3.4.)
GSSG release	Section 3.1.1	Perfused organs.
Aminotriazole inhibition of catalase	Section 3.1.1	Fungi, various bacteria, *Mycoplasmas*.

but in addition can catalyse introduction of halogen atoms onto a wide range of substrates in the presence of hydrogen peroxide and the halide ions, chloride (Cl^-), bromide (Br^-), or iodide (I^-). If SH is the substrate and X^- the halide, these may be written

$$SH + X^- + H_2O_2 + H^+ \rightarrow SX + 2H_2O.$$

Many marine organisms are rich in halogenated compounds and similar enzymes have been isolated from several of these, such as the purple bleeder sponge and the tropical marine sponge. In some cases only one halide can act as a substrate, e.g. a bromoperoxidase has been isolated from several marine organisms.

Ascorbate peroxidase

The chloroplasts of higher plants and the green alga *Euglena* contain no catalase, glutathione peroxidase, or 'nonspecific' peroxidase activities, but they do contain high activities of an ascorbate peroxidase enzyme, which catalyses the overall reaction

$$ascorbate + H_2O_2 \rightarrow H_2O + dehydroascorbate$$

The enzyme purified from *Euglena* is a haem protein inhibited by cyanide and azide and it is likely that the chloroplast enzyme is similar. Disposal of hydrogen peroxide by this enzyme is probably one reason why chloroplasts contain a high internal concentration of ascorbic acid. An ascorbate peroxidase activity has recently been reported in *Trypanosoma cruzi*, which also lacks catalase activity.

Miscellaneous peroxidase activities

It has long been known that myoglobin, haemoglobin, and a complex of haemoglobin with the haemoglobin-binding protein in plasma (*haptoglobin*) display peroxidase activities *in vitro* using hydrogen peroxide and a suitable electron donor. No physiological peroxidase role has ever been ascribed to these proteins but the peroxidase properties of haemoglobin are widely used as the basis of a diagnostic test for gastrointestinal bleeding (*faecal occult blood test*).

3.1.2. Methionine sulphoxide reductase

The action of singlet oxygen or hydroxyl radicals upon the amino acid methionine can cause its oxidation to methionine sulphoxide (Chapter 2). The proteins within the lens of patients suffering from cataract contain a significant amount of methionine sulphoxide and, as we saw in Chapter 2, there is considerable interest in the role of singlet oxygen produced by photosensitization reactions in damaging lens proteins. *E. coli,* yeast, rat tissues, rabbit tissues, the alga *Euglena gracilis,* human and bovine lens, human lung and neutrophils, spinach leaves, and the protozoan *Tetrahymena pyriformis* have all been found to contain an enzyme that reduces methionine sulphoxide back to methionine and hence can re-activate proteins damaged by previous oxidation of their methionine residues. The source of reducing power used by the enzyme is not clear, although there is evidence that NADPH reduces a thiol-containing protein (*thioredoxin*) which then, in the presence of the enzyme, reduces the sulphoxide. The actual importance of this reductase in repairing radical-induced damage to cells cannot yet be evaluated.

3.1.3. Superoxide dismutase

The copper–zinc enzymes

In 1938, T. Mann and D. Keilin in England described a blue-green protein containing copper that they had isolated from bovine blood. They called it *haemocuprein*. In 1953, a similar protein was isolated from horse liver and named *hepatocuprein*. Other proteins of this type were later isolated, such as *cerebrocuprein* from brain. In 1970, it was discovered that the erythrocyte protein contains zinc as well as copper. No enzymic function was detected in any of these proteins, so it was often suggested that they served as metal stores (a few scientists still think that they do, and this question is examined in detail in Section 3.3). However, in 1968 the work of J. M. McCord and I. Fridovich in the USA showed that the erythrocyte protein was able to catalytically remove the superoxide radical and thus they identified its function as a *superoxide dismutase* enzyme. Despite an intensive search, no other substrate on which superoxide dismutase enzymes act catalytically has been discovered, i.e. we may regard them as specific for the superoxide radical.

Copper–zinc-containing superoxide dismutases (CuZnSODs) are highly stable enzymes and thus easily isolated. In purifying this enzyme from erythrocytes, the cells are lysed and haemoglobin removed by treatment with chloroform and ethanol, followed by centrifugation. The enzyme actually enters the organic phase, from which it can be precipitated out by addition of cold propanone (acetone) and then further purified by ion-exchange chromatography. Not many enzymes will tolerate these procedures! Copper–zinc superoxide dismutases are also quite resistant to heating, to attack by proteases, and to denaturation by such reagents as guanidinium chloride, sodium dodecyl sulphate (SDS), or urea.

Subsequent studies have shown that CuZnSODs are found in virtually all eukaryotic cells such as yeasts, plants and animals (Table 3.4) but not generally in prokaryotic cells such as bacteria or blue-green algae. The first exception to this rule to be discovered, in A. M. Michelson's laboratory in France, is the luminescent bacterium *Photobacterium leiognathi,* which contains a CuZnSOD. This organism exists in a symbiotic relationship with the ponyfish, occupying a special gland and imparting a characteristic luminescence to the fish. Comparison of the amino-acid composition of the bacterial enzyme with that of higher organisms shows that it is closely related to fish Cu–ZnSOD enzymes. This might be taken to mean that the bacterium obtained the gene for its CuZnSOD by gene transfer from its host fish, although recent sequencing studies have cast some doubt on this suggestion. The free-living (non-symbiotic) bacterium *Caulobacter crescentus* CB15 also contains a CuZn-SOD, although studies of its amino-acid sequence have shown that it is not closely related to eukaryotic Cu–ZnSODs.

All the CuZnSOD enzymes so far isolated from eukaryotic cells have molecular weights around 32 000 and contain two protein subunits, each of which bears an active site containing one copper ion and one zinc ion (Table 3.4).

Table 3.4. Some systems from which copper–zinc SOD has been purified

Mammalian tissues	Plant tissues
Bovine erythrocytes	*Neurospora crassa*
Human erythrocytes	*Fusarium oxysporum*
Spermatozoa (especially high activity	Green peas
in donkey semen).	Maize seeds
Rat liver	Wheat-germ
Bovine liver	Spinach chloroplasts
Horse liver	Yeast (*Saccharomyces cerevisiae*)
Bovine milk ⎱ (present at low activity	Pea seedlings*
Human milk ⎰ in milk)	Corn seedlings*
Pig liver	Tomatoes*
	Cucumber*
Fish	Green peppers*
Cuttlefish (*Sepia officinalis*)*	**Other organisms**
Ponyfish	
Snapper	Fruit-fly (*Drosophila melanogaster*)
Sea bass	*Photobacterium leiognathi*
Croaker	Chicken liver
Merlin	*Caulobacter crescentus*
Trout	*Trichinella spiralis*
Swordfish	

Unless indicated by a star (*), the enzyme was purified to homogeneity and shown to contain two subunits. All purified enzymes have one ion of copper and one of zinc at each active site.

This list is being added to constantly, perhaps in the hope of finding an enzyme different from the norm.

For all CuZnSODs the reaction catalysed is the same—the dismutation reaction of O_2^- is greatly accelerated (Fig. 3.7)

$$O_2^- + O_2^- + 2H^+ \rightarrow H_2O_2 + O_2 \text{ (ground-state)}$$

Whereas the rate constant for the uncatalysed dismutation reaction depends strongly on the pH of the solution (Chapter 2) and is about $5 \times 10^5 \, M^{-1} \, s^{-1}$ at physiological pH, the reaction catalysed by bovine erythrocyte CuZnSOD is relatively independent of pH in the range 5.3–9.5, the rate constant for reaction of O_2^- with the active site being about $1.6 \times 10^9 \, M^{-1} \, s^{-1}$. Cyanide is an extremely powerful inhibitor of CuZnSODs. These enzymes are also inactivated on prolonged incubation with the compound *diethyldithiocarbamate* $(CH_3CH_2)_2N—\overset{\|}{\underset{S}{C}}—SH$, which

binds to the copper at the active sites and removes this metal from the enzyme. Diethyldithiocarbamate has been used to inhibit CuZnSOD activity in isolated erythrocytes, intestinal cells, and in whole animals, although caution should be exercised in its use since it inhibits a number of other enzymes as well. For example, when 1.5 g of diethyldithiocarba-

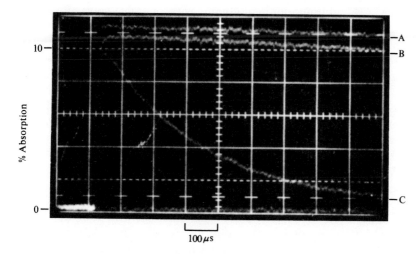

Fig. 3.7. The catalytic action of superoxide dismutase as demonstrated by pulse radiolysis. The oscilloscope traces show the decay at pH 8.8 of O_2^- radical (initial concentration 32 µmoles per litre) as followed by the loss of its absorbance at 250 nm. Trace A, spontaneous dismuatation of O_2^-; C, plus 2 µmoles per litre of SOD; B as C but SOD boiled for 5 min to destroy enzyme activity. Data from G. Rotilio *et al.* (1972) *Biochim. Biophys. Acta* **268**, 605–9, with permission.

mate per kilogram of body weight was injected into mice, the SOD activity, three hours later, of blood had decreased by 86 per cent, that of liver by 71 per cent, and that of brain by 48 per cent.

The copper ions appear to function in the dismutation reaction by undergoing alternate oxidation and reduction, i.e.

$$Enzyme{-}Cu^{2+} + O_2^- \rightarrow E{-}Cu^+ + O_2$$
$$E{-}Cu^+ + O_2^- + 2H^+ \rightarrow E{-}Cu^{2+} + H_2O_2$$
$$\overline{\hspace{3cm}}$$
net reaction: $O_2^- + O_2^- + 2H^+ \rightarrow H_2O_2 + O_2$

The Zn^{2+} does not function in the catalytic cycle but it appears to stabilize the enzyme—this conclusion is drawn from experiments in which the metals are removed from the active sites and replaced either singly or together. Indeed, the regaining of SOD activity on addition of copper ions to the metal-free CuZnSOD enzyme (apoenzyme) has been suggested as a simple method for the measurement of trace amounts of copper. In general, ions of other transition metals, such as Mn^{2+}, cannot replace the copper to yield a functional enzyme, but cobalt, mercury, or cadmium ions can replace Zn(II) in increasing enzyme stability. If the Cu^{2+} is replaced by cobalt ions (Co^{2+}), however, the enzyme can still catalyse O_2^- dismutation, although with a rate constant of only $4.8 \times 10^6 \, M^{-1} s^{-1}$.

The complete amino-acid sequence of CuZnSODs from yeast, human erythrocytes, horse liver, and bovine erythrocytes have been studied and they are all very similar. The three-dimensional structure of the bovine enzyme has been elucidated by X-ray crystallography. Each subunit is composed primarily of eight antiparallel strands of β-pleated sheet structure that form a flattened cylinder, plus three external loops. The copper ion is held at the active site by interaction with the nitrogens in the imidazole ring structures (Fig. 3.8) of four histidine residues (numbers 44, 46, 61, and 118 in the amino acid sequence); whereas the zinc ion is bridged to the copper by interaction with the imidazole of histidine 61 and it also interacts with histidines 69 and 78 and the carboxyl ($-COO^-$) group of aspartate 81. Histidine 61, which interacts with both metals, may be involved in supplying the protons needed for the dismutation reaction. Most of the surface of each protein subunit is negatively charged, repelling O_2^-, except for a positively charged 'track' that leads into the active site. A similar arrangement probably exists in the manganese and iron SODs (see below). Hence O_2^- approaching any other part of a subunit seems to be 'guided' into the the active site. Chemical modifications of these positively charged amino-acid side-chains markedly decrease enzyme activity. Although the two active sites on the enzyme are some distance from each other, the separated SOD subunits themselves catalyse O_2^- dismutation only slowly, if at all.

It is possible to visualise SOD enzymes after electrophoresis on polyacrylamide gels, as explained in the legend to Fig. 3.9. Inhibition by cyanide ion (CN^-) can be used to identify CuZnSOD enzymes. Electrophoresis of some tissue extracts, or even of purified SOD enzymes, has sometimes shown the presence of multiple bands, e.g. cow liver shows seven bands of CuZnSOD activity. Caution must be exercised in

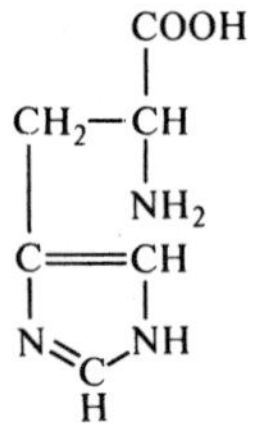

Fig. 3.8. Structure of the amino acid histidine. The ring structure is known as the imidazole ring and contains two nitrogen atoms. Each has five electrons in its outermost shell (see Appendix), three of which are being used in covalent bonding. The remaining two constitute a lone pair (see Appendix) and can interact with metal ions as explained in the text.

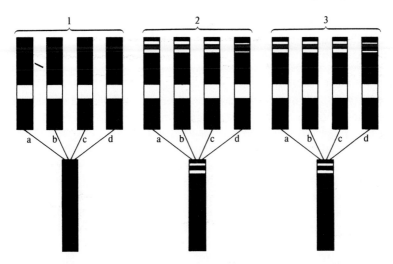

Fig. 3.9. Visualization of SOD on polyacylamide gels. Protein is applied to the gel and electrophoresis carried out. The gel is then soaked in a solution of nitro-blue tetrazolium and exposed to a O_2^--generating system. O_2^- reduces NBT to the blue-coloured formazan (Chapter 2) and so the gel turns blue except at the points where SOD activity is located. The enzyme quickly removes the O_2^-, formazan production is prevented, and so a white 'achromatic zone' is detected (for further details see C. Beauchamp and I. Fridovich (1971) *Anal. Biochem.* **44**, 276–80). The figure shows the pattern obtained in polyacrylamide gel electrophoresis of extracts from (a) brain, (b) heart, (c) liver, and (d) lung of (1) rat, (2) mouse, (3) chicken. Upper panel, no cyanide added. Lower panel, 2 mM CN⁻ ions present, which inhibit CuZnSODs. Figure taken from De Rosa *et al.* (1979) *Biochim. Biophys. Acta* **566**, 32–9, with permission. (The enzyme activity *tetrazolium oxidase*, extensively studied by geneticists, is identical with SOD.)

attributing such multiple bands to the presence of SOD isoenzymes since they might arise by attack on the SOD protein by proteolytic enzymes present in the extract. Storage of purified *Neurospora crassa* CuZnSOD at low temperatures causes it to show multiple bands on subsequent electrophoresis. If, for example, metal ions were lost from the enzyme this should increase its net negative charge. However, purification and analysis of the enzymes has proved the existence of two slightly different forms of CuZnSOD in wheat seeds, and isoenzymes have also been shown to exist in several other organisms. Indeed, an electrophoretic variant of human CuZnSOD known as SOD-2 has been found to occur in Northern Sweden and Northern Finland. Most of the population is homozygous for 'normal' SOD (SOD-1), but there are some heterozygotes with both SOD-1 and SOD-2, and a very few SOD-2 homozygotes. SOD-2 is rarely detected in other populations, except in one of the Orkney islands; perhaps the Vikings may have introduced the SOD-2 gene on one of their rampages! The SOD-2 enzyme has a slightly lower activity than the SOD-1 enzyme, but SOD-2 homozygotes suffer no apparent ill effects from this.

The manganese enzymes

The superoxide dismutase first isolated from *E. coli* proved to be entirely unlike the CuZn enzymes. It was pink rather than blue-green, was not inhibited by cyanide or diethyldithiocarbamate, had a molecular weight of 40 000 rather than 32 000, was destroyed by treatment with chloroform plus ethanol (and hence did not survive the typical purification methods for CuZnSOD) and contained manganese at its active site, this being in the Mn(III) state in the 'resting' enzyme. MnSODs catalyse exactly the same reaction as do the CuZnSODs. At pH 7.0 the rate constants for the two enzymes are similar, but unlike CuZnSODs the rate constant for MnSODs decreases at alkaline pH (e.g. for the *E. coli* enzyme, k_2 at pH 7.8 is 1.8×10^9 but $0.33 \times 10^9 \, \text{M}^{-1} \, \text{s}^{-1}$ at pH 10.2). Thus assays of tissues at high pH for SOD activity underestimate the amount of MnSOD present in relation to CuZnSOD. Manganese SODs are also much more labile to denaturation by heat or chemicals, such as detergents, than are CuZnSODs.

Manganese SOD has since been detected in several other bacteria and also in extracts of animal and plant tissues. For example, the CN^--insensitive SOD bands shown in Fig. 3.9 are attributable to the Mn-enzyme. The activity of the MnSOD in relation to CuZnSOD depends on the tissue and on the species (Fig. 3.9). Mammalian erythrocytes contain no MnSOD, it is about ten per cent of total SOD activity in rat liver but much more than this in human liver. In a normal growth medium, the fungus *Dactylium dendroides* contains 80 per cent of its SOD activity as CuZnSOD, and 20 per cent as MnSOD; but if its copper supply is restricted, more MnSOD is synthesized to maintain the total cellular SOD activity constant. Similarly, a rise in CuZnSOD activity has been observed in the liver of chickens fed on a Mn-restricted diet.

Subcellular fractionation studies upon the liver of rats have shown that most of the CuZnSOD is in the cytosol of the cell, with some activity being present in lysosomes and possibly between the inner and outer mitochondrial membranes. The MnSOD is located in the mitochondrial matrix. Mitochondria isolated from chicken liver, the fungus *N. crassa*, maize seedlings, and spinach leaves also contain MnSOD in the matrix and possibly have CuZnSOD in the intermembrane space. In human and baboon liver, however, there is some MnSOD outside the mitochondria as well as in the matrix, and the 'extra' MnSOD synthesized by copper-depleted *D. dendroides* appears in the cytosol.

MnSODs have been purified from a number of sources (Table 3.5). All

Table 3.5. Some organisms from which manganese SOD has been purified

Organism	Subunit structure	Moles Mn per mole enzyme
Higher organisms		
Maize	Tetramer	2
Luminous fungus (*Pleurotus olearius*)	Tetramer	2
Pea (*Pisum sativum*)	Tetramer	1
Chicken liver	Tetramer	2
Rat liver	Tetramer	4
Human liver	Tetramer	4
Saccharomyces cerevisiae	Tetramer	4
Bacteria		
Rhodopseudomonas spheroides	Dimer	1
E. coli	Dimer	1
Bacillus stearothermophilus	Dimer	1
Mycobacterium phlei	Tetramer	2
Mycobacterium lepraemurium	Dimer	1
Thermus thermophilus	Tetramer	2
Paracoccus denitrificans	Dimer	1–2
Streptococcus faecilis	Dimer	1
Streptococcus mutans	Dimer	1–2
Propionibacterium shermanii	Dimer	3
Bacillus subtilis	Dimer	1
Serratia marcescens	Dimer	1–2

of the MnSODs from higher organisms contain four protein subunits and usually have 0.5 or 1.0 ions of Mn per subunit. Most, but not all, of the bacterial enzymes have two subunits. Removal of the manganese from the active site causes loss of catalytic activity, and it cannot in general be replaced by any other transition-metal ion, including iron, to yield a functional enzyme. Detailed mechanistic studies of MnSODs from *E. coli* and *Bacillus stearothermophilus* show that the manganese undergoes changes of its valency state during catalysis, as would be expected, but the exact nature of these has not been elucidated since it is not a simple two-stage mechanism of the type shown by the CuZnSODs.

The amino-acid sequences of all MnSODs, whether from animals, plants, or bacteria, are extremely similar to each other and totally unrelated to those of the CuZnSODs. This is consistent with the *endosymbiotic theory* for the origin of mitochondria, which suggests that they evolved as a symbiosis between a primitive eukaryote (with CuZnSOD) and a prokaryote (with MnSOD) that eventually became incorporated into the eukaryotic cytoplasm, wrapped in a membrane (the outer mitochondrial membrane) that enclosed some of the CuZnSOD from the eukaryote.

The iron enzyme

From the bacterium *E. coli*, two SOD enzymes can be purified, one of which is the MnSOD as already described. The other was found to be an iron-containing enzyme and similar enzymes were detected subsequently in several other bacteria and in algae. In *E. coli*, both enzymes are found in the cell matrix—an early report that the iron enzyme is located in the periplasmic space (i.e. in-between the cell-wall and cell-membrane) has been retracted, although it is still widely quoted. Iron-containing superoxide dismutases (FeSODs, for short) usually contain two protein subunits, although at least two tetrameric enzymes exist, from *Mycobacterium tuberculosis* and *Methanobacterium bryantii* (Table 3.6). The dimeric enzymes usually contain 1 or 2 moles of iron per mole of enzyme. The iron in the resting state is Fe(III) and it probably oscillates between

Table 3.6. Some organisms from which iron SOD has been purified

Organism	Subunit structure	Moles Fe/per mole enzyme
Bacteria		
E. coli	Dimer	1–1.8
Desulphovibrio desulphuricans	Dimer	1–2
Thiobacillus denitrificans	Dimer	1
Chromatium vinosum	Dimer	2
Photobacterium leiognathi	Dimer	1 (plus some 'non-specifically bound' iron)
Pseudomonas ovalis	Dimer	1–2
Methanobacterium bryantii	Tetramer	2–3
Thermoplasma acidophilum (this enzyme has very low activity and the question has been raised as to whether or not it is truly a FeSOD)	Tetramer	2
Bacillus megaterium	Dimer	1
Mycobacterium tuberculosis	Tetramer	4
Propionibacterium shermanii	Dimer	2
Other organisms		
Mustard (*Brassica campestris*)	Dimer	1–2
Water-lily (*Nuphar luteum*)	Dimer	1
Porphyridium cruentum (red alga)	Dimer	1
Spirulina platensis (blue-green alga)	Dimer	1
Plectonema boryanum (blue-green alga)	Dimer	1
Anacystis nidulans (blue-green alga)	Dimer	1
Euglena gracilis (alga)	Not reported	1
Crithidia fasciculata (trypanosome)	Dimer	2–3

the Fe(III) and Fe(II) states during the catalytic cycle, i.e.

$$Fe^{3+}\text{—enzyme} + O_2^- \rightarrow Fe^{2+}\text{—enzyme} + O_2$$
$$Fe^{2+}\text{—enzyme} + O_2^- + 2H^+ \rightarrow Fe^{3+}\text{—enzyme} + H_2O_2$$

net reaction: $O_2^- + O_2^- + 2H^+ \rightarrow_2 O_2 + O_2$

although this may be an oversimplification of the mechanism in some cases.

The amino-acid sequences of FeSODs are extremely similar to those of MnSODs from all sources, and very different from the sequences of CuZnSODs. The tertiary structure of FeSODs from *Pseudomonas ovalis* and *E. coli* have been determined by X-ray crystallography, and are very different from that of CuZnSOD (see above). Like MnSODs, FeSODs show decreased catalytic activity at high pH values (compared to pH 7) and are not inhibited by CN^-. The rate constants for reaction with O_2^- are slightly lower for FeSODs than for the other types of SOD.

Some bacteria contain both FeSOD and MnSOD, such as *E. coli*, whereas others contain only one enzyme. For example, *Bacillus cereus* contains only FeSOD, and *Streptococcus sanguis*, only MnSOD. However, *Propionibacterium shermanii*, which normally contains a FeSOD, has been reported to produce an MnSOD if it is grown on iron-deficient media. *Photobacterium leiognathi* contains a FeSOD in addition to its CuZnSOD, discussed previously, whereas the non-symbiotic, free-living, strain *Photobacterium sepia* contains a FeSOD but no CuZnSOD. The aerobic bacterium *Nocardia asteroides* has been reported to produce a SOD that contains *both* manganese and iron ions.

No animal tissues have been found to contain FeSOD, but a few higher plants tissues do. Of forty-three plant families investigated by Salin's group in the USA, FeSOD was found in only three, and it has been purified from water-lily and from mustard leaves (Table 3.6). Mitochondria from mustard leaves apparently contain CuZnSOD in the intermembrane space, and MnSOD in the matrix, but the FeSOD appears to be located in the chloroplasts of the plant. There is a great deal of interest in the information about the process of evolution that may be obtained by investigating the SOD enzymes present in 'intermediate' organisms. For example, the bacterium *Paracoccus denitrificans* shares many structural and biochemical features with mitochondria, and it has been proposed that both it and mitochondria might have evolved from a common ancestral bacterium, i.e. *P. denitrificans* resembles the symbiotic bacterium that 'fused' with the primitive eukaryote. Consistent with this, *P. denitrificans* contains a MnSOD. A Cu-protein with some CN^--sensitive SOD activity has also been isolated from *P. denitrificans*, but it should not necessarily be assumed that it is related to a CuZnSOD since a number of other copper proteins react with O_2^- radical (Section 3.3.3).

Assays of superoxide dismutase activity

In investigating the biological importance of SOD, it is obviously necessary to determine the activity of the enzyme in the organism under investigation, and to determine which type of enzyme is present without having to go to the trouble of purification and metal determination. Immunological methods for detecting the amount of CuZnSOD and MnSOD proteins in animal tissues have been developed. Since these two enzymes are very different the antibodies do not cross-react. Table 3.7 shows some results for the amount of CuZnSOD protein in different human tissues. Because of the limited availability of human tissues for assay, these results should be taken as guidelines only, but they do show an especial concentration of SOD in brain, liver, and kidney, which is broadly consistent with the more extensive data from animal studies (Table 3.8).

Direct determination of SOD activity can be carried out by pulse radiolysis (Fig. 3.7), which has been especially useful in investigations of the mechanism of enzyme action. Similarly, the loss of the ultraviolet absorbance of O_2^- when KO_2 is added to an aqueous solution can be observed in a spectrophotometer, although this method can only be used at alkaline pH values when the rate of non-enzymic O_2^- dismutation is low. Any assay carried out at alkaline pH values will underestimate the

Table 3.7. CuZnSOD protein in human tissues

Tissue	CuZnSOD µg per mg protein
Cerebral grey-matter	3.7
Liver	4.71
Erythrocytes	0.52*
Renal cortex	1.93
Renal medulla	1.31
Thyroid	0.38
Testis	2.16
Cardiac muscle	1.82
Gastric mucosa	0.94
Pituitary	0.99
Pancreas	0.39
Lung	0.47

An immunological method was used which measures the enzyme protein rather than enzyme activity. Results were obtained from patients who died after accidents. Data abstracted from Hartz *et al.* (1973) *Clin. Chim. Acta* **46,** 125–32. (*Erythrocyte value per mg of haemoglobin.)

Table 3.8. Superoxide dismutase activities in animal tissues

Animal used	Assay	Tissue	Total SOD activity (units per mg protein)
Mice	Disproportionation of KO_2 in	Pancreatic islets	331
(data from *Biochem. J.*	alkaline solution (one unit of	Liver	660
[1981] **199**, 393–8.)	SOD causes O_2 to decay at the	Kidney	582
	rate of $0.1\,s^{-1}$ in a 3 ml	Erythrocytes	52
	reaction volume).	Heart	390
		Brain	408
		Skeletal muscle	282
Rat	Riboflavin–light–NBT system	Liver	22
(data from *Biochem. J.*	(one unit of SOD inhibits NBT	Adrenal	20
[1975] **150**, 31–9.)	reduction by 50%).	Kidney	13
		Erythrocytes	4
		Spleen	5
		Heart	9
		Pancreas (whole)	1.5
		Brain	3
		Lung	3
		Stomach	7
		Intestine	3
		Ovary	2
		Thymus	1
Rat	Xanthine–xanthine oxidase–	Adipose tissue	11
(data from *FEBS Lett*,	cytochrome c method (see text)		
112, 42–43)			

amount of FeSOD or MnSOD activity in relation to that of CuZnSOD, as explained previously, and appropriate corrections must be introduced. Italian scientists have developed two other methods of measuring SOD activity. One employs an electrolytic cell in which oxygen is reduced to O_2^- at a 'coated dropping mercury electrode'. Very simply, the current flowing in the system depends on the concentration of oxygen available at the electrode. Addition of SOD, by dismutating O_2^- more rapidly, increases oxygen production and hence the flow of current. This method is very useful at alkaline pH values for determination of catalytic rate constants, and is cheaper than pulse radiolysis apparatus. Their other method involves the fact that halide ions will bind to the active site of Mn and CuZnSOD enzymes. Observation of the nuclear magnetic resonance (NMR) behaviour of fluoride ($^{19}F^-$) ions so bound can be used to assess the amount of enzyme present. We do not propose to discuss this method in detail (see the end of the chapter for further references), although it does require access to NMR equipment.

However, most laboratories still use the so-called *indirect assay methods* for SOD activity. In these, O_2^- is generated by some mechanism and allowed to react with a detector molecule. SOD, by removing the O_2^-, inhibits the reaction with the detector. In their original work in 1968 on the SOD activity of the erythrocyte enzyme, McCord and Fridovich used an assay of this type. O_2^- was generated by a mixture of the enzyme xanthine oxidase and its substrate xanthine, and was allowed to reduce cytochrome c, this reduction being accompanied by a change in absorbance at 550 nm. SOD, by removing O_2^-, will inhibit the absorbance change. One unit of SOD activity was defined as that amount that would inhibit cytochrome c reduction by 50 per cent under their assay conditions, and so the units of SOD activity quoted in the literature bear no relation whatsoever to quoted units for other enzymes (1 enzyme unit is normally defined as that amount catalysing transformation of 1 μmole of substrate per minute). Figure 3.10 summarizes the principles of this assay. It may be seen that a decrease in the cytochrome c reduction, looking like a SOD activity, could also be produced by a reagent that inhibits O_2^- generation, i.e. by inhibiting the xanthine oxidase enzyme. Fortunately, this can be checked for by measuring the production of uric acid in the system as an index of xanthine oxidase activity; and such a control should always be done when the assay is being applied to a crude tissue extract. Many such extracts contain cytochrome oxidase, a mitochondrial enzyme complex that re-oxidizes reduced cytochrome c and so interferes with the assay. Chemical modification of the cytochrome

c by attachment of acetyl (CH_3—$\overset{\displaystyle O}{\overset{\displaystyle \|}{C}}$—) groups to some of its amino-acid side-chains prevents it from being a substrate for cytochrome oxidase but still allows it to react with O_2^-. Use of acetylated cytochrome c permits this assay to be used in extracts containing cytochrome oxidase. Other detector molecules can be used instead of cytochrome c, e.g. nitroblue tetrazolium (NBT) which is reduced by O_2^- to a deep-blue coloured formazan (Chapter 2), or adrenalin, which is oxidized by O_2^- to form a pink product known as adrenochrome. The ability of O_2^- to oxidize NADH in the presence of lactate dehydrogenase (Chapter 2), accompanied by a fall in absorbance at 340 nm, has also been used. Table 3.9 summarizes these and some other methods (see also Fig. 3.10). The authors' laboratory routinely uses inhibition of NBT reduction in a xanthine–xanthine oxidase system to assay SOD in tissue extracts. Although this avoids problems with cytochrome oxidase, it must be remembered that the reaction of NBT with O_2^- is very complex (Chapter 2). Although formazan is only sparingly soluble in water, its precipitation can be avoided by keeping absorbance changes fairly low. The compound *luminol* emits light when exposed to O_2^- (again the mechanism is very complex) and has been used as a detector molecule (Table 3.9).

Table 3.9. Indirect methods that have been used to measure SOD activity

Source of superoxide	Detector of superoxide	Reaction measured[1]
Xanthine–xanthine oxidase	Cytochrome c	Reduction, ΔA
	Nitro-blue tetrazolium	Reduction, ΔA
	Luminol	Light-emission
	Adrenalin	Oxidation, ΔA
	NADH + lactate dehydrogenase	Oxidation, ΔA
	Hydroxylamine	Nitrite (NO_2^-) formation (colorimetric method)
	2-Ethyl-1-hydroxy-2,5,5-trimethyl-3-oxazolidine (hydroxylamine derivative)	Oxidation to nitroxide, detected by ESR (Chapter 2)
Autoxidation reactions	Adrenalin	Oxidation, ΔA
	Sulphite	O_2 uptake
	Pyrogallol	O_2 uptake, or ΔA
	6-Hydroxydopamine	Oxidation, ΔA
Directly added $K^+O_2^-$	—	Loss of O_2^-, ΔA in UV
	Nitro-blue tetrazolium	Reduction, ΔA
	Cytochrome c	Reduction, ΔA
	Tetranitromethane	Reduction, ΔA
Illuminated flavins	Nitro-blue tetrazolium	Reduction ΔA O_2 uptake (SOD accelerates)
	Dianisidine	Oxidation, ΔA (*'positive'* assay)
NADH + phenazine methosulphate[2]	Nitro-blue tetrazolium	Reduction, ΔA

[1] ΔA: reaction results in an absorbance change that can be measured using a spectrophotometer.
[2] NOT RECOMMENDED—see text, and also *J. Am. Chem. Soc.* (1982) **104,** 1666.

Since xanthine oxidase is so often used as a source of O_2^- in the laboratory, it is worthwhile saying a little about it. The commercially-available enzyme is usually obtained from cream, and the purification process employed by some manufacturers involves the use of proteolytic enzymes to free the oxidase from the milk fat globule membranes. Sometimes these proteases are still present in the final preparation and this must be carefully checked for. One report of the damaging effects of

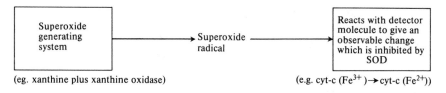

Fig. 3.10. Principle of the indirect assay methods for SOD activity.

O_2^- from a xanthine–xanthine oxidase system upon chloroplast membranes turned out to be an effect of the trypsin contamination of the enzyme preparation.

Xanthine oxidase catalyses oxidation of both hypoxanthine and xanthine as shown below, but it will also act on a number of other substrates such as acetaldehyde (ethanal, CH_3CHO),

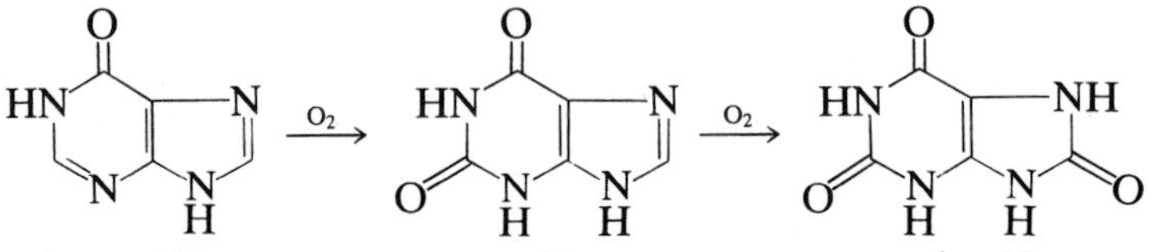

hypoxanthine   xanthine   uric acid

Both hypoxanthine and ethanal can be used instead of xanthine in O_2^--generating systems. A powerful inhibitor of xanthine oxidase is the structurally related compound allopurinol

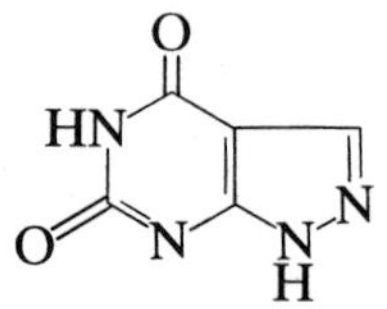

During substrate oxidation, oxygen is reduced both to O_2^- and to hydrogen peroxide at the active site of the enzyme. The ratio of O_2^- to hydrogen peroxide produced by the enzyme increases at alkaline pH values. The purified protein has two active sites, each of which contains one molecule of FAD, one ion of molybdenum, and four ions of iron associated with inorganic sulphur.

Sources of O_2^- other than xanthine oxidase have been used in SOD assays. A mixture of NADH and phenazine methosulphate produces some O_2^-, but we do not recommend it for use because it also directly generates hydroxyl radicals and creates many artefacts (Table 3.9). Illumination of a riboflavin solution in the presence of either EDTA or of the amino acid methionine causes a reduction of the flavin. It then re-oxidizes and simultaneously reduces oxygen to O_2^-, which is allowed to react with a detector molecule such as NBT. SOD will inhibit the formazan production. Flavin photochemistry is extremely complicated, however, and singlet oxygen is also produced (see Chapter 2). In an interesting variation on this assay, an oxygen electrode is used to measure the rate of oxygen consumption during photochemical generation of O_2^-

in the presence of NBT. Reduction of the dye by O_2^- is accompanied by stoichiometric oxygen-production, i.e.

$$NBT + O_2^- \rightarrow NBT^{\cdot} \text{ radical} + O_2$$

On addition of SOD, two O_2^- molecules are required to make one oxygen molecule, and the rate of oxygen uptake increases.

A number of compounds have been shown to oxidize in solution with simultaneous production of O_2^-; these include 6-hydroxydopamine, pyrogallol, the sulphite ion (SO_3^{2-}), and adrenalin (at alkaline pH values). O_2^-, once formed, participates in the oxidation of further molecules, so that addition of SOD greatly slows down the observed rates of oxidation of these compounds. This can be used as an assay for SOD, the rate of oxidation being measured either by an absorbance change (as with adrenalin oxidation to adrenochrome) or by oxygen uptake using an oxygen-electrode (as with pyrogallol or sulphite). The rates of these oxidations are often greatly accelerated by the presence of transition-metal ions, however, and this can cause problems in the assay of crude extracts containing traces of such ions.

Fridovich's laboratory in the USA has developed a 'positive' assay for SOD activity. A solution containing riboflavin and the detector molecule dianisidine is illuminated, whereupon the detector is slowly oxidized, accompanied by an absorbance change at 460 nm. Addition of SOD greatly increases the rate of dianisidine oxidation because it removes O_2^-, which interacts with an intermediate dianisidine radical and thereby decreases the net rate of oxidation. The assay is called 'positive' because addition of the SOD actually causes a reaction to accelerate, instead of inhibiting it. The reaction mechanism is very complicated, however.

Whatever assay is used, it should first be calibrated with pure SOD enzyme, and a known amount of SOD enzyme, added to the crude tissue extract being examined, should be quantitatively detected on subsequent assay. The scientists performing the assay should also think carefully about possible artefacts. To take one example, the compound pamoic acid appeared to inhibit SOD activity in a number of indirect assays, but careful analysis showed that it was actually interfering with the assay mechanism. No one assay will be suitable for all systems, e.g. assays using NBT reduction cannot be applied to eye tissues because of the presence of enzyme systems that reduce NBT directly. Table 3.9 summarizes some of the methods that have been used by different workers; and Table 3.8 shows the activities detected in various animal tissues using some of these assay methods. The units quoted are different for each assay, but comparison of Tables 3.7 and 3.8 shows that the relative amount of SOD activity in different body organs is broadly similar in different mammals.

Distinction between the different SOD types in tissue extracts

We have seen that CuZnSOD is inhibited by cyanide ion (CN^-) whereas FeSOD or MnSOD is not. Inhibition by CN^- can therefore be used to identify CuZnSOD activity in assays of tissue homogenates or on polyacrylamide gels (e.g. Fig. 3.9).

Both CuZnSOD and FeSOD are inactivated on prolonged exposure to hydrogen peroxide, whereas MnSOD is not. The rate of inactivation of the CuZnSOD is faster at higher pH values. It may be due to a reduction of Cu(II) to Cu(I) at the active site by hydrogen peroxide, followed by a reaction of Cu(I) with hydrogen peroxide to generate $OH^\cdot$ (Chapter 2), which then destroys one of the histidine residues essential for the catalytic mechanism. Thus incubation of, say, a bacterial extract with hydrogen peroxide will inactivate FeSOD but not MnSOD and can be used to distinguish the two. Inactivation may be physiologically relevant under certain circumstances (see below).

Another method for distinguishing between different SOD types employs the fact that FeSODs are more sensitive to inhibition by azide. For example, at pH 7.8 azide at a concentration of 10 mmoles per litre inhibits CuZn, Mn and Fe enzymes by about 10%, 30% and 70% respectively. There is some variation however, e.g. the *Methanobacterium bryantii* FeSOD is less sensitive to azide than are other FeSODS. Another approach has been to remove the metals from SOD proteins in bacterial extracts, and then to add either Fe(II) or Mn(II) back to the extract. If a particular band of enzyme activity observed on electrophoresis before metal removal re-appears on addition of, say, Fe^{2+}, then it most likely represented a FeSOD. Despite the close structural similarities between FeSODS and MnSODs, most of them will only work with the correct metal at the active site. There are exceptions to this rule, however, as discussed below (Table 3.14).

3.2. Protection by small molecules

3.2.1. Ascorbic acid (vitamin C)

Pure ascorbic acid is a white crystalline solid, very soluble in water. Plants and most animals can synthesize it from glucose, but humans, other primates, guinea-pigs, and fruit-bats lost one of the necessary enzymes during their evolution, and so require ascorbic acid to be present in the diet, as vitamin C. We rely on the fact that plants can still make it! Ascorbic acid is required *in vivo* as a cofactor for the enzymes proline hydroxylase and lysine hydroxylase involved in the biosynthesis of collagen. Both these enzymes contain iron at their active sites. Ascorbate is also required for the action of the copper enzyme dopamine-β-hydroxylase which converts dopamine into noradrenalin. Deficiency of ascorbate from the human diet causes *scurvy*. Collagen synthesized in the absence of ascorbic acid is insufficiently hydroxylated and does not form fibres properly, giving rise to poor wound-healing and fragility of blood vessels.

The most striking chemical activity of ascorbate is its ability to act as a reducing agent (electron donor). We have already seen (Chapter 2) that its ability to reduce Fe(III) to Fe(II) is important in promoting the uptake of iron in the gut. The observation that dietary ascorbate inhibits the carcinogenic action of several nitroso-compounds fed to animals (Chapter 8) can be attributed to its ability to reduce them to inactive forms. Ascorbate may help to detoxify various organic radicals *in vivo* (e.g. those formed by ionizing radiation—Section 3.4) by a similar reduction process. Indeed, ascorbate is probably required by the above hydroxylase enzymes in order to keep the iron or copper at the active site in the reduced form necessary for hydroxylation to occur.

Donation of one electron by ascorbate gives the semidehydroascorbate radical (Fig. 3.11), which can be further oxidized to give dehydroascorbate. The semidehydroascorbate radical is not particularly reactive and mainly undergoes a disproportionation reaction,

$$2 \text{ semidehydroascorbate} \rightarrow \text{ascorbate} + \text{dehydroascorbate}$$

Dehydroascorbate is unstable and breaks down rapidly in a very complex way, eventually producing oxalic and L-threonic acids. Aqueous solutions of ascorbic acid are stable unless transition-metal ions are present, which catalyse their rapid oxidation at the expense of molecular oxygen. Copper salts are the best catalysts—if you want plenty of vitamin C from your fruit and vegetables, don't cook them in copper pans! Copper-induced oxidation of ascorbate produces hydrogen peroxide and hydroxyl radicals. The reported ability of ascorbate to degrade DNA and damage various animal cells in culture, including cancer cells, can probably be attributed to formation of these species in the presence of traces of copper ions in the solution. Ascorbic acid–Cu^{2+} mixtures inactivate many proteins, probably by formation of hydroxyl radicals.

In Chapter 2 we saw that ascorbate reacts rapidly with O_2^- and $HO_2^{\cdot}$ and even more rapidly with $OH^{\cdot}$ to give semidehydroascorbate. It also scavenges singlet oxygen. The function of ascorbate peroxidase in removing hydrogen peroxide has already been discussed. Hence ascorbate may well help to protect against oxygen-derived species *in vivo*. For example, the lens of the human eye is low in SOD activity but rich in ascorbate, whereas rat lens has more SOD but less ascorbate. In agreement with such a protective role, exposure of animals to elevated

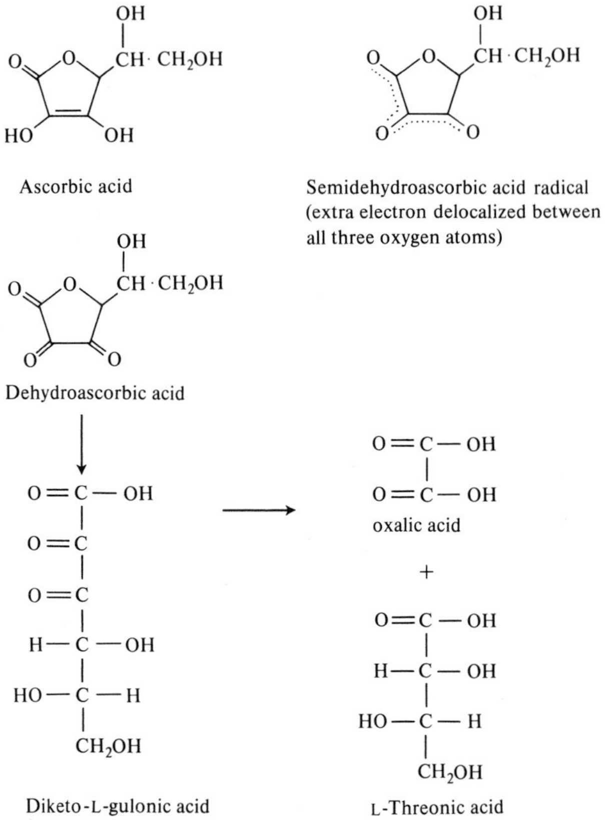

Fig. 3.11. Structure of ascorbic acid and its oxidized forms.

oxygen concentrations, or to ozone, causes a decrease in the ascorbic acid content of the lungs; and administration of ascorbic acid to animals has sometimes been reported to lessen pulmonary damage caused either by ozone or by high-pressure oxygen. In the lung, ascorbate appears to accumulate in the fluid lining the air spaces and may thus act as an extracellular antioxidant, complementing the intracellular SOD and H_2O_2-removing enzymes. Intracellular ascorbate concentrations in several lung cells are in the millimolar range. Treatment with ascorbic acid greatly decreased the incidence of growth abnormalities induced by exposure of a strain of tobacco seedlings to pure oxygen.

Injection of dehydroascorbate into animals affects insulin secretion by the pancreas and induces diabetes. Its decomposition product, oxalic acid, is not very pleasant either. Hence, both animal and plant tissues have evolved mechanisms for converting the oxidized form of ascorbate

Table   3.10. Semidehydroascorbate   reductase
activity in rat tissues

Tissue	Enzyme activity (mean ± SEM)
Adrenal cortex	49.6 ± 2.4
Brain	9.1 ± 0.6
Heart	0
Ileum	3.3 ± 0.3
Kidney	49.3 ± 4.9
Liver	30.9 ± 1.0
Lung	8.9 ± 1.8
Pancreas	16.3 ± 1.1
Skeletal muscle	0
Spleen	6.3 ± 0.3
Testis	11.4 ± 0.3
Thyroid gland	5.8 ± 0.3

Semidehydroascorbate reductase activity was
assayed in homogenates of several rat tissues. The
enzyme reduces semidehydroascorbate (SDA) to
ascorbate at the expense of NADH. Data from
Diliberto *et al.* (1982) *J. Neurochem.* **39**, 563–8.
Enzyme activity is quoted as nanomoles of NADH
oxidized per minute per mg of protein. The high
activity in adrenal gland cortex may be due to the fact
that SDA is formed from ascorbate during the
dopamine-β-hydroxylase reaction (see text).

back to the reduced form. A *dehydroascorbate reductase* enzyme, which catalyses the reaction:

$$\text{dehydroascorbate} + 2\text{GSH} \rightarrow \text{GSSG} + \text{ascorbate}$$

was originally purified from many plant tissues, but later found in several animal and human tissues. Animals also contain a *NADH–semidehydroascorbate reductase* enzyme, which reduces the semidehydroascorbate radical back to ascorbate whilst oxidizing NADH to NAD^+ (Table 3.10). This enzyme has been reported in a few fungi and plant tissues.

Not everything is wonderful about vitamin C, however, whatever the health food shops say! Like O_2^-, ascorbate can reduce Fe(III) ions to Fe(II) and, in the presence of hydrogen peroxide, can stimulate $OH^{\bullet}$ formation by the Fenton reaction. Its overall effect will depend on the concentration of ascorbate present, since it also scavenges $OH^{\bullet}$. Administration of vitamin C to patients with iron-overload has sometimes provoked severe reactions, perhaps due to increased $OH^{\bullet}$ formation *in vivo*. Ascorbate can stimulate iron-dependent peroxidation of membrane lipids (Chapter 4) under certain circumstances.

3.2.2. Glutathione

The role of GSH as a substrate for the H_2O_2-removing enzyme glutathione peroxidase and for dehydroascorbate reductase has already been discussed. In addition glutathione is a scavenger of hydroxyl radicals and singlet oxygen. Since it is present at high concentrations in many cells (Table 3.2) it may help to protect against these species. GSH can reactivate some enzymes that have been inhibited by exposure to high oxygen concentrations. Presumably the oxygen causes oxidation of essential—SH groups on the enzyme, which are regenerated on incubation with GSH. Glutathione is not essential for aerobic life since several strains of aerobic bacteria are known that do not contain it, although they might contain other low-molecular-weight thiol compounds serving a similar purpose. Mutants of *E. coli* unable to synthesize GSH grow normally under air, as do mutants defective in glutathione reductase, although their ability to tolerate elevated oxygen concentrations has not been reported. Deficiencies of GSH synthesis in animal cells have more serious consequences, however, such as haemolysis in erythrocytes. Feeding diets deficient in sulphur-containing amino acids to rats potentiates the toxic effects of elevated oxygen concentrations.

GSH is a cofactor for several enzymes in widely different metabolic pathways, such as glyoxylase, maleylacetoacetate isomerase, prostaglandin endoperoxide isomerase (Chapter 8), and DDT dehydrochlorinase and it may be involved in the synthesis of thyroid hormones. It plays a role in the degradation of insulin in animals and also in the metabolism of herbicides, pesticides, and 'foreign' compounds generally in both animal and plant tissues. For example, corn leaves contain an enzyme which detoxifies the herbicide atrazine by combining it with GSH. The higher the activity of this enzyme, the greater is the resistance of the plant to the herbicide. Many 'foreign compounds' supplied to animals are metabolized in the liver to yield *mercapturic acids* that are excreted. The first stage in this process is conjugation of the compound with GSH by *glutathione-S-transferase* enzymes, as shown in Fig. 3.12. Several enzymes of this type are present in liver, and also in many other animal tissues. Among compounds converted to mercapturic acids in the rat are chloroform, bromobenzene, naphthalene, and paracetamol. The presence of these compounds has the effect of decreasing hepatic GSH concentrations, which reduces the ability of the liver to cope with hydrogen peroxide and other oxygen radicals. Supplying metabolic precursors of glutathione can raise the GSH content of some tissues and protect against these effects. The compound 2-oxothiazolidine-4-carboxylate, a precursor of cysteine *in vivo,* seems especially effective. Injection of methyl esters of GSH, which can cross membranes easily, might also help to raise tissue GSH concen-

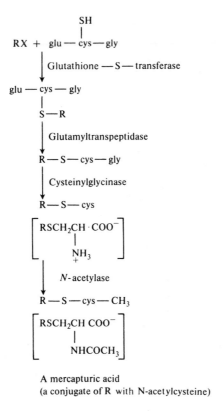

A mercapturic acid
(a conjugate of R with N-acetylcysteine)

Fig. 3.12. Mercapturic acid formation. RX represents the 'foreign' compound.

trations. Professor A. Meister in the USA has proposed a function for GSH in the transport of amino acids across the plasma membrane of animal cells.

If a tissue is exposed to a large flux of hydrogen peroxide and/or hydroxyl radicals, a point might be reached at which the GSH/GSSG ratio cannot be maintained at its normal high value (Table 3.2) and GSSG will accumulate. Unfortunately, GSSG can inactivate a number of enzymes, probably by forming mixed disulphides with them. Mixed disulphides with proteins and with such molecules as coenzyme A accumulate in tissues subjected to 'oxygen radical stress'.

$$\text{enzyme—SH} + \text{GSSG} \rightleftharpoons \text{enzyme—S—S—G} + \text{GSH.}$$

(active)   (inactive)

GSSG has been shown to inhibit protein synthesis in animal cells and the enzymes adenylate cyclase, phosphofructokinase, and phosphorylase

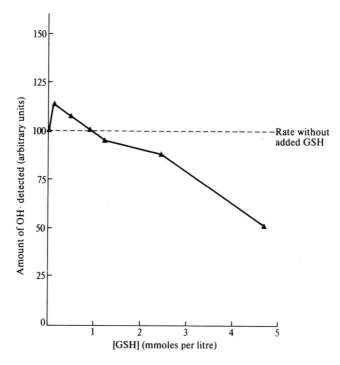

Fig. 3.13. Interaction of reduced glutathione (GSH) with hydroxyl radicals. GSH at various concentrations was added to a system generating hydroxyl radicals (a mixture of hypoxanthine and xanthine oxidase in the presence of an iron salt). At the concentrations present in most body tissues (Table 3.2) GSH decreased the amount of OH˙ detected because it can scavenge this radical. At low concentrations, however, GSH stimulated OH˙ formation because it can interact with Fe^{2+} salts and hydrogen peroxide (produced by the xanthine oxidase) to itself produce OH˙. Data from Rowley and Halliwell (1982) *FEBS Lett.* **138,** 33–6. (Other thiol compounds showed similar effects.)

phosphatase. This is presumably why cells keep their GSSG concentrations low under normal conditions; and it shows that too much oxygen can turn a protective mechanism into a damaging one. GSH can interact with hydrogen peroxide and iron salts to *produce* hydroxyl radicals, but this would only be significant *in vivo* if GSH concentrations were very low (Fig. 3.13).

3.2.3. Uric acid

Professor Bruce Ames in the USA has pointed to the fact that uric acid, present in human blood plasma at concentrations in the range 0.12–0.45 mmoles per litre, is a powerful scavenger of singlet oxygen and of hydroxyl radicals. Japanese scientists had previously established the ability of uric acid to inhibit lipid peroxidation. The extracellular fluids of the body contain only low activities of superoxide dismutase and catalase and Ames has therefore suggested that uric acid may, like ascorbate, contribute an anti-oxidant defence. Glucose, present in the human blood-stream at 4.5 mM, will also scavenge hydroxyl radicals (Chapter 2). The radical produced by reaction of uric acid with OH˙ reacts quickly with oxygen to form a peroxy radical ($R—O_2˙$). This peroxy radical is less reactive than OH˙, although it can damage at least one enzyme.

3.3. The superoxide theory of oxygen toxicity

The discovery of SOD in aerobic cells led directly to the proposal, from Fridovich's laboratory in the USA, that the superoxide radical is a major factor in oxygen toxicity and that the SOD enzymes are an essential defence against it. Even if it turned out to be untrue, this *superoxide theory of O_2 toxicity* would still have been extremely valuable because of the great amount of experimental work that it provoked, which has led to many important discoveries. Indeed, in the past few years more papers have been published on SOD than on any other type of enzyme! There is much truth in the theory, however; although, like all good theories, it has required modification in the light of new experimental facts. Let us evaluate the current status of the theory.

3.3.1. Is superoxide formed in vivo? (inside living tissue)

Most oxygen taken up by higher aerobes is reduced to form water by the operation of the mitochondrial electron transport chain. A few enzymes, such as glycollate oxidase, reduce oxygen directly to hydrogen peroxide as described in Chapter 2. However, a number of enzymes have been discovered that actually reduce oxygen to O_2^-, which is released into free solution. Table 3.11 describes some of these. Xanthine oxidase and the oxidase action of peroxidase have already been discussed. However, work by Italian scientists has shown that 'xanthine oxidase' from liver, and probably in most other animal tissues and in plants, is in fact a xanthine dehydrogenase enzyme *in vivo,* which transfers electrons from the substrate onto NAD^+ rather than onto oxygen. Xanthine dehydrogenase becomes converted to xanthine oxidase during its purification, either because of attack by proteolytic enzymes or by oxidation of some thiol (—SH) groups. Thus the metabolism of xanthine or hypoxanthine by this enzyme, which is located in the cytosol of animal and plant tissues, would not normally produce O_2^- *in vivo*. However, there is evidence for some xanthine oxidase activity *in vivo* in thyroid gland and intestinal cells. For

Table 3.11. Enzymes that generate the superoxide radical

Enzyme	Location	Comments
Peroxidases (nonspecific)	Plants and bacteria	O_2^- produced during the oxidase reaction (see Section 3.1.1).
Xanthine oxidase	intestine, ischaemic tissues	see Section 3.3.1. Forms both O_2^- and H_2O_2 at the active site.
Nitropropane dioxygenase	*Hansenula mrakii* (a yeast)	catalyses oxidation of 2-nitropropane into acetone. O_2^- produced and involved in the catalytic mechanism. SOD inhibits the oxidation.
Indoleamine dioxygenase	Most animal tissues, especially small intestine, not liver. Activity of enzyme in lung increases during virus infection or after injection of bacterial endotoxin in (30–100 fold) but is not increased by exposure to elevated oxygen concentrations.	Cleaves the indole ring of tryptophan and several related compounds such as serotonin 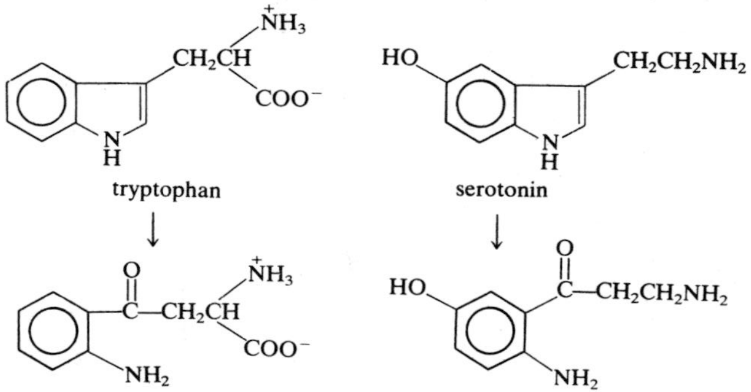 O_2^- produced and involved in the catalytic mechanism, SOD inhibits the oxidation. Inhibition of the SOD in isolated rabbit intestine cells by diethyldithiocarbamate increased tryptophan degradation, as did addition of xanthine.
Tryptophan dioxygenase	Liver	Same reaction as above but specific for tryptophan.
Galactose oxidase	Fungi	Copper enzyme. O_2^- produced and involved in the catalytic mechanism. Oxidizes a —CH$_2$OH group of the sugar galactose to —CHO.
Aldehyde oxidase	Liver	Contains molybdenum, iron. Produces free O_2^-. Broad substrate specificity.

example, addition of xanthine to whole intestinal cells increases indoleamine dioxygenase activity *in vivo,* an effect suppressed by the xanthine oxidase inhibitor allopurinol (see Table 3.11). Thyroid xanthine oxidase may provide the hydrogen peroxide needed by the peroxidase that adds iodine onto the thyroid hormones (Section 3.1.1).

In intestine deprived of its oxygen supply (this state is called *ischaemia*) there is a rapid conversion of xanthine dehydrogenase to oxidase activity, presumably by proteolytic attack. Deprivation of oxygen causes extensive degradation of ATP in the cells and accumulation of hypoxanthine. Obviously, ischaemia itself causes damage to the tissue, but a great deal more damage occurs when the oxygen supply is restored, an effect known to physiologists as *reperfusion damage*. J. M. McCord in the USA has shown that reperfusion damage in cat intestine can be decreased by including SOD in the reperfusion fluid. He has suggested that, on re-admission of oxygen, the accumulated hypoxanthine is oxidized by the xanthine oxidase generated during ischaemia and the excessive O_2^- production causes further damage (Fig. 3.14). Ischaemia and reperfusion damage are of the greatest importance in the pathology of heart disease (e.g. blockage of a coronary artery can shut off the blood supply to part of the heart muscle) and of stroke (cutting off the blood supply to the brain), and it is interesting to speculate that SOD may be therapeutically useful if injected under the right conditions. *In vitro* experiments have shown that SOD, hydrogen-peroxide metabolising enzymes and the iron-chelating agent desferrioxamine (Chapter 2) can decrease reperfusion damage in animal hearts.

Several biologically important molecules also oxidize in the presence of oxygen to yield O_2^-; these include glyceraldehyde, the reduced forms of riboflavin and its derivatives FMN and FAD, adrenalin, tetrahydropteridines, and thiol compounds such as cysteine. Tetrahydropteridines act as cofactors for several oxygenase enzymes, including those that catalyse hydroxylation of the aromatic amino acids, phenylalanine and tyrosine.

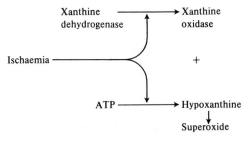

Fig. 3.14. A suggested mechanism for damage during intestinal and cardiac ischaemia. For details, see *Gastroenterology* (1981) **81,** 22–9.

Oxidation of adrenalin (a very complex reaction) and of photochemically reduced flavins has been employed in SOD assays, as discussed previously. The reduced forms of several biological flavoproteins and flavoenzymes have also been shown to release O_2^- in the presence of oxygen. Oxidation of molecules such as adrenalin and thiol compounds (cysteine, GSH) *in vitro* is greatly accelerated by the presence of transition metals such as iron and copper. Indeed, Swedish scientists have pointed out that several of the components present in the growth media used to culture anaerobic bacteria oxidize to produce O_2^- and hydrogen peroxide in the presence of oxygen and could thus contribute to the toxic effects seen when the bacterial cultures are exposed to oxygen.

In general it is difficult to assess the contribution of these oxidation reactions to O_2^- formation *in vivo,* but there is one such reaction that is of great significance to the cells in which it occurs—that of oxyhaemoglobin. The haemoglobin molecule has four protein subunits, two α amino-acid chains and two β chains. Each chain carries a haem group to which the oxygen reversibly attaches. The iron in the haem ring of deoxyhaemoglobin is in the Fe(II) state, but when oxygen attaches to it, an intermediate structure results, in which an electron is delocalized between the iron ion and the oxygen (see Appendix), i.e.

$$Fe^{2+}\!-\!O_2 \rightleftharpoons Fe^{3+}\!-\!O_2^-$$

The bonding is intermediate between Fe(II) bonded to O_2, and Fe(III) bonded to the superoxide radical. Every so often a molecule of oxyhaemoglobin undergoes decomposition and releases O_2^- (this is a gross oversimplification of the actual mechanism by which O_2^- is released but it is sufficient for our purpose):

$$haem\!-\!Fe^{2+}\!-\!O_2 \rightarrow O_2^- + haem\!-\!Fe^{3+}$$

The product with Fe(III) present in the haem ring is unable to bind oxygen and is thus biologically inactive; it is known as *methaemoglobin.* It has been estimated that about three per cent of the haemoglobin present in human erythrocytes undergoes such oxidation every day, and so these cells are exposed to a constant flux of O_2^-. Mature mammalian erythrocytes contain no nuclei or mitochondria, nor can they synthesize proteins or membrane lipids. Since they have to survive for an average of 120 days in the circulation of humans, they must carefully protect themselves against O_2^- using CuZnSOD, catalase, glutathione peroxidase, and pentose phosphate pathway enzymes as described previously (and summarized in Fig. 3.15). A *methaemoglobin reductase* enzyme is additionally present to reduce Fe(III) haem to Fe(II) haem and reactivate the haemoglobin. Haemoglobin oxidation is speeded up by the presence of transition-metal ions, especially copper, and of nitrite ion, NO_2^-. The

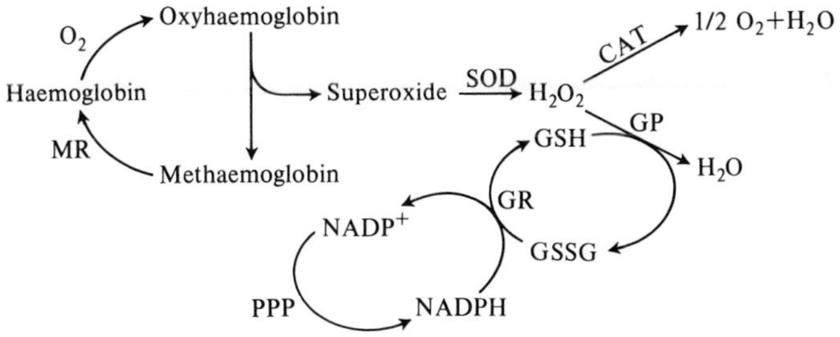

Fig. 3.15. Protection of erythrocytes against damage resulting from the oxidation of haemoglobin. MR, Methaemoglobin reductase; SOD, Cu–Zn superoxide dismutase; CAT, catalase; GP, glutathione peroxidase; GR, glutathione reductase; PPP, pentose phosphate pathway (first enzyme: glucose-6-phosphate dehydrogenase).

presence of large amounts of nitrate (NO_3^-) in the water supply of rural areas, due to excessive use of inorganic fertilisers, can cause problems in young bottle-fed babies: the NO_3^- in the water used to make up feeds is reduced by gut bacteria to NO_2^-, which is absorbed and causes excessive methaemoglobin formation that leads to problems in tissue oxygenation. Several mutant haemoglobins oxidize much more rapidly than normal, as do the isolated α- or β-chains that accumulate during thalassaemias (Chapter 2). Oxymyoglobin and the oxygenated form of the leghaemoglobin present in the root nodules of leguminous plants (Chapter 1) can also release O_2^- ions.

O_2^- formation has additionally been reported from 'negative air ionizers' (supposed to improve your health and prevent allergies), during exposure of the amino acid tryptophan (Table 3.11) to near-UV light (which may be of relevance in the eye, since lens proteins contain tryptophan) and from light-exposed *pheomelanin*. Melanin is a pigment formed by the oxidation and polymerization of the aromatic amino acid tyrosine,

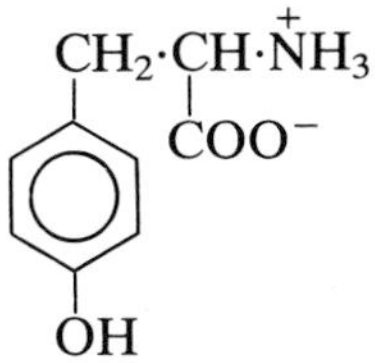

and the brown or black melanins (*eumelanins*) of human skin afford protection against the ultraviolet component of sunlight. The melanin polymer contains a number of unpaired electrons left over from the polymerization process and these can be detected by ESR. Hence one could regard melanin as a large free-radical. The movement of these

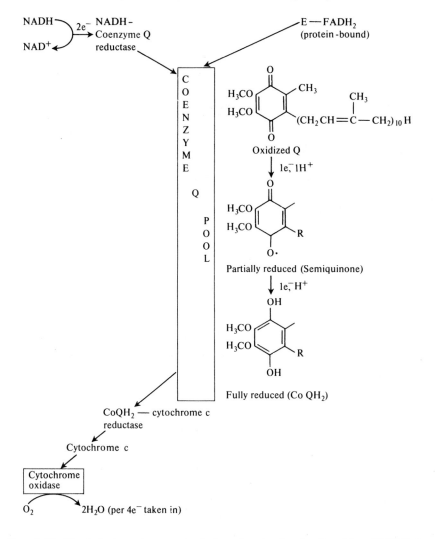

Fig. 3.16. The electron transport chain of animal mitochondria. NADH is oxidized to NAD^+ by a multienzyme complex known as 'NADH–coenzyme Q reductase' and the two electrons are eventually passed onto coenzyme Q. The reductase complex contains flavoproteins (FMN at active site) and non-haem-iron proteins. Coenzyme Q accepts the electrons to form both semiquinone and fully reduced forms. It can also accept electrons from various reduced flavoproteins generated by the Krebs cycle and β-oxidation of fatty acids. From Q the electrons pass through another multienzyme complex ('coenzyme Q cytochrome c reductase', which contains cytochromes b) onto cytochrome c. Cytochromes are haem proteins which accept electrons by allowing Fe(III) at the centre of the haem ring to be reduced to Fe(II), i.e. they can accept one electron at a time per molecule. Different cytochromes, designated by small letters, contain different

unpaired electrons between different energy levels helps to absorb the UV radiation. The slightly different, red-brown or yellow pigment found in the skin and hair of fair-skinned red-headed humans is *pheomelanin*. On exposure of this to light it is photodegraded and some O_2^- is formed. Eumelanin also absorbs oxygen upon illumination, but is not degraded.

Probably the most important sources of O_2^- *in vivo* in most aerobic cells are the electron transport chains of mitochondria and endoplasmic reticulum. The chloroplast electron transport chain is discussed in Chapter 5. The most important function of animal mitochondria is the oxidation of NADH and of $FADH_2$ produced during the Krebs cycle, β-oxidation of fatty acids, and other metabolic pathways. Oxidation is achieved by an electron transport chain located in the inner mitochondrial membrane and described in Fig. 3.16. The energy released by these oxidations is used to drive ATP synthesis. For every four electrons fed into the cytochrome oxidase complex (Fig. 3.16) a molecule of oxygen is reduced to two molecules of water. Cytochrome oxidase releases no detectable oxygen radicals into free solution, i.e. it has evolved so as to keep all the radical intermediates of oxygen-reduction firmly bound to the proteins. The exact mechanism of the reduction is not yet known, but scientists in Sweden and the USA have identified a number of partially-reduced oxygen intermediates tightly bound to the cytochrome oxidase complex. Some other components of the electron transport chain are less well-designed, however, and 'leak' a few electrons onto oxygen whilst passing the great bulk of them on to the next component in the chain. As would be expected from the chemistry of oxygen (Chapter 1), this leakage produces a univalent reduction to give O_2^-. The main sites of leakage seem to be the NADH–coenzyme Q reductase complex, and the reduced forms of coenzyme Q itself. Mitochondria isolated from several animal tissues have been shown to produce hydrogen peroxide *in vitro*, most, if not all of which arises by the dismutation of O_2^- by mitochondrial SOD activity. Mitochondria possess MnSOD in the matrix, and may well have CuZnSOD in the space between the inner and outer membranes, as explained previously. The rate of O_2^-, and hence

---

proteins and different haem groups. Finally, reduced cytochrome c is re-oxidized by the multienzyme complex, 'cytochrome c oxidase' which contains cytochrome a, cytochrome a_3 and copper. For every four electrons taken in by this complex, one oxygen molecule is fully reduced to two molecules of water. The enzyme dihydro-orotic acid dehydrogenase, which catalyses a step in pyrimidine synthesis, feeds electrons into the electron transport chain at several points in the region of CoQ. The ubisemiquinone radical can be detected in respiring mitochondria by observing its ESR signal.

hydrogen peroxide, production by mitochondria is increased at elevated oxygen concentrations. For example, in slices of rat lung exposed to air, about nine per cent of total oxygen uptake could be attributed to O_2^- formation, the rest being due to cytochrome oxidase activity. In an atmosphere containing 85 per cent oxygen, however, O_2^- formation accounted for 18 per cent of the total oxygen uptake.

Mitochondria from some trypanosomes and from plant tissues (e.g. spinach leaves, mung beans, potato tubers, and Jerusalem artichokes) have also been shown to produce hydrogen peroxide, probably via O_2^- produced from CoQ and from NADH–coenzyme Q reductase, as in animal mitochondria. The electron transport chains located in the plasma membranes of several aerobic bacteria, e.g. *E. coli* and *Paracoccus denitrificans* have been shown to produce O_2^-, as have rabbit spermatozoa (although its exact origin is unknown).

The endoplasmic reticulum of many animal and some plant tissues contains a special cytochrome known as *cytochrome P_{450}.* Several slightly different forms of P_{450} exist. This name was given because the reduced form of the cytochrome complexes with carbon monoxide to produce a species that absorbs light strongly at 450 nm. Cytochrome P_{450} is involved in the oxidation of a wide range of substrates at the expense of molecular oxygen. One atom of the oxygen enters the substrate and the other forms water, such a reaction being known as a *mono-oxygenase* or *mixed-function oxidase* reaction. The functioning of cytochrome P_{450} requires a reducing agent (RH_2), and the overall reaction catalysed can be represented by the following equation, in which AH is the substrate:

$$AH + O_2 + RH_2 \rightarrow A \cdot OH + R + H_2O.$$

Liver endoplasmic reticulum is especially rich in P_{450}, which metabolizes a large number of chemicals. Some of these can increase synthesis of one or more forms of the cytochrome when fed to animals. One such inducer is the barbiturate *phenobarbital,* hydroxylation of which increases its solubility and facilitates its excretion from the body. Other substrates include insecticides such as heptachlor and aldrin, hydrocarbons such as benzpyrene, and drugs such as phenacetin, amphetamine, and paracetamol. Usually the product of reaction with P_{450} is less toxic than the starting material, but this is not always the case: there is evidence that it is the hydroxylated products of paracetamol (Chapter 6) and of carcinogenic hydrocarbons such as benzpyrene (Chapter 8) that actually do the cellular damage that these compounds cause.

Often the initial product undergoes further reactions, e.g. *deamination* (removal of an amino group, —NH_2) in the case of amphetamine. Two hydroxylated intermediates may exist, i.e.

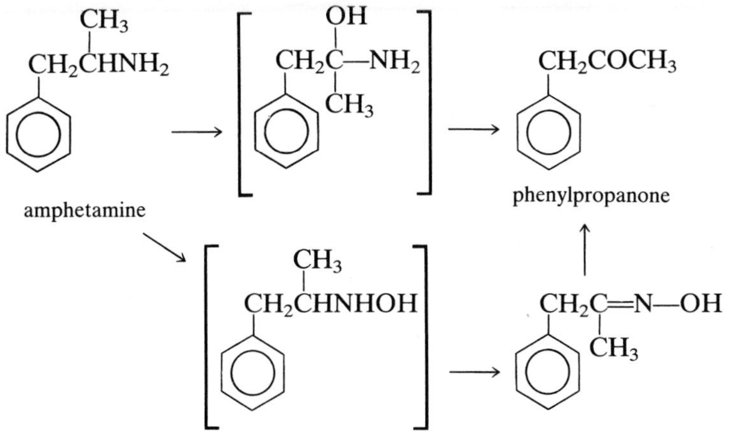

N-demethylation (removal of a methyl group, CH_3—) produces the aldehyde formaldehyde, HCHO. Metabolism of the carcinogen *dimethylnitrosamine* in this way produces methyl ions (CH_3^+) which can attack DNA

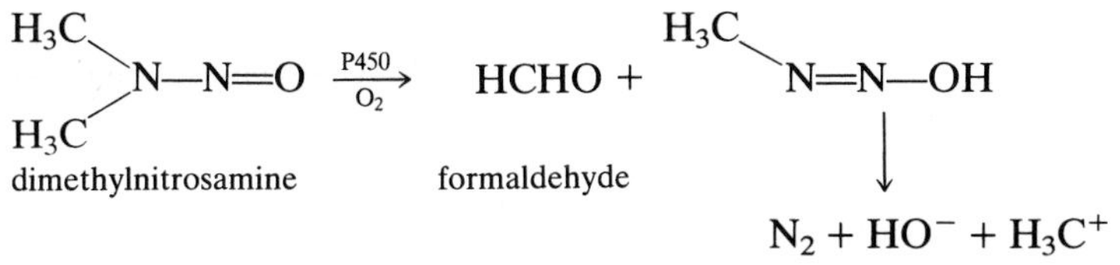

Removal of hydrocarbon groups attached to oxygen atoms (*O-dealkylation*) can also occur, e.g.

$$R\!-\!O\!-\!CH_3 \xrightarrow[O_2]{P_{450}} [R\!-\!O\!-\!CH_2\!-\!OH] \longrightarrow ROH + HCHO$$

In the liver the electrons required by the P_{450} system are donated by NADPH via a flavoprotein enzyme *NADPH-cytochrome P_{450} reductase*. Figure 3.17 shows a possible mechanism for substrate hydroxylation by P_{450}, but the nature of the actual hydroxylating species at the active site of this protein is not yet clear (see figure legend). Adrenal cortex mitochondria contain a cytochrome P_{450} which is involved in the hydroxylation of cholesterol to give the adrenal steroid hormones (e.g. aldosterone, hydrocortisone, and corticosterone) but the electrons it requires are donated by a non-haem-iron protein known as *adrenodoxin*.

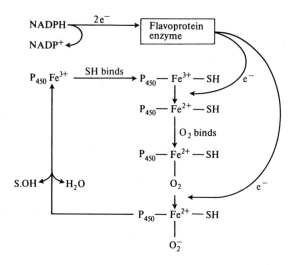

Fig. 3.17. Mechanism for substrate hydroxylation by cytochrome P_{450} in liver. SH represents the substrate molecule. The mechanism of P_{450} hydroxylation appears similar in other tissues or bacteria, but the source of reducing power used is different. The one shown is for liver endoplasmic reticulum. The mechanism by which the P_{450}—Fe(II)—O_2^- complex produces a hydroxylating species is not known. It is possible that there is a loss of water to give P_{450}—$[FeO]^{3+}$, in which the iron has a nominal valency of 5. It may be that the actual valency state of the iron is +4 (i.e. a ferryl species, FeO^{2+}) and the extra oxidizing capacity is located elsewhere on the molecule, as in compound I of horseradish peroxidase.

A flavoprotein enzyme transfers electrons from NADPH to adrenodoxin. Cytochrome P_{450} is found in some bacteria, e.g. in *Pseudomonas putida* it serves to hydroxylate camphor. Here electrons are supplied by the non-haem-iron protein *putidaredoxin,* which is kept reduced at the expense of NADH by a flavoprotein enzyme. The hydroxylated camphor can then be metabolized by the cells to provide energy.

The reduced forms of adrenodoxin, putidaredoxin, and NADPH-cytochrome P_{450} reductase from liver, are capable of reducing oxygen to give O_2^-. In addition, there is some evidence that the oxygenated intermediates of cytochrome P_{450} itself (Fig. 3.17) can decompose in a minor side-reaction and release O_2^-. Indeed, isolated microsomal fractions from various tissues have been shown to produce hydrogen peroxide rapidly *in vitro* in the presence of NADPH, presumably via dismutation of O_2^-. The rate of hydrogen peroxide production is increased at elevated oxygen concentrations. Increasing the amount of P_{450} and its reductase by pretreatment of animals with phenobarbital increases the rates of H_2O_2-production by liver microsomes later isolated from the animals.

In perfused rat liver, however, the basal rate of GSSG release as an index of H_2O_2-production (Section 3.1.1) is much smaller than expected from the rates observed with microsomes *in vitro,* and it does not increase on pretreatment of the animals with phenobarbital. Hence H_2O_2-formation and, by implication, O_2^- formation by the endoplasmic reticulum *in vivo* does not occur as rapidly as would be expected from experiments on microsomes. Perhaps during the fragmentation and membrane vesicle formation that occurs on cell disruption to produce microsomes (Chapter 2) the arrangement of the components of the P_{450} system within the membrane is altered so that electrons escape more easily to oxygen. This should be borne in mind in experiments with isolated microsomal fractions. The oxygen concentration adjacent to the endoplasmic reticulum *in vivo* must also be much lower than that seen by microsomes incubated *in vitro.*

Liver endoplasmic reticulum also contains an enzyme system, *desaturase,* that introduces carbon–carbon double bonds into fatty acids. The system requires oxygen, NADH, or NADPH, and a special cytochrome known as cytochrome b_5. Electrons from NAD(P)H are transferred to b_5 by a flavoprotein enzyme, and reduced b_5 then donates electrons to the desaturase enzyme. Both cytochrome b_5 and the flavoprotein can leak electrons onto oxygen to make O_2^-, and this might be an additional source of O_2^- *in vivo.*

The membrane surrounding the cell nucleus also contains an electron transport chain, of unknown function, that can 'leak' electrons to give O_2^-, at a rate increasing with oxygen concentration, in the presence of NADH or NADPH. It resembles the microsomal electron transport system and may be of especial importance *in vivo* because the oxygen radicals that it generates are close to the cell's DNA.

Contemplation of all the information given above shows that production of O_2^- would be very likely to occur within all aerobic cells, but it is very difficult to assess the actual amount produced. In extracts of the bacterium *Streptococcus faecalis* incubated with NADH, 17 per cent of the oxygen consumed was used to form O_2^-, although this figure obviously need not apply to intact cells. It is probably an overestimate, since disruption of cellular structures probably allows electrons to escape to oxygen more easily, as in the case of liver microsomes. Oxygen uptake associated with formation of O_2^- and hydrogen peroxide can be easily demonstrated in whole leaves or illuminated chloroplasts (Chapter 5) and the study of low-level chemiluminescence (Chapter 2) gives evidence for oxygen radical reactions in bacteria, animal organs both isolated and *in situ,* whole cells of the soil amoeba *Acanthamoeba castellanii,* and phagocytic cells (Chapter 7). Pretreatment of guinea-pigs with diethyldithiocarbamate to inactivate

CuZnSOD (Section 3.1.3) decreases the rate at which liver catalase activity is inhibited by aminotriazole in these animals. Inhibition of catalase by aminotriazole requires hydrogen peroxide, so it may be concluded that inhibition of CuZnSOD had decreased H_2O_2-production and therefore that this enzyme must have been acting on O_2^- *in vivo*.

Thus our answer to the question posed at the head of this section would be "Yes, but we don't know how much!".

3.3.2. Is superoxide a damaging species?

Sources of the O_2^- radical, such as photochemical or enzymic systems, have been observed to inactivate enzymes, cause erythrocyte haemolysis, kill bacteria, degrade DNA, and destroy animal cells in culture. Table 3.12 summarizes some of these effects. In organic solvents O_2^- is a powerful nucleophile and base (Chapter 2). The interior of biological membranes is hydrophobic, being similar in chemical nature and viscosity to a 'light oil' (Chapter 4) and O_2^- produced in this environment could be extremely damaging, e.g. destroying phospholipids by a nucleophilic attack upon the carbonyl groups of the ester bonds linking the fatty acids to the glycerol 'backbone' of the phospholipids. Much of the O_2^- generated within cells comes from membrane-bound systems, and it is certainly possible that some of it is formed in the membrane interior, especially as oxygen is more soluble in organic solvents than it is in water (Chapter 1).

However, all the effects listed in Table 3.12 were produced by O_2^- in aqueous solution, in which O_2^- mainly acts as a reducing agent and undergoes the dismutation reaction to form hydrogen peroxide (Chapter 2). It seems unlikely that O_2^- alone could do the damage, and in many cases protection is seen not only with SOD but also with catalase (Table 3.12). It was therefore suggested that O_2^- and hydrogen peroxide can interact to form the highly reactive product hydroxyl radical according to the overall equation

$$H_2O_2 + O_2^- \rightarrow O_2 + OH^- + OH^\cdot$$

Indeed, scavengers of hydroxyl radical such as mannitol, formate, and thiourea (Chapter 2) can often protect against damage induced by O_2^--generating systems (Table 3.12). Formation of $OH^\cdot$ in a wide range of systems generating O_2^- has been detected by its ability to hydroxylate aromatic compounds, degrade tryptophan, give a characteristic ESR signal with spin-traps, oxidize methional into ethene and dimethylsulphoxide into methane, and decarboxylate benzoic acid (see Chapter 2 for a detailed discussion of these methods).

The above reaction was first postulated by F. Haber and J. Weiss in

1934 and has become known as the *Haber–Weiss reaction*. Unfortunately the second-order rate constant for the reaction in aqueous solution has since been shown to be virtually zero, and it certainly could not occur at the low steady-state concentrations of O_2^- and hydrogen peroxide present *in vivo*. Several scientists, including J. M. McCord in the USA, and the authors, have shown that the OH˙ formation can be accounted for if the Haber–Weiss reaction is catalysed by traces of transition-metal ions. Since iron salts are available *in vivo* (Section 2.4.2), a possible mechanism can be written as follows, although recent work by the authors has shown it to be a gross oversimplification:

$$Fe^{3+} + O_2^- \rightarrow Fe^{2+} + O_2 \qquad (O_2^- \text{ reducing the iron salt})$$

$$Fe^{2+} + H_2O_2 \rightarrow Fe^{3+} + OH˙ + OH^- \qquad \text{(Fenton reaction)}$$

Net: $O_2^- + H_2O_2 \xrightarrow[\text{catalyst}]{\text{Fe-salt}} O_2 + OH˙ + OH^-$

(sometimes called an *iron-catalysed Haber–Weiss reaction*)

The authors' laboratory showed that the iron chelator desferrioxamine binds Fe(III) tightly so that it cannot be reduced by O_2^-, and hence it inhibits OH˙ formation. Desferrioxamine is virtually specific for the chelation of Fe^{3+} ions. (It can bind aluminium (Al^{3+}) ions and indeed it has been used to decrease aluminium toxicity in patients undergoing prolonged blood dialysis. Al^{3+} does not catalyse OH˙ formation, however.) The less specific chelating agents DETAPAC and phenanthroline (Chapter 2) also inhibit the iron-catalysed Haber–Weiss reaction, whereas EDTA does not, since Fe(III)–EDTA chelates are reduced by O_2^- quite quickly. Use of the above chelating agents has shown that an iron-catalysed Haber–Weiss reaction accounts for the OH˙ production in a number of systems, including the degradation of DNA and of synovial fluid by a xanthine–xanthine oxidase system. In addition, it has been shown that iron-saturated lactoferrin (i.e. with two moles of Fe^{3+} per mole of protein) can cause formation of hydroxyl radicals in O_2^--generating systems. It is possible that transferrin can act similarly, although there is considerable debate on this point. Certainly, partially iron-loaded proteins are not effective catalysts of the Haber–Weiss reaction, and these are the major species present normally (Chapter 2) except, of course, in iron-overload conditions. A further suggestion is that denatured haemoglobin may catalyse OH˙ formation. Such denatured haemoglobin might be formed during the decomposition of unstable haemoglobins.

Hydroxyl radicals, once generated, will react with the molecules in their immediate surroundings. Superoxide and hydrogen peroxide,

Table 3.12. Some deleterious effects of systems generating the superoxide radical

Source of O_2^-	System studied	Damage	Comments
Heart-muscle submitochondrial particles	activity of NADH-CoQ reductase	activity lost	damage prevented by SOD. The O_2^- generated by the complex inactivates it unless SOD is present. Catalase not protective.
Illuminated FMN	Bacteria	loss of viability	protection by SOD.
Xanthine + xanthine oxidase	human synovial fluid	degradation—loss of viscosity and lubricating power	SOD, catalase and iron chelators protect
Xanthine + xanthine oxidase	bacteriophage R17	inactivation	SOD protects partially.
Illuminated FMN	ribonuclease	loss of activity	SOD protects partially.
Illuminated FMN	calf myoblast cells	growth abnormality, some cell death	SOD protects partially.
Hypoxanthine + xanthine oxidase	rat brain membrane, Na^+, K^+ ATPase	inactivation	SOD protects partially.
Acetaldehyde + xanthine oxidase	erythrocyte membranes	lysis	SOD protects.
Acetaldehyde + xanthine oxidase	arachidonic acid	oxidation	both SOD and catalase protect.
Hypoxanthine + xanthine oxidase	DNA	degradation, single-strand breaks, attack on sugar moiety	SOD, OH· scavengers and catalase protect. Iron salts needed.

Autoxidation of dihydroxyfumarate	rat thymocytes	inhibition of Na^+-dependent amino acid uptake	SOD protects, but not catalase.
Hypoxanthine + xanthine oxidase	cheek pouch of living hamster (perfused with O_2^--generating system)	Increased permeability of blood vessels, leakage of contents	SOD protects.
Autoxidation of dialuric acid	E. coli	loss of viability	both SOD and catalase protect.
Acetaldehyde + xanthine oxidase	Staphylococcus aureus	loss of viability	SOD, OH˙ scavengers and catalase protect. Traces of iron chelates needed for killing.
Xanthine + xanthine oxidase	rat lung in vivo (instilled into lungs)	acute lung inury, oedema.	SOD protects, but not catalase.
Xanthine + xanthine oxidase	rat heart ornithine decarboxylase	inactivation	SOD protects, also mannitol (OH˙ scavenger).
Xanthine + xanthine oxidase	rat heart mitochondria	lowered P/O ratios, and lower respiratory control	SOD and catalase protect.
Xanthine + xanthine oxidase	rat heart or liver mitochondria	inhibition of net Ca^{2+} uptake	protected by SOD or mannitol (OH˙ scavenger).
Xanthine + xanthine oxidase	dog heart sarcoplasmic reticulum	decreased Ca^{2+} uptake	some protection by SOD and mannitol.

however, are much less reactive and can diffuse away from their sites of formation, leading to OH˙ generation in different parts of the cell whenever they meet a 'spare' transition-metal ion. In a sense, therefore, they can be damaging *because* of their poor reactivity.

Identification of OH˙ as the reactive oxidizing radical species produced in the Fenton reaction was at first fiercely contested, although it is now generally accepted. For example, formation of a *ferryl radical* in which the iron has a valency of +4 was suggested to occur instead

$$Fe^{2+} + H_2O_2 \rightarrow FeOH^{3+} \ (or\ FeO^{2+}) + OH^-$$

Ferryl radical may well exist at the active site of peroxidase compounds I and II (Section 3.1.1) and possibly in cytochrome P_{450} (legend to Fig. 3.17) and it clearly has extensive substrate-oxidizing and -hydroxylating properties. It is not therefore surprising that this controversy has spilled over into discussions of the iron-catalysed Haber–Weiss reaction, in which 'OH'' is produced by the Fenton reaction. However, if the reactive species produced in O_2^--generating systems is not OH˙, then it must still attack spin traps such as DMPO to produce the correct 'OH'' signal, react with scavengers of OH˙ with the same rate-constants as does OH˙ itself and require both O_2^- and hydrogen peroxide for its formation. Ferryl radicals have not yet been demonstrated to do any of these things. This is not to say, however, that OH˙ is the *only* oxidizing radical found in systems containing O_2^-, hydrogen peroxide, and iron salts. This point is well illustrated by studies on the initiation of lipid peroxidation, discussed in Chapter 4.

In the iron-catalysed Haber–Weiss reaction, the only role apparent for O_2^- is to reduce Fe(III) to Fe(II), which is consistent with its known chemistry in aqueous solution. Several scientists have argued that this could not happen *in vivo* because the concentration of other biological reducing agents, such as GSH, NADH, NADPH, or ascorbate, would be much greater than that of O_2^-. However, work in the authors' laboratory has shown that GSH or NAD(P)H do not prevent O_2^--dependent formation of OH˙ *in vitro*. Ascorbate at physiological concentrations can partially replace O_2^- in reducing Fe(III), but it is rapidly oxidized by O_2^- and by OH˙ so that the OH˙ production eventually becomes O_2^--dependent. Figure 3.18 shows an experiment that illustrates this point. What happens *in vivo* will therefore depend on the ascorbate concentration.

Mixtures of Cu^{2+} salts and hydrogen peroxide in the presence of a reducing agent also produce free hydroxyl radicals. In this case, O_2^- is a more effective reducing agent than is ascorbate. The binding of Cu^{2+} ions to its physiological ligands such as histidine or albumin does not completely prevent its reduction by O_2^- (Table 3.13) and it is likely that

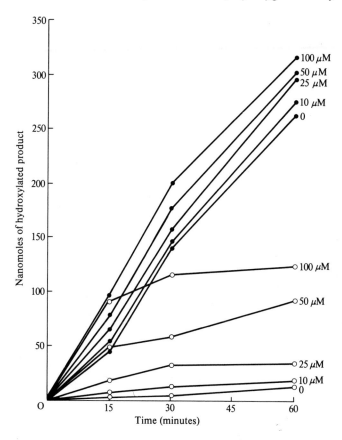

Fig. 3.18. Effect of ascorbic acid on hydroxyl radical production by a mixture of hypoxanthine and xanthine oxidase. OH· production by a mixture of hypoxanthine and xanthine oxidase is completely inhibited by SOD, catalase, or the iron chelator desferrioxamine. Addition of ascorbate to a hypoxanthine–xanthine oxidase system at concentrations below $50 \, \mu mol \, l^{-1}$ increases OH· production slightly but it is still largely inhibited by SOD, indicating that O_2^- is still the major reductant of Fe(III). At $50 \, \mu mol \, l^{-1}$ ascorbate, SOD only inhibited OH· formation after 15 minutes had elapsed. Presumably during this period ascorbate was reducing Fe(III) itself until its concentration had fallen so low that O_2^- once more became the major reductant. At $100 \, \mu mol \, l^{-1}$ ascorbate, SOD only inhibited after 30 minutes. Ascorbate concentrations in body fluids are usually $50 \, \mu mol \, l^{-1}$ or less (very much less in many diseases). On the other hand, lung cells contain millimolar concentrations of internal ascorbate (Section 3.2.1) and so the action of O_2^- in reducing Fe(III) would be difficult to envisage unless it happens at a specific site in the cell. Closed circles, no SOD added; open circles, SOD present. Data from Rowley and Halliwell (1983) *Clin. Sci.* **64,** 649.

Table 3.13. Simple compounds that react with superoxide

Compound	Comments
Fe^{3+}—EDTA	both Fe^{3+}—EDTA and Fe^{2+}—EDTA react with O_2^- probably by the reactions: $$Fe^{3+}\text{—EDTA} + O_2^- \rightarrow O_2 + Fe^{2+}\text{—EDTA}$$ $$Fe^{2+}\text{—EDTA} + O_2^- \rightarrow [O_2^-\text{—}Fe^{2+}\text{—EDTA}] \text{ complex.}$$ The reaction mechanism is complicated by the generation of OH˙ from reaction of Fe^{2+}—EDTA with H_2O_2. Probably overall not a significant catalyst of O_2^- dismutation. 'Unchelated' iron salts react more slowly with O_2^- (exact rate-constant not known).
Tetrakis-(4-N-methyl) pyridyl porphine-Fe^{3+}	catalyses O_2^- dismutation with about 2–3% of the rate of the SOD enzyme. The manganese and cobalt complexes are also effective but the copper complex is not. 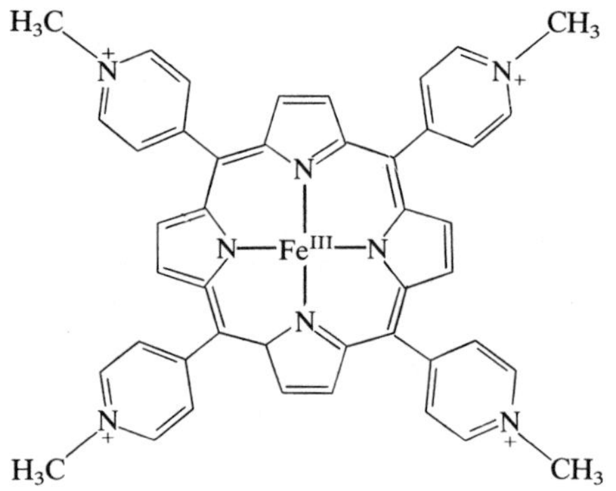

Mn^{2+}-polyphosphate Mn^{2+}-lactate Mn^{2+}-succinate Mn^{2+}-malate	slowly catalyse O_2^- dismutation but Mn(II) in the absence of a complexing agent is a poor catalyst. EDTA largely abolishes catalysis by Mn(II) (compare iron, above). May be an important catalyst of O_2^- dismutation in some *Lactobacillaceae* (see text). The most active complex, Mn-lactate, has about one-sixty-fifth the rate constant for O_2^- that MnSOD has.
Copper complexes	catalyse O_2^- dismutation more rapidly at low pH than equimolar amounts of CuZnSOD. Much less effective when the Cu is bound to proteins such as albumin. Some Cu–amino acid complexes (e.g. with lysine or with histidine) are still quite good catalysts at pH 7.4 as is a complex of Cu with the anti-inflammatory drug *indomethacin*. Many simple copper complexes are unstable, however, and problems can be caused due to the O_2^--scavenging ability of free Cu(II) ions present in the solution. EDTA abolishes reaction of O_2^- with Cu salts but has no effect on CuZnSOD.

hydroxyl radicals are still formed and immediately attack the binding molecule (Chapter 2).

Inspection of Table 3.12 shows that the iron-dependent or copper-dependent Haber–Weiss reactions do not explain everything, as there are several damaging effects prevented by SOD but not by catalase. A further illustration of this comes from the organism *Streptococcus sanguis*, whose growth appears to be independent of the availability of iron. It contains no haem compounds and thus is lacking in catalase and peroxidase activities. It is damaged by exposure to a O_2^--generating system but damage is not prevented by $OH^{\cdot}$ scavengers.

Dismutation of O_2^- catalysed by the SOD enzymes produces hydrogen peroxide and oxygen in the ground state. An additional explanation that has been offered to explain the toxicity of O_2^--generating systems is that they produce singlet oxygen $^1\Delta g$, either in the non-enzymic dismutation reaction of O_2^- or in the metal-catalysed Haber–Weiss reaction (which would account for inhibition by both SOD and catalase). If singlet oxygen is produced in either of these reactions, however, the amount is very small and difficult to detect. Indeed, O_2^- itself is a quencher of singlet oxygen by electron transfer, i.e.

$$O_2^- + O_2^* \rightarrow O_2^- + O_2$$

Finally, it should be noted that the protonated form of O_2^-, $HO_2^{\cdot}$, is a more reactive radical than is O_2^- itself. Even though only small amounts of $HO_2^{\cdot}$ are present at physiological pH (Chapter 2) they still might exert some biological effects.

The answer to the question posed at the head of this section is therefore "Yes, O_2^- is damaging", as shown in Table 3.12. Much of this damage may be done by O_2^-dependent generation of more reactive species such as $OH^{\cdot}$, but this still means that removal of O_2^- by SOD would be a valuable protective mechanism.

3.3.3. Is the true function of superoxide dismutase that of removing superoxide radicals?

The specificity of SOD for reaction with O_2^- has frequently been used to probe for the involvement of this radical in biochemical systems (see table 3.12). Yet there have been several claims that SOD is not specific for reactions with O_2^-. An early suggestion that SOD could catalyse the conversion of singlet oxygen back to the ground state has not been substantiated. The SOD *proteins* can, of course, react with singlet oxygen because they often contain histidine, tryptophan, and methionine residues which react with this species (Chapter 2). For the same reason the SOD proteins react with hydroxyl radicals. The same is true of any other protein, however, including heat-denatured SOD, the apoenzyme (the protein with metals removed from its active sites), catalase, and serum albumin. Thus if a large quantity of SOD or other protein is added to a system producing OH$^\bullet$ or singlet oxygen then there will be scavenging. To prove an involvement of O_2^- one must keep the molar concentrations of SOD very low and carry out a control using heat-denatured enzymes, apoenzymes, or another protein, such as albumin, at the same concentration. A similar point applies to the use of catalase inhibition as a means of detecting the involvement of hydrogen peroxide in a reaction.

It should perhaps be mentioned here that inhibition by SOD must be interpreted with caution in systems containing quinones and semi-quinones. As we saw in Chapter 2, many semiquinones react reversibly with oxygen

$$\text{semiquinone} + O_2 \rightleftharpoons \text{quinone} + O_2^-$$

The equilibrium will tend to be dragged over to the right because of the non-enzymic dismutation of O_2^-. Addition of SOD, by removing O_2^- much more rapidly, can further decrease the concentration of semi-quinone present. As Winterbourn in New Zealand has shown, a reaction that is actually caused by the semiquinone might be mistakenly attributed to O_2^- as a result of the inhibition on addition of SOD.

Some diphenols can bind to the active site of CuZnSOD and partially reduce the copper to the Cu(I) form, becoming themselves simultaneous-ly oxidized to the semiquinone form. However, this is a reaction seen with many copper proteins. SOD does not *catalyse* the oxidation of diphenols but merely reacts stoichiometrically with them (i.e. only one mole of diphenol reacts per active site). Wardman and Fielden in England have studied the interaction of CuZnSOD with a range of radical species, including semiquinones and radicals derived from FMN, NAD, aromatic nitro compounds, and organic peroxy radicals (Chapter 4), but no catalysis of radical removal was evident in any case. On the basis of present evidence we are able to conclude that the SOD enzymes are specific for O_2^- as substrates and therefore that their use as 'probes' is valid if suitable controls are employed.

A number of simple metal complexes are able to react with O_2^- *in vitro* and in some cases to catalyse its dismutation. Table 3.13 describes some of these. It should perhaps be pointed out here that the use of 'indirect' assay systems for SOD activity (Table 3.9) to study these low-molecular-weight SOD 'mimics' should in general be avoided in favour of pulse radiolysis or stopped-flow methods using $K^+O_2^-$. We recommend this because many of the complexes interfere with O_2^--generating systems (e.g. Cu^{2+} ions inhibit xanthine oxidase and stimulate oxidation of adrenalin, sulphite, or pyrogallol) or with the detector molecules (e.g.

Fe^{2+}—EDTA reduces cytochrome c directly). The high dismutase activity of Cu^{2+}-complexes *in vitro* has led a few scientists to speculate that the biological role of SOD is not the dismutation of O_2^- but something else, perhaps just a metal-storage protein, and that the dismutase activity of the enzyme is merely a consequence of the fact that it happens to contain copper. If this were so, however, then one might expect some other copper proteins to dismutate O_2^-. At physiological pH, the rate constant for O_2^- dismutation by each active site of CuZnSOD is about 1.6×10^9 $M^- s^{-1}$. Some other copper proteins do react with O_2^- but the rate constants are much lower e.g.

cytochrome c oxidase, $k = 2 \times 10^7 M^{-1} s^{-1}$
caeruloplasmin, $k = 7 \times 10^5 M^{-1} s^{-1}$
galactose oxidase, $k = 3 \times 10^6 M^{-1} s^{-1}$.

These reactions with O_2^- are not catalytic; the O_2^- is merely reducing Cu^{2+} at the active site to Cu^+ and the reaction then stops. The ability of Cu^{2+} ions to react with O_2^- is decreased by the presence of proteins, such as albumin, whereas that of CuZnSOD is not. Indeed, this enzyme shows an arrangement of charged residues that seem designed to guide O_2^- into the active site (Section 3.1.3) and evidence for a similar 'electrostatic facilitation' of O_2^--binding has been presented for FeSOD and MnSOD enzymes. Finally, it must be noted that the latter two enzymes are far more effective in catalysing O_2^- dismutation than are equivalent amounts of free or complexed Mn or Fe salts under any conditions (Table 3.13).

3.3.4. Evidence bearing on the superoxide theory of oxygen toxicity

According to the superoxide theory of oxygen toxicity as originally formulated, the damaging effects of elevated O_2 are due to an increased production of O_2^- and radicals derived from it; and SOD is essential in allowing organisms to survive in the presence of any oxygen at all. The classical way of proving such an assertion would be to obtain a mutant strain of an aerobe that lacked SOD activity and could not grow in the presence of oxygen. Indeed, there was an early report from Fridovich's laboratory of a mutant strain of *E. coli* that could not make SOD at high growth temperatures. At moderate temperatures it grew normally under both aerobic and anaerobic conditions, but at high temperatures it refused to grow aerobically, although it still grew under anaerobic conditions. Unfortunately, its catalase and peroxidase activities were not examined, and the mutant has since been lost. More recently, oxygen-intolerant mutants of *E. coli* K12 have been isolated and examined more thoroughly. In one type of mutant, peroxidase and catalase were deficient, whereas, in a second type, peroxidase, catalase, and MnSOD activities were all decreased. Some mutant strains eventually regained the capacity for growth in the presence of oxygen; these revertants were of two types. In one group, some or all of the missing enzyme activities had returned. In the second group the enzyme defects were still present, but the cells no longer respired. As they did not take up oxygen, they presumably no longer produced O_2^- or hydrogen peroxide and thus did not need protective enzymes.

Apart from the cases just mentioned, SOD activity has been found to be present in practically all aerobes examined, even in species such as *Bacillus popilliae* or *Acholeplasma laidlawii* that contain no catalase or peroxidases. In the murine leprosy bacillus *Mycobacterium lepraemurium* SOD constitutes about ten per cent of the total cell protein. We have already seen that if the fungus *Dactylium dendroides* is prevented from synthesizing CuZnSOD by removing copper from the growth medium, it steps up synthesis of the manganese enzyme so as to maintain total cellular SOD activities unchanged.

The first question about the validity of the superoxide theory came from studies upon anaerobes, it being suggested that organisms living without oxygen would not make O_2^- and thus would not need a SOD. Hence the presence of a SOD activity in anaerobes might be taken to suggest that its true biological function is not that of removing O_2^-. Original reports from Fridovich's laboratory showed that the anaerobes they studied were indeed lacking in SOD activity. Since then, there have been several reports that a few anaerobes do contain SOD, the first coming from Professor Morris's laboratory in Wales. Table 3.14 summarizes some of these reports. It may be seen that when SOD is present, it is usually (not always!) the FeSOD in small amounts (compare the activities with those of aerobically grown *E. coli* in Table 3.14).

Does the presence of SOD in a few anaerobes mean that SOD does not function to remove O_2^-? Not necessarily, since, as we saw in Chapter 1, many so-called anaerobes can survive brief exposure to oxygen, albeit with growth inhibition. It is reasonable to suppose that SOD is present within them in order to permit survival during such exposures. Consistent with this interpretation, growth of *E. coli* under strictly anaerobic conditions for several generations causes loss of the MnSOD but FeSOD remains. On re-exposure to oxygen, the MnSOD is promptly re-synthesized. The FeSOD presumably allows survival whilst more SOD is made. Some scientists have found a correlation between oxygen tolerance and the SOD content of anaerobes, but others have not. This is perhaps not surprising, since the activities of catalase, peroxidase, and NADH oxidase enzymes (Section 3.1.1) will also be important. Even if SOD is absent, an NADH oxidase, by reducing oxygen to water, can remove oxygen from the immediate environment of the bacteria and permit

Table 3.14. Anaerobic bacteria that contain SOD activity

Anaerobe studied	Type of SOD	Activity (units per mg protein)
Chlorobium thiosulphatophilum	FeSOD	14
Chromatium	FeSOD	0.6
Desulphovibrio desulphuricans	FeSOD	0.6
Clostridium perfringens	FeSOD	15.6
Bacteroides distasonis	FeSOD	0.1–0.4 (most strains)
	FeSOD	3.2–3.9 (strain ATCC8503)
Actinomyces naeslundii	MnSOD	—
Propionibacterium shermanii	FeSOD (but produces MnSOD if grown under Fe-restricted conditions)	
Bacteroides fragilis	FeSOD	—

SOD activity can vary with the growth medium and with position in the cell cycle, so do not take the values too literally. For comparison, aerobically grown *E. coli* contained 44 units SOD per mg protein under the same assay conditions. Most anaerobes examined contain no SOD activity at all.

survival. Some anaerobes, such as *Bacteroides fragilis*, synthesize more SOD if oxygen is present.

The experiments with mutants described above suggest that SOD is important in allowing survival in oxygen, but they do not prove it to be essential because of simultaneous alterations in catalase and peroxidase activities. Another approach to finding evidence for the superoxide theory is *induction experiments*—exposure of organisms to elevated oxygen concentrations should cause them to form more O_2^- *in vivo* and this might perhaps lead to synthesis of more SOD if insufficient is present to cope with the increased O_2^- generation. Much evidence of this type has been accumulated. For example, elevated oxygen concentrations increase the total SOD activity in *E. coli B* cells due to increased synthesis of the MnSOD. *E. Coli B* cells grown under an atmosphere of 100 per cent oxygen were much more resistant to the toxic effects of high-pressure oxygen than were cells grown under air. By contrast, exposure of *Bacillus subtilis* to elevated oxygen concentrations increases its catalase activity but not that of SOD, and this organism is equally sensitive to high-pressure oxygen, whether grown previously under air or under 100 per cent oxygen. Exposure of anaerobically grown *E. coli K12* cells to oxygen causes induction of catalase and peroxidase activities, but much greater oxygen concentrations are required than those necessary to increase synthesis of MnSOD. This organism has two forms of catalase, one of which is always present and the other is only synthesized in the presence of oxygen. This effect may be due to formation of hydrogen peroxide *in vivo* since addition of hydrogen peroxide can induce catalase synthesis in *E. coli* cultures under either aerobic or anaerobic conditions.

It is possible to increase the SOD activity of *E. coli* cells by supplying them with a compound that increases intracellular O_2^--generation; such compounds include the antibiotic streptonigrin, the herbicide paraquat (Chapter 6), juglone, menadione, pyocyanine, methylene blue, and phenazine methosulphate. Strains of *E. coli* with elevated SOD activity are resistant to the toxic effects of these compounds and, *vice versa,* strains with elevated SOD due to treatment with these compounds are less sensitive to toxic oxygen effects. Many of these drugs induce H_2O_2-degrading enzymes as well, perhaps as a result of H_2O_2-generation by the increased SOD activity. These compounds appear to act by being reduced within the cell, followed by a reaction of the reduced forms with oxygen to make O_2^-. The toxicity of streptonigrin to *E. coli* is aggravated by the presence of iron salts and inhibited by desferrioxamine, consistent with it being mediated by an iron-catalysed Haber–Weiss reaction. Further discussion of the toxicity of paraquat can be found in Chapter 6.

The SOD activity present within an aerobe should depend not only on the oxygen concentration but also on what percentage of this oxygen is used to make O_2^-. For example, when *E. coli K12* is grown in continuous culture both its capacity for respiration and its SOD activity increase in parallel with growth rate, whereas peroxidase and catalase activities do not. A sudden rise in the availability of carbon source for the bacteria does not immediately result in an increased growth rate: there is a lag period before the rate increases, which correlates with the time taken for the cells to raise their SOD activity to a level characteristic of the greater rate of growth. Here the SOD level is probably responding to the respiratory activity of cells i.e. as they grow more quickly, more oxygen is used and more O_2^- is formed *in vivo*. Consistent with this interpretation is the observation that when *E. coli* is grown in an aerated rich growth medium containing glucose, the cells take up only small amounts of oxygen as they first use up the glucose, and release lactate and other organic acids into the medium. When the glucose is exhausted, the cells increase their respiratory capacity and oxidize other components of the medium, this increase being accompanied by a rise in SOD levels.

Induction experiments in which SOD activity was correlated with oxygen exposure have been carried out also in *Streptococcus faecalis,* vibrios, the macroaerophile *Campylobacter sputorum,* the nitrogen-fixing bacterium *Rhizobium japonicum,* and the yeast *Saccharomyces cerevisiae.* A mutant strain of the alga *Chlorella* with increased SOD activity was much more resistant to elevated oxygen concentrations and to streptonigrin than was the wild-type; and oxygen has also been observed to increase SOD activity in potato tubers and in the blue-green alga *Anabaena cylindrica.* The oxygen produced by the photosynthetic activity of symbiotic algae has been observed to increase SOD activity in their host, the sea anemone *Anthopleura elegantissima.* Anemones that have symbiotic algae have much greater SOD activities than those which do not.

As for animal tissues, if adult rats are placed in pure oxygen gas they rapidly develop symptoms of lung damage and usually die after 60–72 h. If, however, they are exposed to gradually increasing oxygen concentrations they can adapt to survive in 100 per cent oxygen and the gain of the ability to survive is strongly correlated with an increased content of SOD in the lungs. Catalase, glutathione reductase, and glutathione peroxidase activities are also increased in the lungs, as is the amount of glutathione present. If rats are pretreated with diethyldithiocarbamate, which inhibits CuZnSOD both *in vitro* and *in vivo,* the toxic effects of high O_2 tensions are enhanced; although care must be exercised in interpreting this observation since diethyldithiocarbamate inhibits several other enzymes as well. The decrease in SOD activity is followed by a loss of glutathione peroxidase activity in the lung. Such changes also occur in lung homogenates treated with diethyldithiocarbamate *in vitro* but the loss of glutathione peroxidase can be prevented by adding SOD. This indicates that *in vitro,* and presumably *in vivo* as well, the glutathione peroxidase is inactivated by O_2^- (or, more likely, by $OH^\cdot$ derived from it) as a consequence of the inhibition of SOD. It must be pointed out, however, that lung is a very complex tissue containing many different cell types. Damage caused by exposure to oxygen may well destroy some cell types and encourage others to proliferate, so that changes in enzyme activities as assayed in homogenates of whole lungs may simply be due to changes in cell populations, or might well underestimate very large enzyme changes taking place in only a few cell types. For example, sublethal oxygen exposure causes a proliferation in rat lungs of the cells known as 'granular pneumocytes' or 'type II cells'. Cultures of such cells *in vitro* showed increased SOD activity if they were isolated from rats previously exposed to elevated oxygen concentrations, but activities of glutathione peroxidase or glucose-6-phosphate dehydrogenase were not altered.

Newborn rats are very much more resistant to oxygen toxicity than are adults. Autor in the USA has presented evidence that this occurs because the SOD activity of their lungs is induced much more rapidly under high O_2 than it is in adults. The same phenomenon is seen in newborn rabbits. Catalase and glutathione peroxidase activities increase as well. If induction of these three enzymes is prevented (e.g. by injection of protein synthesis inhibitors), newborn rats become highly sensitive to oxygen. The increase in SOD activity in the lungs of newborn rats occurs both in the pulmonary macrophages (an increase in MnSOD) and in the extracellular fluid of the lung (an increase in CuZnSOD). Pulmonary macrophages are discussed further in Chapter 7. Newborn rats kept in pathogen-free environments have lower SOD activities in lung than normal, and they are more sensitive to oxygen-poisoning. Treatment of adult rats with bacterial endotoxin enhances their resistance to oxygen and induces an increase in SOD, catalase, and glutathione peroxidase activities in the lungs. This might be because bacteria infecting mammals are dealt with by polymorphonuclear leukocytes, which produce O_2^- and other radicals after engulfment of the bacteria (Chapter 7). The presence of bacterial endotoxin could be a 'signal' to the organism to produce more SOD in order to help protect the host tissues against O_2^- from the leukocytes. Simultaneous administration of diethyldithiocarbamate to the animals abolishes the protective effect of the endotoxin, even though it does not prevent the increase in catalase and glutathione peroxidase activities. Unlike adult rats, adult mice cannot be 'adapted' to survive in 100 per cent oxygen by pre-exposure to 60 or 80 per cent oxygen, and there is no increase in their lung SOD activity under such conditions. Similarly, injection of endotoxin does not increase SOD activity or confer protection against 100 per cent oxygen.

Other evidence for induction of SOD in animal tissues has been obtained. The compound 2,4-dinitrophenol is an uncoupler, i.e. it breaks the link between operation of the mitochondrial electron transport chain and the synthesis of ATP (Fig. 3.16) and allows electron transport to proceed at a rapid, uncontrolled rate. Treatment of rats with dinitrophenol caused a marked increase in MnSOD activity in their liver mitochondria, presumably due to increased electron flow and O_2^- generation in these organelles. Petkau *et al.* in Canada studied breast cancers induced by treatment of female rats with a carcinogen. SOD activity at the centre of the tumour, which had a restricted oxygen supply, was less than at the outer edges. If the rats were exposed to 85 per cent oxygen for 5 days, SOD activities in both tumour regions increased but the activity ratio of centre to margin remained about the same. Several other observations are consistent with an important role of SOD in animal tissues. For example, the SOD activity of erythocytes from a wide range of animals is virtually constant when expressed per unit haemoglobin, whereas catalase and glutathione peroxidase activities vary widely. As we saw in Chapter 1, the swim-bladders of certain fish have to tolerate high oxygen concentrations. The SOD activity of the swim bladder of the toadfish *Opsanus tau* was found to be higher than that of any other body tissue examined, but its catalase and glutathione peroxidase activities are fairly low.

The above results on the correlation between SOD activity and resistance to oxygen toxicity are fully consistent with the idea that SOD plays an important protective role but they do not, of course, *prove* rigourously that it does. Since catalase and peroxidase activities often increase as well, they also must be important. Indeed, it has been observed that both CuZnSOD and FeSOD are inactivated on prolonged exposure to hydrogen peroxide *in vitro* although MnSOD is not. This has been described previously as a means of distinguishing between the different enzymes. Inactivation can happen *in vivo* as well. For example, a decrease of SOD activity in the presence of H_2O_2-generating drugs has been observed in erythrocyte suspensions treated with aminotriazole to inhibit catalase. Feeding aminotriazole to rabbits inhibits catalase activity in the lens of the eye and causes loss of CuZnSOD activity. When wild-type strains of blue-green algae are brightly illuminated under 100 per cent oxygen (as can sometimes happen in their normal environment, e.g. in Israeli ponds) they are quickly killed. Strains of *Plectonema boryanum* sensitive to this 'photo-oxidative' death contained FeSOD as the major cellular SOD activity, but a resistant strain of this blue-green algae contained mainly MnSOD. Under photo-oxidative conditions, FeSOD activity is rapidly lost, presumably due to the action of hydrogen peroxide produced during photosynthesis, whereas MnSOD is not.

We have already seen that inhibition of SOD in rat lung can lead to inactivation of glutathione peroxidase, which further illustrates the interdependence of these enzymes. The superoxide radical can also partially inhibit catalase and horseradish peroxidase activities by reducing the Fe(III) iron in the haem ring of the active enzymes to the Fe(II) form or the Fe(II)–O_2 form, neither of which can degrade hydrogen peroxide. The survival time of rats exposed to pure oxygen was increased by about 70 per cent when liposomes containing both catalase and SOD were injected intravenously before and during O_2 exposure; liposomes containing either SOD or catalase alone had much less effect. It follows that SOD is not the *sole mechanism* that operates against oxygen toxicity; rather it is a very important mechanism that operates in conjunction with a wide range of others. Which is the most important will depend on the organism, the nature of the oxidative challenge, the tissue in question, and even the time of day which can affect, among other things, GSH concentrations. For example, rabbit bone marrow contains little SOD although it has substantial catalase activity. When cotton plants were exposed to elevated O_2, there was a significant rise in the glutathione reductase activity of leaves, but no change in SOD. Further, SOD may not be the only mechanism by which O_2^- is removed in aerobes. Ascorbate reacts rapidly with O_2^- and it may play an important role in this respect in isolated chloroplasts (Chapter 5) and in the lens of the human eye. Human lens is poor in SOD but rich in ascorbate, whereas rat lens has more SOD but less ascorbate. As a final example of the complexity of higher organisms, however, it must be pointed out that exposure of rats to oxygen concentrations *lower* than normal (10 per cent oxygen) for several days also increases the SOD activity of the lung; and rats so pretreated can then survive prolonged exposure to 100 per cent oxygen.

The biological role of SOD has been further questioned as a result of the discovery of aerobic organisms that contain none of this enzyme. A few *Leptospira* strains and three virulent strains of the aerobic gonococcus *Neisseria gonorrhoeae* were found to contain no SOD activity, although they were exceptionally rich in catalase. *Mycoplasma pneumoniae* contains O_2^--generating systems, but neither SOD- nor catalase-activities. However, if one accepts that the metal-ion-catalysed Haber–Weiss reaction is a major explanation of the toxicity of O_2^--generating systems, it is obvious that protection against OH˙ formation could be achieved either by efficient removal of O_2^- or by efficient removal of hydrogen peroxide. Both are not necessarily required. Hence one can find organisms with SOD but no catalase or peroxidase activities, such as *Bacillus popilliae*. The gonococcus, by contrast, has invested in exceptionally high catalase activities and can do without SOD (the inhibition of catalase by O_2^- mentioned above is only partial). Consistent with this argument, catalase-negative strains of *Listeria monocytogenes* have increased SOD compared with catalase-positive strains. Perhaps *Mycoplasma pneumoniae,* a pathogen of the human respiratory tract, contains no metal complexes that catalyse OH˙ formation. Addition of *M. pneumoniae* to isolated human cells causes membrane damage, an effect prevented by simultaneous addition of SOD. Cells taken from trisomy 21 patients, which contain elevated SOD concentrations (Chapter 8), show much less damage by *M. pneumoniae*. Given that no protection mechanism is 100 per cent efficient *in vivo* and that FeSOD and CuZn-SOD are inactivated by hydrogen peroxide, most organisms have probably evolved to contain both SOD and H_2O_2-removing systems to make doubly sure that OH˙ is not generated to an unacceptable extent under normal conditions.

Several aerotolerant strains of *Lactobacillaceae* contain SOD activity, but a few do not, such as *Lactobacillus plantarum*. This organism accumulates manganese ions from its growth medium to an internal concentration of 25 mmoles per litre or more! If accumulation is prevented by removing the metal ions from the medium the organism will not grow in the presence of oxygen. Since simple complexes of manganese salts with biological components slowly catalyse O_2^- dismutation (Table 3.13), Fridovich has suggested that these complexes function to remove O_2^- *in vivo*. Strains without high internal Mn or SOD are unable to survive in air, consistent with this argument. *L. plantarum* also possesses an H_2O_2-degrading enzyme that contains manganese ions.

The existence of a few aerobic organisms without SOD clearly means that this enzyme is not essential to all aerobic life. The massive accumulation of evidence reviewed above, however, shows that SOD is a major protector against O_2^- toxicity and, if it is not present, something else must take over its job.

3.4. Oxygen radicals and radiation damage

It has been known for a long time that the damaging effects of ionizing radiation on cells are aggravated by the presence of oxygen (Chapter 1). Much of the initial damage done is due to formation of the hydroxyl radical, $OH^{\cdot}$, which can react with other cellular components to produce organic radicals (Chapter 2).

Often these organic radicals can be 'repaired' by reaction with ascorbic acid or glutathione. If $R^{\cdot}$ is used to denote them, the 'repair' can be represented by the equations

$$R^{\cdot} + GSH \rightarrow RH + GS^{\cdot}$$

$$2GS^{\cdot} \rightarrow GSSG$$

$$R^{\cdot} + \text{ascorbate} \rightarrow RH + \text{semidehydroascorbate}$$

We have already seen that mechanisms exist *in vivo* for removing GSSG and semidehydroascorbate. However, the presence of oxygen may 'fix' the damage by forming other radicals that cannot be repaired, e.g.

$$R^{\cdot} + O_2 \rightarrow RO_2^{\cdot} \text{ (peroxy radical)}$$

Strains of bacteria or human cell lines that cannot synthesize GSH are equally sensitive to ionizing radiation under both aerobic and anaerobic conditions; they do not show the normal protection seen when oxygen is removed. Similarly, treatment of human lymphoid cells with *buthionine sulphoximine,* an inhibitor of GSH synthesis, increases their sensitivity to irradiation. Indeed GSH, its precursors and other thiol compounds such as cysteamine, have been used as radioprotectors. Both GSH and ascorbate can scavenge the hydroxyl radical directly although, at least for GSH, the 'repair' reaction seems to be more important *in vivo*.

In the presence of oxygen, the hydrated electrons formed by ionizing radiation can produce O_2^- (Chapter 2) which, if metal ions are present, can give more $OH^{\cdot}$ by a catalysed Haber–Weiss reaction, and thus increase the damage. Consistent with this, the presence of Fe(III) salts increased the damaging effects of ionizing radiation upon macrophages. Addition of SOD to the growth medium could partially protect *Acholeplasma laidlawii* or *E. coli* cells against damage by ionizing radiation, an effect also seen with several animal cell lines in culture. Treatment of Chinese hamster cells or human lymphocytes with diethyldithiocarbamate (DIECA) increased their sensitivity to radiation, consistent with a protective role of CuZnSOD *in vivo*. In such experiments, it is essential to remove the inhibitor completely before irradiating because, being a thiol compound, DIECA is an efficient radioprotector! Several radioresistant bacterial strains such as *Arthrobacter radiotolerans* and *Micrococcus radiodurans* contain exceptionally high activities of SOD. Canadian scientists have shown that injection of CuZnSOD into the bloodstream of mice reduces the lethal effects of X-irradiation, whereas injection of inactivated enzyme does not. The effects depend very much on the time and dose of SOD administered, and on the intensity of radiation used; but SOD appears to be particularly protective towards the bone marrow. The mechanism of this radio-protective effect of SOD remains to be established.

3.5. Further reading

Ames, B. N., *et al.* (1981). Uric acid provides an antioxidant defence in humans against oxidant- and radical-caused aging and cancer: a hypothesis. *Proc. nat. Acad. Sci. USA* **78**, 6858.

Archibald, F. S., and Fridovich, I. (1982). The scavenging of O_2^- by manganous complexes in vitro. *Arch. Biochem. Biophys.* **214**, 452.

Autor, A. P. (ed.) (1982). *Pathology of oxygen.* Academic Press, New York.

Biaglow, J. E. *et al.* (1983). The role of thiols in cellular response to radiation and drugs. *Radiat. Res.* **95**, 437.

Brigelius, R., *et al.* (1975). Superoxide dismutase activity of Cu(tyr)$_2$ and Cu, Co erythrocuprein. *Hoppe–Seyler's Z. physiolog. Chem.* **356**, 739.

Brigelius, R. *et al.* (1983). Identification and quantitation of glutathione in hepatic protein mixed disulphides and its relationship to glutathione disulphide. *Biochem. Pharmacol.* **32**, 2529.

Britton, L., and Fridovich, I. (1977). Intracellular location of the superoxide dismutases of *E. coli*—a re-evaluation. *J. Bact.* **131**, 815.

Brot, N. and Weissbach, H. (1983). Biochemistry and physiological role of methionine sulphoxide residues in proteins. *Archs Biochem. Biophys.* **223**, 271.

Chance, B., Sies, H., and Boveris, A. (1979). Hydroperoxide metabolism in mammalian organs. *Physiol. Rev.* **59**, 527.

Chedekel, M. R., *et al.* (1978). Photodestruction of pheomelanin: role of O$_2$. *Proc. natn. Acad. Sci. USA* **75**, 5395.

Clark, I. A., and Hunt, N. A. (1983). Evidence for reactive oxygen intermediates causing hemolysis and parasite death in malaria. *Infection and Immunity,* **39**, 1.

Cudd, A., and Fridovich, I. (1982). Electrostatic interactions in the reaction mechanism of bovine erythrocyte superoxide dismutase. *J. biol. Chem.* **257**, 11443.

De Mello, M. P., *et al.* (1980). Excited indole-3-aldehyde from the peroxidase-catalysed aerobic oxidation of indole-3-acetic acid. Reaction with an energy transfer to tRNA. *Biochem.* **19**, 5270.

Diplock, A. T. (1981). Metabolic and functional defects in selenium deficiency. *Philos. Trans. R. Soc.* **B294,** 105.

Gilbert, H. F. (1982). Biological disulphides: the third messenger? *J. biol. Chem.* **257,** 12086.

Halliwell, B. (1981). Free radicals, oxygen toxicity and aging. In *Age pigments,* (ed. R. S. Sohal) p. 1. Elsevier/North Holland, Amsterdam.

Halliwell, B. (1983). Superoxide dismutase and the superoxide theory of oxygen toxicity. A critical appraisal. In *Copper proteins,* Vol. II, (ed. R. Lontie) p. p. 63. CRC Press, Boca Raton, Florida, USA.

Harris, J. I., *et al.* (1980). Structural comparisons of superoxide dismutases. *Eur. J. Biochem.* **106,** 297.

Hassan, H. M., Dougherty, H., and Fridovich, I. (1980). Inhibitors of superoxide dismutases; a cautionary tale. *Arch. Biochem. Biophys.* **199,** 349.

Kirby, T. W., and Fridovich, I. (1982). A picomolar spectrophotometric assay for superoxide dismutase. *Analyt. Biochem.* **127,** 435.

Kittridge, K. J. and Willson, R. L. (1984). Uric acid substantially enhances the free radical-induced inactivation of alcohol dehydrogenase. *FEBS Lett* **170,** 162.

Kosower, N. S., and Kosower, E. M. (1978). The glutathione status of cells. *Int. Rev. Cytology* **54,** 109.

Marklund, S., Beckman, G., and Stigbrand, T. (1976). A comparison between the common type and rare genetic variant of human cupro-zinc superoxide dismutase. *Eur. J. Biochem.* **65,** 415.

Martin, J. R., and Fridovich, I. (1981). Evidence for a natural gene transfer from the ponyfish to its bioluminescent bacterial symbiont. *Photobacter leiognathi. J. biol. Chem.* **256,** 6080.

Meister, A. (1983). Selective modification of glutathione metabolism. *Science* **220,** 472.

Oberley, L. W. (ed.) (1982). *Superoxide dismutase,* Vols. I and II. CRC Press, Florida.

Paine, A. J. (1978). Excited states of O_2 in biology; their possible involvement in cytochrome P_{450}-linked oxidations as well as in the induction of the P_{450} system by many diverse compounds. *Biochem. Pharmacol.* **27,** 1805.

Pasternack, R. F., Banth, A., Pasternack, J. M., and Johnson, C. S. (1981). Catalysis of the dismutation of O_2^- by metalloporphyrins. *J. inorg. Biochem.* **15,** 261.

Rao, K. K., and Cammack, R. (1981). Evolution of ferredoxin and superoxide dismutases in micro-organisms. In *Molecular and cellular aspects of microbial evolution.* (eds Carlile *et al.*) p. 175. Cambridge University Press, England.

Reiter, B. (1979). The lactoperoxidase–thiocyanate–H_2O_2 antibacterium system. In *Oxygen free radicals and tissue damage.* p. 285. CIBA Foundation Symposium 65 (new series). Excerpta Medica, Amsterdam.

Rigo A., *et al.* (1979). Nuclear magnetic relaxation of ^{19}F as a novel assay method of superoxide dismutase. *J. biol. Chem.* **254,** 1759.

Seib, P. A., and Tolbert, B. M. (eds) (1982). *Ascorbic acid: chemistry, metabolism and uses.* Advances in Chemistry, Series 200, American Chemical Society, Washington DC, USA.

Sinet, P. M., Heikkila, R. E., and Cohen, G. (1980). Hydrogen peroxide production by rat brain *in vivo. J. Neurochem.* **34,** 1421.

Shlafer, M., Kane, P. F., and Kirsh, M. M. (1982). Superoxide dismutase plus catalase enhances the efficiency of hypothermic cardioplegia to protect the globally ischaemic, reperfused heart. *J. thorac. cardiovasc. Surg.* **83,** 830.

Singh, A., and Singh, H. (1982). Time-scale and nature of radiation biological damage. Approaches to radiation protection and post-irradiation therapy. *Prog. Biophys. mol. Biol.* **39,** 69.

Smith, A. M., Morrison, W. L., and Milham, P. J. (1982). Oxidation of indole-3-acetic acid by peroxidase: involvement of reduced peroxidase and compound III with superoxide as a product. *Biochem.* **21,** 4414.

Steinman, H. M. (1982). Copper–zinc superoxide dismutase from *Caulobacter crescentus* CB15. *J. biol. Chem.* **257,** 10283.

Tainer, J. A., *et al.* (1982). Determination and analysis of the 2Å structure of copper–zinc superoxide dismutase. *J. mol. Biol.* **160,** 181.

Turrens, J. F., Crapo, J. D. and Freeman, B. A. (1984). Protection against oxygen toxicity by intravenous injection of liposome-entrapped catalase and superoxide dismutase. *J. Clin. Invest.* **73,** 87.

Winterbourn, C. C., (1983). Lactoferrin-catalysed hydroxyl radical production. Additional requirement for a chelating agent. *Biochem. J.* **210,** 15.

Wislocki, P. G., Miwa, G. T., and Lu, A. Y. H. (1980). Reactions catalysed by the cytochrome P_{450} system. In *Enzymatic basis of detoxication,* Vol. I. Academic Press, New York.

Yamazaki, I. and Yokota, K. (1973). Oxidation states of peroxidase. *Mol. cell. Biochem.* **2,** 39.

4 Lipid peroxidation: a radical chain reaction

Lipid peroxidation has been broadly defined by A. L. Tappel in the USA as the "oxidative deterioration of polyunsaturated lipids", i.e. lipids that contain more than two carbon–carbon double covalent bonds ($>C=C<$). Oxygen-dependent deterioration, leading to *rancidity,* has been recognized as a problem in the storage of fats and oils since antiquity and is even more relevant today with the popularity of 'polyunsaturated' margarines and cooking oils, and the importance of paints, plastics, lacquers, and rubber, all of which can undergo oxidative damage. The first attempts to study this problem began in 1820 when de Saussure, using a simple mercury manometer, observed that a layer of walnut oil on water exposed to air absorbed three times its own volume of air in the course of eight months. This initial lengthy period was followed by a second phase of rapid air-absorption, the oil taking up sixty times its own volume of air in ten days. During the following three months, the rate of air uptake gradually diminished, so that the oil had eventually taken up 145 times its own volume of oxygen. Parallel with these changes, the oil became viscous and evil-smelling. Commenting on these experiments a few years later, the famous chemist Berzelius suggested that oxygen absorption might account for not only the autoxidation of oil exposed to air but also a host of similar phenomena. In particular, Berzelius suggested that autoxidation might be involved in the spontaneous ignition of wool after its lubrication with linseed oil, a common cause of disaster in textile mills at that time. Interestingly, Berzelius also discovered the element selenium, of vital importance as a protective mechanism as discussed later in this chapter.

The sequence of reactions which is now recognized as the basis of lipid peroxidation was worked out in detail by Farmer and others at the British Rubber Producers Association research laboratories in the 1940s. The relevance of these reactions to biological systems was not appreciated until later, however.

The membranes surrounding cells and cell organelles contain large amounts of polyunsaturated fatty-acid side-chains. What therefore stops us from going rancid ourselves?

4.1. Membrane structure

The major constituents of biological membranes are lipid and protein, the amount of protein increasing with the number of functions the membrane performs. In the nerve myelin sheath, which serves merely as an insulator of the nerve axon, only 20 per cent of the dry weight of the membrane is protein, but most membranes have 50 per cent or more protein, and the highly complicated inner mitochondrial membrane (Chapter 3) and chloroplast thylakoid membrane (Chapter 5) have 80 per cent protein. Some proteins are loosely attached to the surface of membranes (*extrinsic proteins*), but most are tightly attached (*intrinsic proteins*), being partially embedded in the membrane or, in some cases, located in the membrane interior or completely traversing the membrane. Lipid peroxidation can damage membrane proteins as well as the lipids.

Membrane lipids are generally *amphipathic* molecules, i.e. they contain hydrocarbon regions that like to stay together and have little affinity for water, together with polar parts that like to associate with water. In animal cell membranes the dominant lipids are *phospholipids* based on glycerol (Fig. 4.2) but some membranes, particularly cell surface membranes, contain significant proportions of sphingolipids and the hydrophobic molecule *cholesterol* (Fig. 4.1). The commonest phospholipid in animal cell membranes is *lecithin* (phosphatidylcholine). By contrast, the membranes of subcellular organelles such as mitochondria or nuclei rarely contain much sphingolipid or cholesterol. The fatty-acid side-chains of membrane lipids in animal cells have unbranched carbon

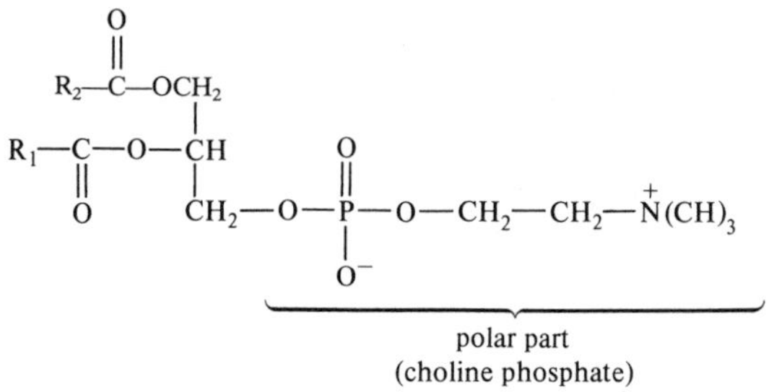

Fig. 4.1. Lipid molecules found in membranes. R_1, R_2, etc. represent long, hydrophobic, fatty-acid side-chains (for structures, see Fig. 4.2). In most lipids these are joined by *ester* bonds to the alcohol glycerol (Fig. 4.2), i.e.

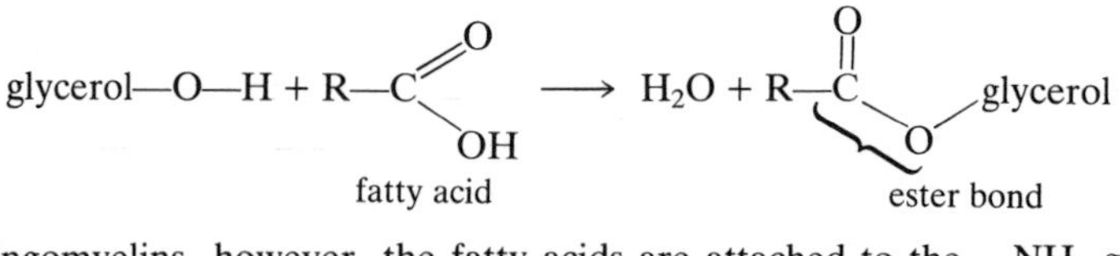

In sphingomyelins, however, the fatty acids are attached to the $—NH_2$ group of sphingosine (Fig. 4.2). All these lipid molecules contain a polar (hydrophilic) part that can interact with water, but in cholesterol this is very small (only an $—OH$ group) so that, overall, cholesterol is a very hydrophobic molecule.

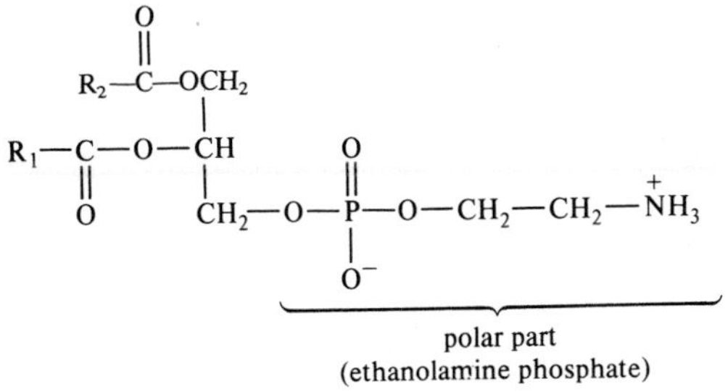

Phosphatidylethanolamine

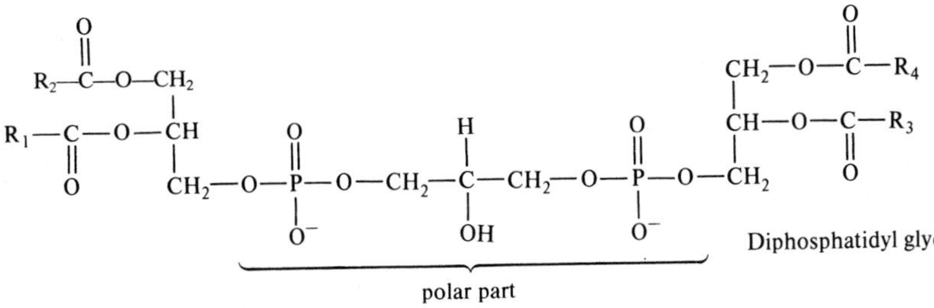

polar part
(ethanolamine phosphate)

Diphosphatidyl glycerol

polar part

Sphingomyelin

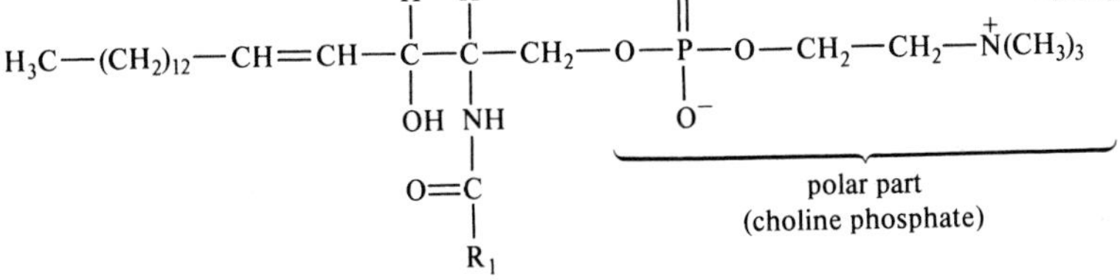

polar part
(choline phosphate)

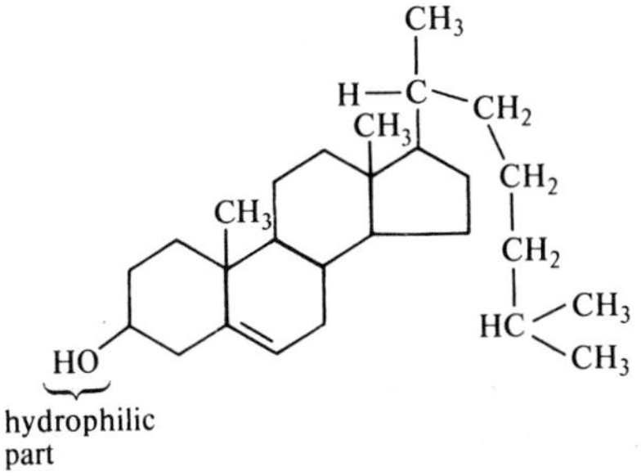

Cholesterol

hydrophilic
part

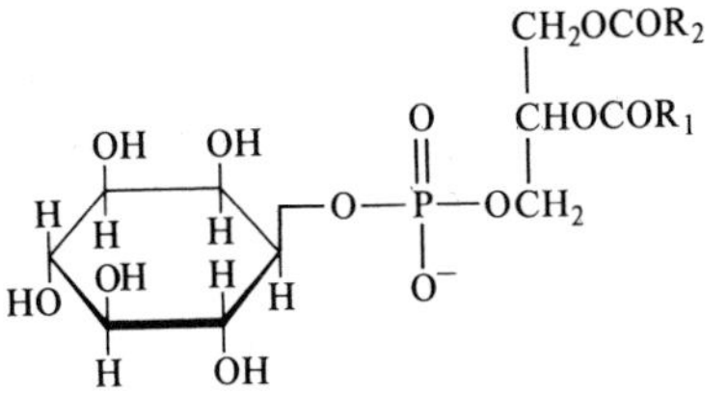

Phosphatidylinositol
(polar part is the sugar inositol)

Fig. 4.1. (*Continued*)

$CH_3 \cdot (CH_2)_{14} \cdot COOH$ Palmitic acid (C_{16}, saturated)

$CH_3 \cdot (CH_2)_{16} \cdot COOH$ Stearic acid (C_{18}, saturated)

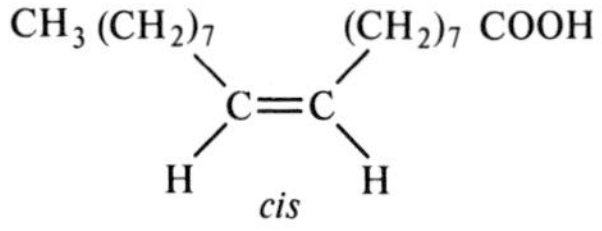

Oleic acid (C_{18}, one double bond)

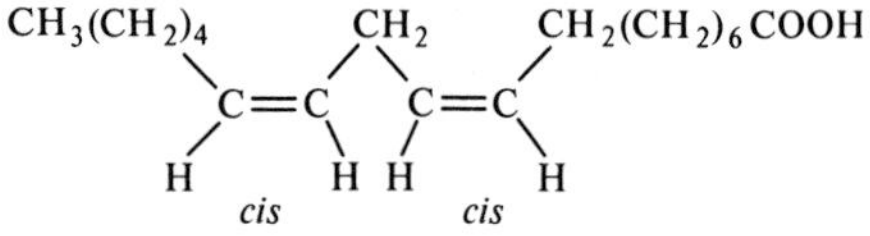

Linoleic acid
(C_{18}, two double bonds, at carbons 9 and 12. The carbon of the —COOH group is counted as carbon number 1).

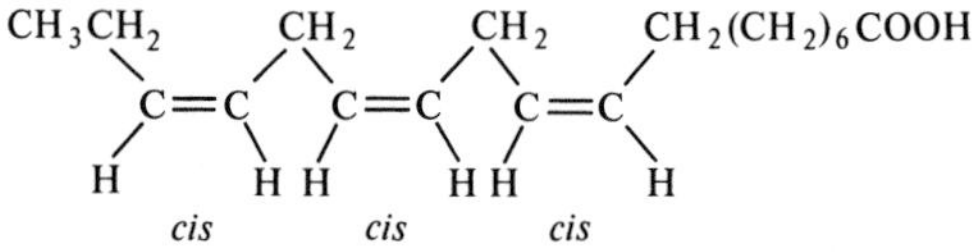

Linolenic acid
(C_{18}, polyunsaturated. Double bonds at C9, 12, 15).

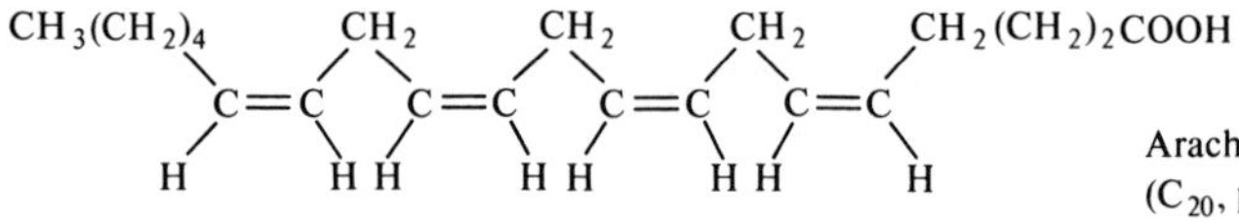

Arachidonic acid
(C_{20}, polyunsaturated. Double bonds at C5, 8, 11, 14)

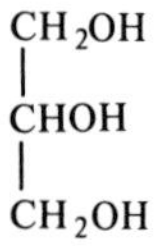

CH_2OH
|
$CHOH$
|
CH_2OH

Glycerol

$H_3C—(CH_2)_{12}—CH=CH—\overset{\overset{\displaystyle H}{|}}{C}—\overset{\overset{\displaystyle H}{|}}{C}—CH_2OH$
          $\underset{OH}{}$ $\underset{NH_3^+}{}$

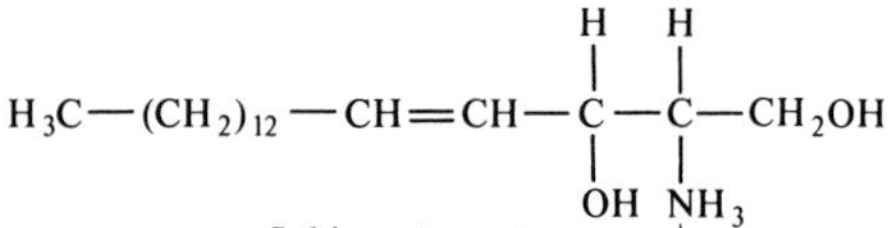

Sphingosine

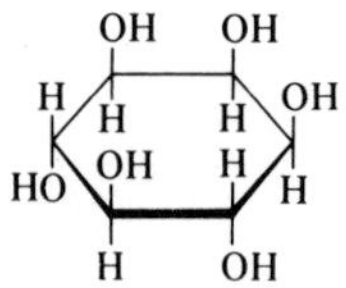

Inositol

Fig. 4.2. Fatty acids and other 'building blocks' of the membrane lipids.

chains and contain even numbers of carbon atoms, mostly in the range 14–24, and the double bonds are of the *cis*-configuration. A double bond in a carbon chain prevents rotation of the groups attached to the carbon atoms forming it, so that they are forced to stay on one side of the double bond or the other. For example if the two hydrogen atoms of the acid $CH_3(CH_2)_6CH{=}CH(CH_2)_6COOH$ are on the same side of the double bond the *cis*-configuration results, i.e.

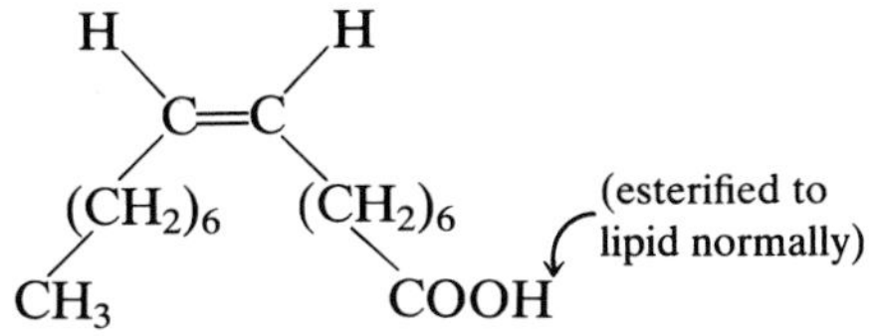

H   H
　╲   ╱
　 C＝C
　╱   ╲
$(CH_2)_6$ $(CH_2)_6$ (esterified to
　│   │   lipid normally)
CH_3 COOH

rather than the *trans*-arrangement

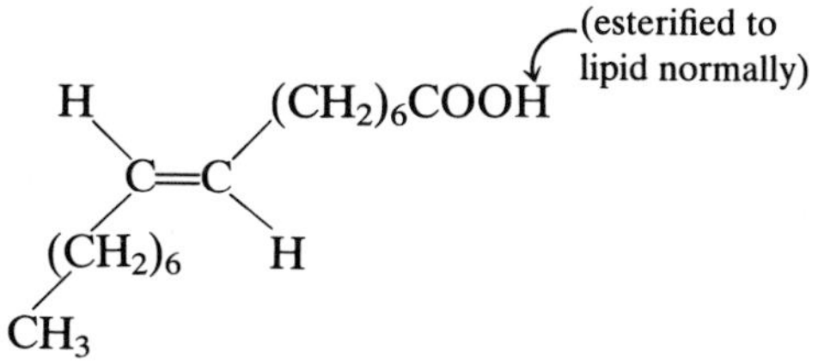

　　　　　　　　　　　　　　(esterified to
　　　　　　　　　　　　　　 lipid normally)
H $(CH_2)_6COOH$
　╲   ╱
　 C＝C
　╱   ╲
$(CH_2)_6$ H
　│
CH_3

Thus the membrane fatty-acid side-chains have 'kinks' in them whenever a *cis* double bond occurs. Phospholipids contain a number of unsaturated and polyunsaturated fatty-acid side-chains, the structures of some of which are summarized in Fig. 4.2.

The lipid composition of bacterial membranes depends very much on the species and even on the culture conditions and stage in the growth cycle. Membrane fractions usually contain 10–30 per cent lipid. In gram-positive bacteria (i.e. those that take up *Grams stain*, used by microscopists) phosphatidylglycerol is present, but phosphatidylethanolamine (Fig. 4.1) is more common in gram-negative species. *Mycoplasmas* contain a lot of cholesterol in their membranes.

As the number of double bonds in a fatty-acid molecule increases, its melting point drops. For example, stearic acid (Fig. 4.2) is solid at room temperature whereas linoleic acid is a liquid. Fatty acids with zero, one, or two double bonds are more resistant to oxidative attack than are the polyunsaturated fatty acids; yet polyunsaturated fatty-acid side-chains are present in many membrane phospholipid molecules.

Since membrane lipids are amphipathic molecules, on exposure to water they will tend to aggregate with their hydrophobic regions clustered together away from the water and their hydrophilic regions in contact with it so that, for example, charged atoms (see the structures in Fig. 4.1) can be stabilized by hydration (Appendix, Section A.2.3). How this arrangement is achieved depends on the relative amounts of lipid and water present. When phospholipids are shaken or sonicated in aqueous solution they form *micelles*, but as more phospholipids are added *liposomes* result, bags of aqueous solution bounded by a *lipid bilayer*. Figure 4.3 shows the structures involved. Liposomes can be surrounded by a single lipid bilayer (*unilamellar*) or several bilayers, as shown in the electron micrograph of a liposome preparation in Fig. 4.4. The interior of liposomes contains a portion of the aqueous solution in which they were made, so they have sometimes been used to study membrane permeability, and as 'parcels' for transporting drugs to target tissues.

There is considerable evidence that the lipid bilayer (Fig. 4.3) is the basic structure of all cell and organelle membranes, proteins being inserted in different parts of the bilayer (as explained previously and summarized diagrammatically in Fig. 4.5). In each half of the lipid bilayer, protein and lipid molecules can diffuse extremely quickly—indeed a lipid molecule can get from one end of a bacterial cell to the other in one or two seconds! However, exchange of lipid molecules between the two halves of the bilayer is rare. This *membrane fluidity* is due to the presence

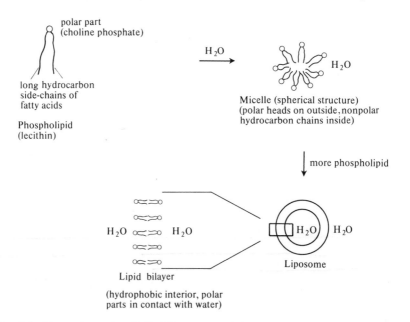

Fig. 4.3. Formation of a lipid bilayer on mixing phospholipids with aqueous solutions.

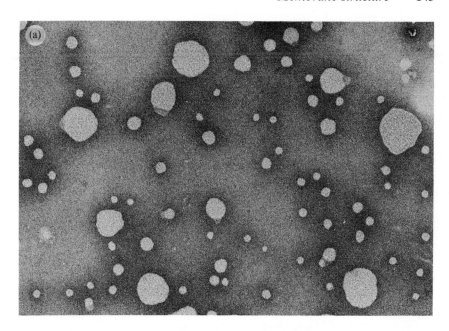

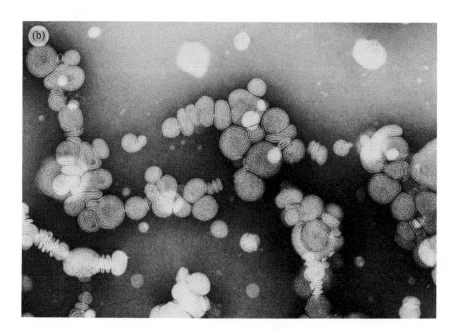

Fig. 4.4. Electron micrograph of a liposome preparation. (a) Unilamellar; (b) multilamellar liposomes.

Electron micrograph of a cell membrane

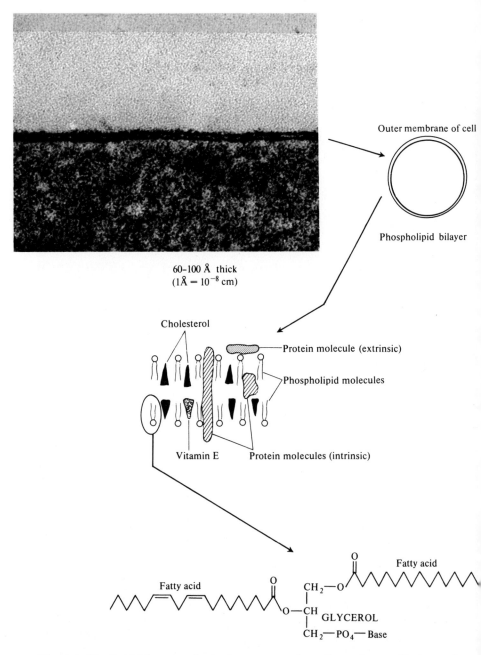

Outer membrane of cell

Phospholipid bilayer

60–100 Å thick
$(1\text{Å} = 10^{-8}\text{ cm})$

Cholesterol

Protein molecule (extrinsic)

Phospholipid molecules

Vitamin E

Protein molecules (intrinsic)

Fatty acid

Fatty acid

CH_2—O

O—CH GLYCEROL

CH_2—PO_4—Base

Fig. 4.5. The lipid bilayer as the basic structure of a cell membrane. Cholesterol is found in cell plasma membranes but not usually in organelle membranes. The membrane shown is that of a sheep red blood cell as seen under the electron microscope.

of unsaturated and polyunsaturated fatty-acid side-chains in many membrane lipids, which lower the melting point of the membrane interior so that it effectively gains the chemical nature and viscosity of a 'light oil'. Damage to polyunsaturated fatty acids tends to reduce membrane fluidity, which is known to be essential for the proper functioning of biological membranes.

4.2. The peroxidation process

Initiation of a peroxidation sequence in a membrane or polyunsaturated fatty acid is due to the attack of any species that has sufficient reactivity to abstract a hydrogen atom from a methylene (—CH_2—) group. Since a hydrogen atom has only one electron (see Appendix), this leaves behind an unpaired electron on the carbon, —ĊH—. The presence of a double bond in the fatty acid weakens the C—H bonds on the carbon atom adjacent to the double bond and so makes H˙ removal easier. The carbon radical tends to be stabilized by a molecular rearrangement to produce a *conjugated diene* (Fig. 4.6), which then easily reacts with an oxygen molecule to give a *peroxy radical,* R—OȮ˙. The presence of these radicals has been detected during peroxidation of rat liver microsomes by the spin-trapping method (Chapter 2). Peroxy radicals can abstract a hydrogen atom from another lipid molecule—this is the *propagation stage* of lipid peroxidation and so, once the process is initiated, it tends to continue (i.e. it is a *chain reaction*). The peroxy radical combines with the hydrogen atom that it abstracts to give a *lipid hydroperoxide,* R—OOH. A probable alternative fate of peroxy radicals is to form *cyclic peroxides* (Figs 4.6 and 4.7). Pure lipid hydroperoxides are fairly stable molecules at physiological temperatures, but in the presence of transition-metal complexes, their decomposition is catalysed. Many metal complexes that can do this are present *in vivo* and they include simple complexes of iron salts with phosphate compounds (Chapter 2) as well as non-haem-iron proteins. Both Fe(III) and Fe(II) salts are effective, although the latter much more so. Free haem is effective, as are haemoglobin, myoglobin, cytochromes (including cytochrome P_{450}), and horseradish peroxidase. Ferritin catalyses hydroperoxide decomposition to an extent proportional to the amount of iron it contains. By contrast, transferrin and lactoferrin do not stimulate hydroperoxide decomposition until they are fully iron-saturated (which does not happen *in vivo* except in iron-overload; see Chapter 2) and then they are only weakly effective. Commercially available ATP, ADP, and phosphate salts for phosphate buffers are always contaminated with traces of iron salts (Table 4.1 summarizes some results on this obtained in the authors laboratory). Treatment of solutions of reagents with *Chelex* resin, which binds both

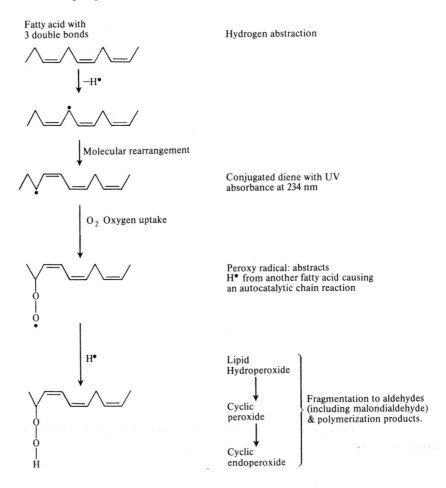

Fig. 4.6. Initiation and propagation reactions of lipid peroxidation. The peroxidation of a fatty acid with three conjugated double bonds is shown.

iron and copper salts, usually decreases the observed rate of peroxidation of membrane lipids in the absence of added metals *in vitro*. Various chelating agents (Chapter 2) have often been used to assess the participation of metal ions in lipid peroxidation *in vitro*. EDTA or DETAPAC can either stimulate or inhibit peroxide decomposition catalysed by iron salts depending on the ratio of the concentration of chelator to that of iron salt (high chelator/iron ratios usually inhibit, lower ones usually stimulate) whereas desferrioxamine or rhodotorulic acid inhibit at all concentrations tested. Injection of iron salts into animals greatly increases the rate of peroxidation observed *in vivo* and tissues

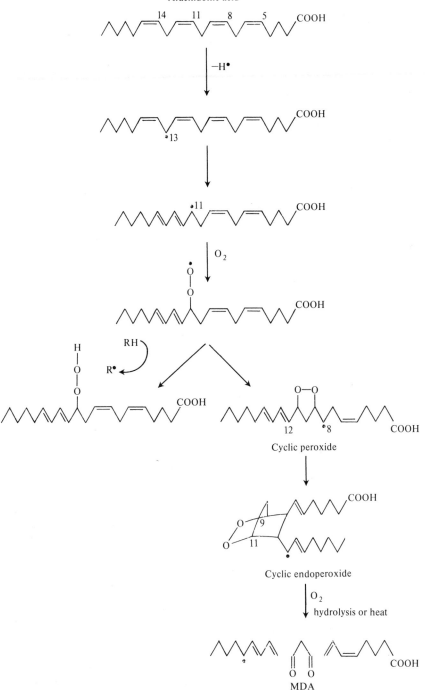

Fig. 4.7. Formation of lipid hydroperoxides and cyclic peroxides from arachidonic acid. MDA, malondialdehyde (malonaldehyde).

Table 4.1. Iron contamination of laboratory reagents

Solution analysed	Iron concentration (μ moles l^{-1})
5.8 M Hydrochloric acid	1–2
'Saline buffer' (67.5 mM Na$_2$HPO$_4$ + 4 mM KCl adjusted to pH 7.4 with HCl)	10
Old saline buffer (stored in laboratory for several weeks in a flask covered with 'parafilm')	18
50 mM EDTA	8
0.5 M Sodium formate	9
0.5 M Urea	6
0.5 M Thiourea	3
20 mM Ascorbic acid	4

Solutions were made up in double-distilled water that itself contained no iron. The iron content was measured by atomic absorption analysis (data from Wong *et al.* (1981) *J. inorg. Biochem.* **14**, 127–34). The presence of these traces of iron is of great significance in studies of lipid peroxidation and of the iron-catalysed Haber–Weiss reaction (Chapter 3).

from iron-overloaded humans or animals show a rapid rate of peroxidation *in vitro* in the absence of added iron salts.

A reduced iron compound can react with a lipid hydroperoxide in a similar way to its reaction with hydrogen peroxide—it causes fission of an O–O bond to form an *alkoxy radical*

$$H\text{—}O\text{—}O\text{—}H + Fe^{2+}\text{—complex} \rightarrow Fe^{3+}\text{—complex} + OH^-$$
$$+ OH^\cdot (\text{Fenton reaction})$$

Hydrogen Peroxide (annotation above H—O—O—H)

$$R\text{—}O\text{—}O\text{—}H + Fe^{2+}\text{—complex} \rightarrow Fe^{3+}\text{—complex} + OH^- + R\text{—}O^\cdot.$$
lipid hydroperoxide   alkoxy radical

With an Fe(III)-compound a peroxy radical, $ROO^\cdot$, will form

$$R\text{—}O\text{—}O\text{—}H + Fe^{3+}\text{—complex} \rightarrow R\text{—}OO^\cdot + H^+ + Fe^{2+}\text{—complex}.$$

Peroxy radicals seem to be less reactive than alkoxy radicals. In both cases, however, the formation of radical products will stimulate the chain reaction of lipid peroxidation by causing more initiation (Fig. 4.6). For example, if Fe(II) reacts with a hydroperoxide on the fifth carbon from the methyl end of the fatty acid, pentane gas can be produced. This can happen with linoleic acid and arachidonic acid (Fig. 4.2)

$$CH_3(CH_2)_4\overset{\overset{\displaystyle H}{|}}{\underset{\underset{\displaystyle OOH}{|}}{C}}-R + Fe^{2+} \longrightarrow Fe^{3+} + OH^- + CH_3(CH_2)_4\overset{\overset{\displaystyle H}{|}}{\underset{\underset{\displaystyle O^\bullet}{|}}{C}}-R$$

(R, rest of molecule)   alkoxy radical

$\Bigg\downarrow$ β-scission reaction

$$CH_3(CH_2)_3CH_3 \xleftarrow[\substack{\text{from another}\\\text{fatty-acid side-}\\\text{chain}}]{\text{abstracts } H^\bullet} CH_3(CH_2)_3\dot{C}H_2 + \overset{\overset{\displaystyle H}{|}}{\underset{\underset{\displaystyle O}{\|}}{C}}-R$$

pentane   pentane radical

Ethane (CH_3—CH_3) and ethene (H_2C=CH_2) gases are produced in similar reactions from linolenic acid. *β-scission* is a well-known reaction of radicals, especially alkoxy radicals. A whole series of complex reactions leads not only to hydrocarbons but also to the production of compounds containing the carbonyl group $>$C=O, especially aldehydes (RCHO) (Table 4.2). One such product that has been isolated by Austrian scientists from rat liver microsomes undergoing lipid peroxida-

Table 4.2. Formation of carbonyl compounds from fatty acids and microsomes during lipid peroxidation in the presence and absence of iron salts

System	$FeSO_4$ present	Malonaldehyde formed (nmol mg^{-1})	Other carbonyl compounds formed (nmol mg^{-1})
Oleic acid	no	—	—
	yes	0	62
Linoleic acid	no	0	98
	yes	12	317
Linolenic acid	no	—	—
	yes	13	562
Arachidonic acid	no	19	588
	yes	45	1728
Microsomes	no	5	10
	yes	60	104

Data were abstracted from the article by H. Esterbauer in McBrien and Slater (1982); see reading list. Peroxidation was carried out in Tris/HCl buffer pH 7.4 at 37 °C for 18 h in shaken flasks in the presence or absence of added Fe(II) ions (20 μM $FeSO_4$). ADP–Fe(II)-stimulated peroxidation of microsomes was carried out for 30 min at 37 °C. (Also see *Biochem. J:* (1982) **208**, 129–40.)

tion is 4-hydroxy-2,3-*trans*-nonenal

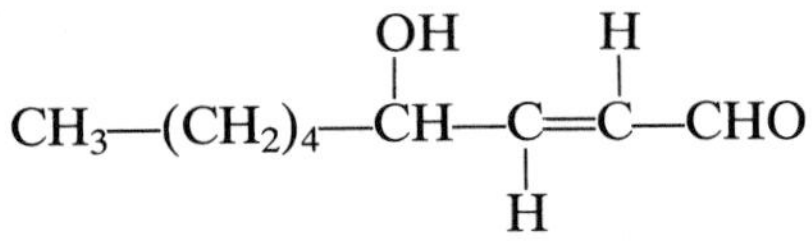

$$CH_3-(CH_2)_4-\overset{\overset{\displaystyle OH}{|}}{CH}-C=\overset{\overset{\displaystyle H}{|}}{C}-CHO$$
$$\underset{\underset{\displaystyle H}{|}}{}$$

Malonaldehyde (sometimes called malondialdehyde) is also formed in small amounts; it has the structure

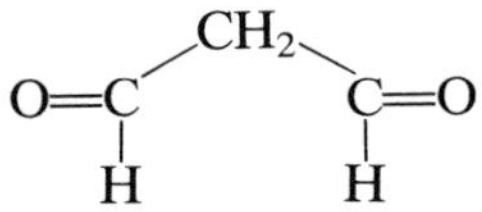

$$O=\overset{\overset{\displaystyle CH_2}{\diagup \diagdown}}{\underset{\underset{\displaystyle H}{|}}{C}} \quad \overset{}{\underset{\underset{\displaystyle H}{|}}{C}}=O$$

   In most membranes, the propagation reactions of lipid peroxidation will not proceed very far before they meet a protein molecule, which can then be attacked and damaged. Aldehydes can react with —SH groups on proteins, e.g. for hydroxy-*trans*-nonenal

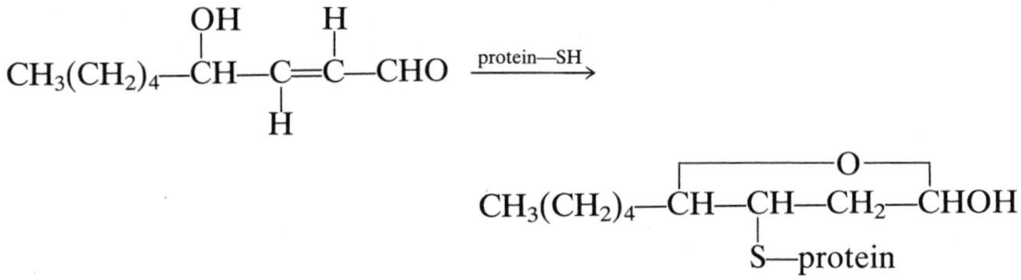

$$CH_3(CH_2)_4-\overset{\overset{\displaystyle OH}{|}}{CH}-C=\overset{\overset{\displaystyle H}{|}}{C}-CHO \quad \xrightarrow{\text{protein—SH}}$$
$$\underset{\underset{\displaystyle H}{|}}{}$$

$$CH_3(CH_2)_4-\overset{\overset{\displaystyle \lceil \quad\quad\quad O\quad \rceil}{}}{CH}-\overset{\overset{}{}}{CH}-CH_2-CHOH$$
$$\underset{\underset{\displaystyle S\text{—protein}}{|}}{}$$

In addition, aldehydes can attack amino groups on the protein molecule to form both intramolecular cross links and also cross-links between different protein molecules, e.g. for malonaldehyde

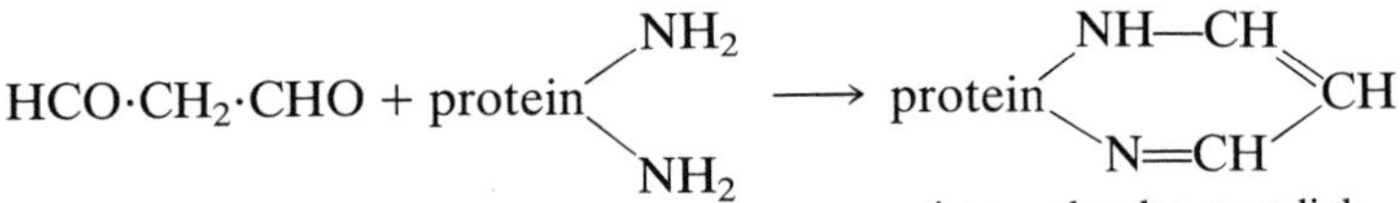

$$HCO\cdot CH_2\cdot CHO + protein \overset{\diagup NH_2}{\underset{\diagdown NH_2}{}} \longrightarrow protein \overset{\diagup NH-CH}{\underset{\diagdown N=CH}{\diagdown}} CH$$

intramolecular cross-link

$$HCO\cdot CH_2\cdot CHO + 2 \text{ protein—NH}_2 \longrightarrow$$

$$protein—NHCH=CH—CH=N—protein$$

intermolecular cross-link

Treatment of membranes with malonaldehyde *in vitro* has frequently been observed to cross-link and aggregate membrane proteins. Enzymes that require —NH$_2$ or —SH groups for their activity are usually inhibited during lipid peroxidation, e.g., the glucose-6-phosphatase enzyme found in liver microsomal fractions is inhibited as its —SH groups are attacked. Hydroxy-*trans*-nonenal and other low-molecular-weight products of lipid peroxidation have been shown to inhibit protein synthesis and to interfere with the growth of bacteria and animal cells in culture. Several of the lipid peroxidation products show prostaglandin-like activity (prostaglandins are discussed further in Chapter 7), probably due to the formation of cyclic endoperoxides (Fig. 4.7). Peroxidation of liver or erythrocyte membranes is known to cause formation of high-molecular-weight protein aggregates within the membrane, probably due to radical processes. The surface receptor molecules that allow cells to respond to hormones can be inactivated during lipid peroxidation (Fig. 4.8), as are enzymes such as glucose-6-phosphatase or the Na$^+$K$^+$ATPase involved in maintenance of correct ion balance within cells. In general, the overall effects of lipid peroxidation are to decrease membrane fluidity, increase the 'leakiness'

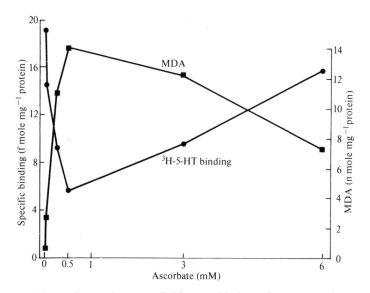

Fig. 4.8. Effect of ascorbate on lipid peroxidation. Cortex membranes were prepared from rat brains by centrifugation—brain fractions usually contain significant amounts of endogenous iron. The effect of ascorbic acid on the rate of lipid peroxidation, measured by the TBA test and expressed as the amount of malonaldehyde produced (Section 4.6.9) is shown. The binding of tritium [^{3}H]-labelled 5-hydroxytryptamine to receptors on the membrane was studied— loss of binding means that the receptors have been damaged. Data from Muakkassah-Kelly *et al.* (1982) *Biochem. Biophys. Res. Commun.* **104,** 1003–10, with permission.

of the membrane to substances that do not normally cross it (such as Ca^{2+} ions), and inactivate membrane-bound enzymes. Continued fragmentation of fatty-acid side-chains to produce aldehydes and hydrocarbons such as pentane will eventually lead to complete loss of membrane integrity. Rupture of, say, the membranes of lysosomes in this way can spill hydrolytic enzymes into the rest of the cell to cause further damage. Lipid peroxidation of erythrocyte membranes causes them to lose their ability to change shape and squeeze through the smallest capillaries, and it will eventually lead to haemolysis. It has been suggested that the loss of viability of mammalian sperm on prolonged incubation at 37 °C is due to the accumulation of products of lipid peroxidation, and the loss of the germinating ability of soybean seeds, stored under warm, damp conditions, has been attributed to the same reason. In some bacteria the DNA is close to or attached to the cell membrane and it can be damaged during peroxidation.

Studies on a mutant of *E. coli* that could not synthesize fatty acids, and thus incorporated into its membrane lipids whatever acids it was given in the growth medium, showed that the toxicity of hyperbaric oxygen became greater as the percentage of polyunsaturated fatty acids in the membrane was increased. It must be remembered that oxygen is quite soluble in organic solvents (Chapter 1) and thus will concentrate in the hydrophobic interior of membranes. Injection of peroxidized lipids into experimental animals produces deleterious effects, e.g. damage to the heart in rats and 'fatty liver' in both rats and rabbits. Exposure of animals to elevated oxygen concentrations causes increased peroxidation of their cell membranes.

4.3. Enzymic and non-enzymic lipid peroxidation

Scientists often loosely refer to iron salts and other iron complexes as 'initiating' peroxidation, but in many cases what is really happening is that they are causing the decomposition of preformed lipid peroxides, as explained above, to generate alkoxy or peroxy radicals that are the true initiators. Copper salts also stimulate peroxidation, probably by a similar mechanism. (Indeed, it has been suggested that the copper wire in one form of the intrauterine contraceptive device stimulates lipid peroxidation in the cervical mucus, which can damage spermatozoa). Iron(II) salts, but not iron(III) salts, form oxygen radical species in the presence of oxygen that also initiate peroxidation (see below). All commercially available fatty acids and membrane lipids contain significant amounts of preformed lipid peroxides—how they get there will be considered later. Indeed, the authors have obtained more than one commercial batch of arachidonic acid in which very large amounts of hydroperoxides were present. Lipid peroxidation is of the greatest importance in the pathology of iron-overload diseases and Wilson's disease (Chapter 2).

Iron(II) and its complexes stimulate peroxidation more than does Fe(III). This may be ascribed to the greater solubility of iron(II) salts in aqueous solution, the greater reactivity of alkoxy as opposed to peroxy radicals and the ability of iron(II) salts to form oxygen radical species. The rate of peroxidation of purified membrane lipids or microsomal fractions in the presence of added Fe(III) complexes can be greatly increased by addition of ascorbic acid, a biological reducing agent (Chapter 3). For a given concentration of iron salt, low concentrations of ascorbate stimulate peroxidation, probably by reducing Fe(III) to Fe(II), whereas high concentrations inhibit the peroxidation (Fig. 4.8). It has been suggested that high concentrations of ascorbate may reduce some of the lipid peroxy radicals directly to hydroperoxides, and thus interfere with the chain reaction. Mixtures of GSH and iron salts have been reported to stimulate lipid peroxidation, but the general view is that GSH is protective *in vivo* because it is a substrate for glutathione peroxidase (see below), as well as being able to react directly with various aldehydes produced during peroxidation and thus protect the —SH groups of membrane proteins (see above). Consistent with this view, treatment of rats or mice with reagents that decrease liver GSH concentrations seems to increase their susceptibility to peroxidation.

Peroxidation of cell or organelle membranes stimulated by Fe(II), or Fe(III) plus ascorbate, *in vitro* does not require the activity of any enzymes. Another way of stimulating lipid peroxidation *in vitro* is to add lipid hydroperoxides or artificial organic hydroperoxides such as *tert*-butyl hydroperoxide or cumene hydroperoxide (as used for Fig. 4.9). The decomposition of these to alkoxy or peroxy radicals accelerates the chain reaction of lipid peroxidation. Decomposition is again facilitated by metal ions and their complexes, e.g. by methaemoglobin and cytochrome P_{450}. Indeed, cytochrome P_{450} will catalyse a hydroperoxide-dependent hydroxylation of certain substrates, such as cyclohexane or toluene. This activity is sometimes known as the 'peroxygenase' action of cytochrome P_{450}. It probably has no physiological significance, but studies of it have given interesting information about the mechanism of hydroxylation by P_{450}. If SH is the hydroxylatable substrate and XOOH the peroxide, peroxygenase activity can be represented by the equation

$$SH + XOOH \rightarrow S{-}OH + X{-}OH$$

$$\text{peroxide} \qquad \begin{array}{c}\text{hydroxylated} \quad \text{alcohol} \\ \text{substrate}\end{array}$$

The decomposition of cumene and *tert*-butyl hydroperoxides in the

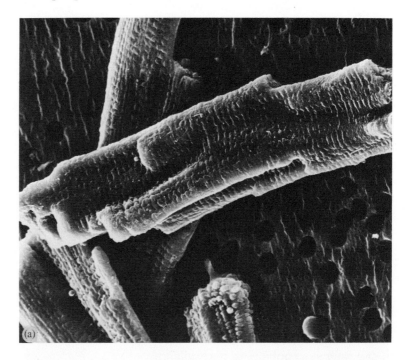

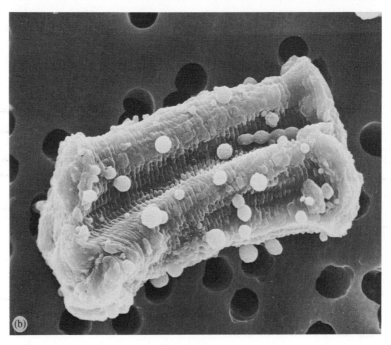

presence of metal ions or their complexes can be written

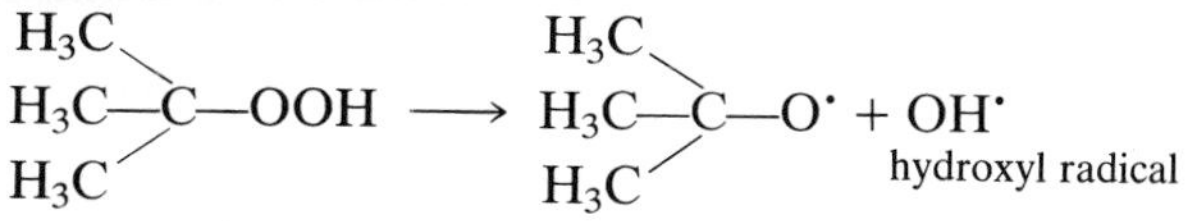

$$
\begin{array}{c}
H_3C \\
H_3C\text{—}C\text{—OOH} \\
H_3C
\end{array}
\longrightarrow
\begin{array}{c}
H_3C \\
H_3C\text{—}C\text{—O}^{\bullet} + OH^{\bullet} \\
H_3C
\end{array}
\quad \text{hydroxyl radical}
$$

tert-butyl hydroperoxide

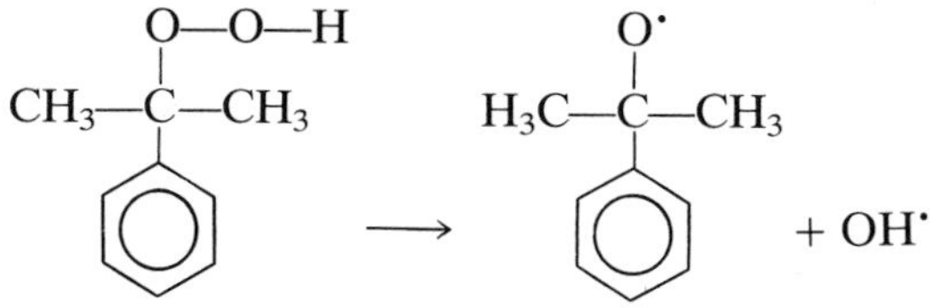

$$
\begin{array}{c}
O\text{—}O\text{—}H \\
| \\
CH_3\text{—}C\text{—}CH_3 \\
\bigcirc
\end{array}
\longrightarrow
\begin{array}{c}
O^{\bullet} \\
| \\
H_3C\text{—}C\text{—}CH_3 \\
\bigcirc
\end{array}
+ OH^{\bullet}
$$

cumene hydroperoxide

Note that hydroxyl radicals are also produced.

Microsomal fractions from several animal tissues (e.g. liver, kidney, and skin) undergo lipid peroxidation in the presence of NADPH and Fe(III) salts (often added as complexes with ADP, pyrophosphate, or even EDTA, reagents which help to keep the iron in solution). An antibody raised against the microsomal flavoprotein NADPH–cytochrome P_{450} reductase (Chapter 3) inhibits this peroxidation by more than 90 per cent. During it, cytochromes b_5 and P_{450} are attacked, the haem groups being degraded. Apparently the P_{450} reductase enzyme, as well as reducing cytochrome P_{450}, can donate electrons to some Fe(III)-complexes and so generate Fe(II), which stimulates peroxidation. Although it requires an enzyme, this mechanism is in principle no different from non-enzymic peroxidation in that the enzyme is serving merely to reduce the iron salts. Nuclear membranes contain an electron transport chain (Chapter 3) and undergo peroxidation on incubation with NADPH and Fe(III) salts *in vitro*; and a similar phenomenon has been shown with the inner mitochondrial membrane in the presence of NADH or NADPH. Here iron is reduced by the NADH–coenzyme Q reductase complex.

Fig. 4.9. Lipid peroxidation as a mechanism of injury to heart cells. Isolated beating cardiac myocytes were prepared from rat hearts. (a) Shows their normal appearance as seen by scanning electron microscopy. (b) The cells shown had been treated with cumene hydroperoxide for 30 minutes. Note the contraction and formation of 'blebs' of membrane on the cell surface. The cumene hydroperoxide stimulated peroxidation of the membrane lipids, as followed by diene conjugation. Photograph by courtesy of Dr A. A. Noronha-Dutra.

4.4. Initiation of lipid peroxidation: the role of oxygen-derived species

Unlike ground-state oxygen, singlet oxygen $^1\Delta g$ can react rapidly with compounds containing $\diagup C{=}C\diagdown$ bonds to give hydroperoxides (Chapter 2). The lifetime of singlet oxygen in the hydrophobic interior of membranes is much greater than it is in aqueous solution. Illumination of unsaturated fatty acids in the presence of sensitizers of singlet-oxygen formation, such as chlorophyll, rose bengal, methylene blue, bilirubin, or porphyrins, induces rapid peroxidation. Similar effects have been seen with intact membranes. For example, the rod outer segments in the retina of the frog are rich in polyunsaturated lipids and form peroxides upon illumination, probably due to singlet-oxygen formation sensitized by retinal (this is further discussed in Chapter 5). Illumination of phospholipid liposomes containing the porphyrin haematin (Chapter 2) together with cholesterol not only peroxidizes the fatty-acid side-chains but also converts some of the cholesterol into its 5-α-hydroperoxide (Chapter 2), probably by reaction with singlet oxygen. Illumination of erythrocytes in the presence of porphyrins or bilirubin causes peroxidation, aggregation of membrane proteins, inactivation of enzymes and transport carriers, and eventually haemolysis. Fungi of the genus *Cercospora* produce a toxin known as *cercosporin* that can attack plant cells in the light, but not in the dark. It has been suggested that cercosporin sensitizes formation of singlet oxygen, which then initiates lipid peroxidation in the plant cell, producing disruption of membranes and leakage of ions.

It has been further suggested that singlet oxygen may be *formed* during the complex degradation reactions undergone during lipid peroxidation and might then contribute to the chain reaction by causing more initiation. The evidence for this singlet-oxygen formation is dubious, however, since it was obtained by using 'singlet-oxygen scavengers' that are by no means specific for reaction with this species. Indeed, it has recently been suggested that DABCO, diphenylfuran, and β-carotene (Chapter 3) can react with at least one organic peroxy radical (the trichloromethylperoxy radical, $CCl_3O_2^{\cdot}$). Nevertheless, something with an oxidizing capacity comparable to singlet oxygen does appear to be produced during lipid peroxidation in some systems and it is probably best to refer to it as a 'singlet-oxygen-like factor'.

A number of scientists have observed that O_2^--generating systems can stimulate the peroxidation of fatty acids or of membranes. For example, Fridovich and Porter in the USA observed that a mixture of xanthine oxidase and its substrate ethanal (chapter 3) could peroxidize arachidonic acid. This was inhibited by mannitol (a $OH^{\cdot}$ scavenger), superoxide dismutase, catalase, or the iron-chelator DETAPAC, which suggests that peroxidation was initiated by hydroxyl radicals produced by an iron-catalysed Haber–Weiss reaction (Chapter 3), i.e.

$$Fe^{3+} + O_2^- \rightarrow Fe^{2+} + O_2$$

$$Fe^{2+} + H_2O_2 \rightarrow Fe^{3+} + OH^{\bullet} + OH^-$$

$$Lipid—H + OH^{\bullet} \rightarrow Lipid^{\bullet} + H_2O$$

Hydroxyl radicals are known to be capable of abstracting hydrogen atoms from membrane lipids and causing formation of conjugated dienes (Fig. 4.6). The rate constant for the reaction of $OH^{\bullet}$ with artificial lecithin bilayers is about $5 \times 10^8 \, M^{-1} \, s^{-1}$, and the reaction causes increased leakiness of the bilayer. Exposure of cell membranes, fatty acids, or unsaturated food oils to ionizing radiation, which generates $OH^{\bullet}$ (Chapter 2), cause rapid peroxidation. Indeed, this is a problem in the use of irradiation to sterilize packaged foods. The superoxide radical formed during irradiation can give rise to more $OH^{\bullet}$ as described above.

Damage to the erythrocyte membrane by xanthine oxidase plus ethanal has also been observed, and may be due not only to $OH^{\bullet}$ but also possibly to the formation of singlet oxygen in O_2^--generating systems (Chapter 3). The superoxide radical itself is insufficiently reactive to abstract $H^{\bullet}$ from membrane lipids; in any case it would not be expected to enter the interior of membranes because of its charged nature (Chapter 2). In agreement with this, O_2^- does not cross biological membranes easily, the one exception to this rule being the erythrocyte membrane. Here it can make use of the 'anion channel' in the membrane through which chloride (Cl^-) and bicarbonate (HCO_3^-) ions normally pass. It is possible, however, that the protonated form of O_2^-, $HO_2^{\bullet}$, which is more reactive, could attack fatty acids directly: evidence for oxidation of linoleic acid to its hydroperoxide by $HO_2^{\bullet}$ has been presented by Bielski in the USA. $HO_2^{\bullet}$, being uncharged, should enter membranes more readily. Another suggestion, by Pryor in the USA, is that O_2^-, or perhaps $HO_2^{\bullet}$, could react with preformed lipid hydroperoxides in the membrane to give alkoxy radicals

$$Lipid—OOH + O_2^- \rightarrow Lipid—O^{\bullet} + OH^- + O_2$$

'Non-enzymic' lipid peroxidation observed *in vitro* when Fe(II) chelates, or Fe(III) plus ascorbate, are added to cell membranes or to liposomes, is not significantly inhibited by catalase, SOD, or scavengers of $OH^{\bullet}$, even though $OH^{\bullet}$ radicals can be detected in the reaction mixture (e.g. by spin-trapping, aromatic hydroxylation, or the deoxyribose method; see Chapter 2). Hence the major initiator of peroxidation in these systems is probably neither free $HO_2^{\bullet}$ nor free $OH^{\bullet}$. The peroxidation of microsomal fractions and other cell membranes in the presence of NADPH and Fe(III) chelated to ADP is also unaffected by SOD, which is perhaps not surprising since if NADPH–cytochrome P_{450} reductase can reduce the Fe(III)–ADP complex, then there is no need for O_2^- to do it. Formation of both OH˙ and O_2^- in this system has again been detected, however (Chapter 3), although added scavengers of OH˙ do not inhibit peroxidation.

If OH˙ radicals can be detected in these various reaction mixtures, why is it that addition of scavengers of these radicals does not usually inhibit the peroxidation process? Firstly, it is possible that those OH˙ radicals important in initiation are formed by metal ions tightly bound to the membranes. Such 'site-specific' production of OH˙ could not be prevented by scavengers in bulk solution. Secondly, once the peroxidation process is underway the major initiators may be alkoxy or peroxy radicals formed by decomposition of lipid hydroperoxides; OH˙ radicals may not be necessary under these conditions. Remember that many lipids already contain substantial amounts of pre-formed hydroperoxide. A third possibility is that some other species initiates peroxidation; suggestions by S. D. Aust in the USA include perferryl, ferryl and Fe(II)–Fe(III)–oxygen radical complexes. With present experimental methodology, it is impossible to distinguish between these possibilities.

Exposure of animals and humans to ozone (O_3) damages the lungs and the damage is associated with lipid peroxidation. Although ozone is not a radical itself (Chapter 1), it can react with a wide variety of organic molecules, including membrane lipids, to produce radical species, and it can thus stimulate lipid peroxidation. For example, exposure of erythrocytes to ozone inactivates membrane-bound enzymes and causes formation of protein aggregates. Reduction of fluidity has been observed in ozone-treated plant cell membranes.

Finally, a mixture of the enzyme lactoperoxidase with hydrogen peroxide and iodide (I^-) ions can peroxidize membranes, perhaps due to the formation of an iodine radical that initiates peroxidation by abstracting hydrogen atoms. Whether or not such reactions play any role in the anti-bacterial action of lactoperoxidase (Chapter 2) has yet to be established.

4.5. Peroxidation of other molecules

Membrane lipids are not the only molecules found in cells that contain many $\diagup C{=}C\diagdown$ bonds. Several others are present, and they can be peroxidized under appropriate conditions. For example, the aldehyde retinal is probably oxidized as it sensitizes singlet-oxygen formation (Chapter 2). Retinal is obtained in the body by the oxidation of *retinol*, which has a —CH_2OH group in place of the —CHO in retinal. Retinol is essential in the human diet, usually being known as *vitamin A*, one of the fat-soluble vitamins. Like retinal, vitamin A can undergo peroxidation—

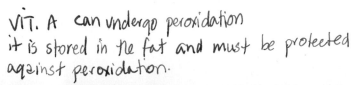

exposure of it to Fe(II) salts *in vitro* causes rapid oxidation to a mixture of products. Any excess vitamin A absorbed from the gut is stored in the fat deposits of adipose tissue, but it must be protected against peroxidation. Similarly, several antifungal antibiotics such as candicidin, nystatin, and amphotericin contain conjugated double bonds and are easily oxidized, resulting in loss of their antifungal activity. Peroxidation of the inner mitochondrial membrane can cause oxidative degradation of coenzyme Q, which has a polyunsaturated side-chain (Chapter 3).

The human bloodstream contains several *lipoproteins*—complexes of protein molecules with lipid. For example, the *pre-β-lipoproteins* carry fats made in the liver to the adipose tissue. Lipoproteins contain phospholipids, which are susceptible to peroxidation reactions that can, in turn, damage the proteins. Products of lipid peroxidation can be shown to accumulate in plasma stored at 4 °C, and peroxidation of the lipids alters the biological properties of the lipoproteins. Even in fresh plasma, some peroxidation of lipoproteins is evident and it is perhaps fortunate that plasma contains no 'free' iron-salts to catalyse hydroperoxide decomposition (Chapter 2). The most sensitive to peroxidation are the *low-density lipoproteins* which are involved in the transport of cholesterol around the body. The cholesterol they contain is mainly esterified (using the —OH group) to linoleic acid.

The low-density lipoproteins bind to receptors on the surface of cholesterol-requiring cells and then enter these cells, being degraded in lysosomes and the cholesterol thereby released. In the disease *familial hypercholesterolaemia* the receptors are absent or malfunctioning so that cholesterol in low-density lipoproteins accumulates in the blood and causes the formation of *atherosclerotic plaques* in blood vessels. Deposition of these plaques interferes with the smooth flow of blood and can cause blood clotting. If the blood supply to a part of the brain or the heart is shut off in this way, *ischaemia* (Chapter 3) results, causing a stroke or heart attack respectively. Highly peroxidized or malonaldehyde-treated low-density lipoproteins do not interact with receptors properly, which might conceivably lead to similar pathological consequences. Hence peroxidation of lipoproteins must be carefully controlled. Oxidation products of cholesterol, also formed during lipid peroxidation, have been shown to be damaging to arterial walls.

4.6. Measurement of lipid peroxidation

A number of techniques are available for measuring the rate of peroxidation of membrane lipids or fatty acids. Each technique measures something different, however, and no one method by itself can be said to be an accurate measure of 'lipid peroxidation'.

1. 4.6.1. Uptake of oxygen   *measurement of O_2 uptake Rate*

Since peroxidation is accompanied by the uptake of oxygen in the formation of peroxy radicals (Fig. 4.6), and also in subsequent decomposition reactions, measurement of the rate of oxygen uptake in an oxygen-electrode is a useful overall index of the progress of peroxidation. Indeed, de Saussure used this method in his pioneering studies on walnut oil, although only a simple mercury manometer was available to him.

2. 4.6.2 Titrimetric analysis using iodine release *measure the formation of I_2 from the rxn between I^- and ROOH*

Lipid hydroperoxides are capable of oxidizing iodide (I^-) ions into iodine (I_2), which may be estimated by titration with sodium thiosulphate

$$ROOH + 2I^- + 2H^+ \rightarrow I_2 + ROH + H_2O$$

Although this method can be used in studies upon purified lipids, it is rarely of value in biological systems because of the presence of many other oxidizing agents that can cause iodine production from I^-, such as hydrogen peroxide. The amounts of peroxide present at a given time will obviously depend not only on the rate of initiation of peroxidation but also on how quickly they are decomposing to give other products.

3 4.6.3. Diene conjugation *formation of conjugate Dienes; these absorb U.V light*

The oxidation of unsaturated fatty-acid side-chains is accompanied by the formation of conjugated dienes (Fig. 4.6), which absorb ultraviolet light in the wavelength range 230–235 nm (Fig. 4.10). This UV-absorption reaches a plateau just before oxygen uptake is observed. The method is extremely useful in studies upon pure lipids and it measures an early stage in the peroxidation process, but it often cannot be used directly on biological materials because many of the other substances present, such as haem proteins, chlorophylls, purines, and pyrimidines, absorb strongly in the ultraviolet, and create such a high background that spectrophotometric measurements become grossly inaccurate. Another source of error is that the polyunsaturated fatty acids themselves absorb UV light at only slightly lower wavelengths than the conjugated dienes (Fig. 4.6). The breakdown of lipid hydroperoxides produces several carbonyl compounds that absorb in the ultraviolet, and this must be borne in mind.

4 4.6.4. Measurement of hydrocarbon gases

This technique, developed by Riely, Cohen, and Lieberman in the USA, is based on the formation of hydrocarbon gases such as pentane and

Measurement of hydrocarbon gases by the technique of gas-liquid chromatography.

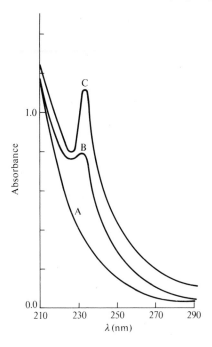

Fig. 4.10. Lipid peroxidation followed by the formation of conjugated dienes. Ethyl linoleate (an ester of linoleic acid and the alcohol ethanol) was purified, and its light absorption plotted at different wavelengths (plot A). Plot B shows the sample after oxidation by air at 30 °C for 8 h; and plot C, a sample in which peroxidation had been speeded up by addition of nitrogen dioxide (see Chapter 6). In both cases the 'shoulder' of UV-absorbance due to conjugated diene formation is clearly visible. By courtesy of Dr W. Pryor.

ethane during lipid peroxidation (Section 4.2). These gases can be easily measured by the technique of gas–liquid chromatography. It must be emphasized that they are minor end-products of peroxidation, and their formation depends on the presence of transition-metal ions to decompose hydroperoxides. Hence an increased rate of gas production might reflect increased availability of such metal ions rather than increased initiation of peroxidation. Some of the hydrocarbon gases are metabolized in the liver, introducing a further source of error. For example, drugs that affect liver metabolism might be thought to be changing the rate of lipid peroxidation *in vivo*. Bearing this in mind, the great value of the technique is that formation of these gases can be measured in the expired breath of whole animals and of humans, i.e. this is potentially an assay for lipid peroxidation *in vivo*. The expired breath is passed through silica gel at low temperature to adsorb and concentrate the hydrocarbons, which are then desorbed and assayed in a gas chromatograph. The greatest care

Fig. 4.11. Hydrocarbon gases. Control experiments for gas production by gut bacteria are essential.

must be taken in experiments of this kind to control for hydrocarbon production from the bacteria always present on the skin (and fur) and in the gut. It is quite possible to ignite the flatulence from cows because of the rich variety of hydrocarbons present (Fig. 4.11)! 'Normal' air in large cities is contaminated with hydrocarbons from motor vehicles, creating a high 'background', so that the air breathed by animals or human subjects must first be purified. In fact, the major hydrocarbon in expired human breath is isoprene, which is probably formed during biosynthesis of cholesterol. Increased production of pentane from human volunteers during severe exercise has been reported, and injection of cumene hydroperoxide into animals increases their ethane production. *Tert*-butyl hydroperoxide has been shown to increase ethane production in the isolated rat liver.

4.6.5. Measurement of other end-products of peroxidation

There are many 'end-products' of peroxidation in addition to hydrocarbon gases, and these could be measured as well. Thin-layer chromatography has allowed separation and identification of a wide variety of carbonyl compounds and other products; and carbonyl compounds can also be separated and assayed by gas–liquid chromatography, sometimes after reaction with *2,4-dinitrophenylhydrazine* to form products known as *2,4-dinitrophenylhydrazones*. Table 4.2 (Section 4.2) shows the formation of carbonyl compounds, including malonaldehyde, during peroxidation of

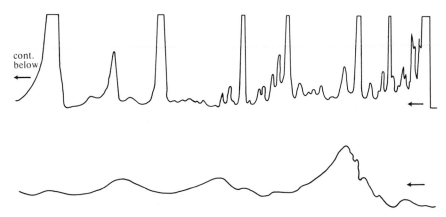

cont.
below
←

Fig. 4.12. Secondary oxidation products formed during the autoxidation of linoleic acid. Separation of carbonyl compounds by gas–liquid chromatography following the autoxidation of linoleic acid for five days. Many of the compounds arise from peroxidic precursors which fragment during the heating conditions of chromatography, especially if the column contains traces of transition-metal ions.

fatty acids. A huge variety of different compounds is formed, as can be seen in Fig. 4.12 for peroxidizing linoleic acid.

6.) 4.6.6. Loss of fatty acids

Lipid peroxidation results in the ultimate destruction of unsaturated fatty-acid side-chains, so it is possible to measure the overall rate of the process by measuring the loss of each fatty acid. The membrane under study must be disrupted and the lipids hydrolysed to release the fatty acids, which can then be chemically converted into volatile products (e.g. by formation of esters with methanol) and separated by gas–liquid chromatography. Great care must be taken to avoid peroxidation of the fatty acids during the hydrolysis and extraction procedures, e.g. by carrying out the reactions under nitrogen gas. Additional information can be gained by separating the different classes of membrane lipids (Fig. 4.1) before hydrolysis to release the fatty acids. Table 4.3 shows the effect of lipid peroxidation on the erythrocyte membrane. The rapid loss of polyunsaturated fatty acids is obvious.

7.) 4.6.7. Light emission

As we saw in Chapter 3, Professor Chance's group in the USA has pioneered the study of chemiluminescence as an index of radical reactions taking place in intact animal organs, either perfused or *in situ*.

Table 4.3. Loss of fatty acids during peroxidation of the red blood cell membrane

Number of carbon atoms in fatty acid	number of $>\!C\!=\!C\!<$ bonds in fatty acid	Percentage of total fatty acids in normal membrane	Percentage after lipid peroxidation
16	0	21	21
18	0	14	14
18	1	12	11
18	2	10	8
20	4	15	5
22	0	3	3
22	4	2	1

Stimulation of lipid peroxidation in isolated perfused animal lung or liver by supplying *tert*-butylhydroperoxide is accompanied by increased light emission, a phenomenon also seen in isolated liver microsomes treated with this peroxide. Hence one source of light *in vivo* is the peroxidation of membrane lipids, the light probably arising both from the formation of something resembling singlet oxygen during the peroxidation (Section 4.4) and also from the formation of carbonyl compounds in their excited states (Chapter 3). Measurements of light emission in isolated liver cells correlate well with other techniques of measuring lipid peroxidation, such as ethane production, which makes it a potentially useful assay method. It must be realized that other chemical reactions can produce light however.

4.6.8 Measurement of fluorescence

Fluorescent pigments partly derived from lipids are known to accumulate in tissues as a function of age. These so-called *age-pigments* are complexes of lipid oxidation products with protein, and more shall be said about them in Chapter 8.

The reaction of carbonyl compounds, such as malonaldehyde, with side-chain amino-groups of proteins, free amino-acids, or even nucleic-acid bases produces products known as *Schiff bases*.

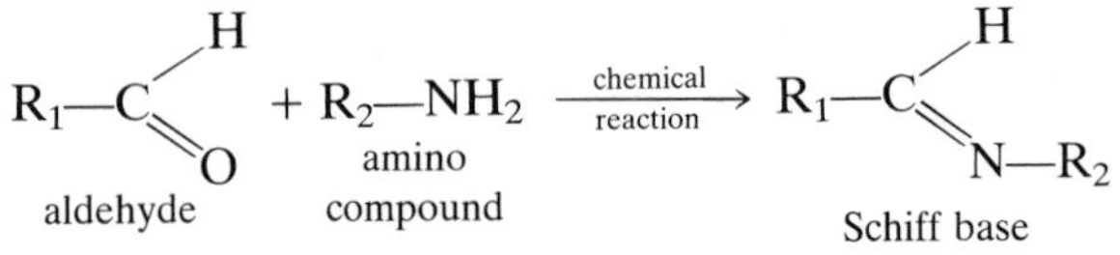

Malonaldehyde, with two carbonyl groups, can cross-link two amino

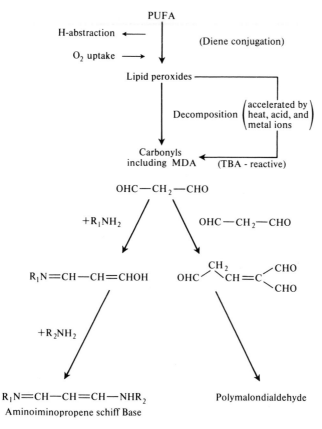

Fig. 4.13. Formation of fluorescent products during lipid peroxidation.

compounds to produce fluorescent molecules that have the general formula R_1—N=CH—CH=CH—NH—R_2 where R_1 and R_2 are the compounds to which the amino groups are attached. The products are called *aminoiminopropene Schiff bases* (Fig. 4.13) and seem to be the precursors of the age-pigments. Such fluorescent Schiff bases can be easily extracted from biological systems for analysis. Fluorescence is a very sensitive method for measuring peroxidation and correlates well with other methods, but it is, of course, a measurement of a late stage in the peroxidation process. The exact wavelength at which the products fluoresce depends on the nature of the side-chains R_1 and R_2. Phosphatidylethanolamine (Fig. 4.1) and the closely related lipid phosphatidyserine

$$\underset{\substack{| \\ \text{R—CH}_2\text{—CH—COO}^-}}{\overset{+}{\text{NH}_3}}$$

(instead of $\text{R—CH}_2\text{—CH}_2\text{—}\overset{+}{\text{NH}_3}$) both have free amino-groups and can also react with carbonyls to give fluorescent products.

A common property of aldehydes is their *polymerization*—several molecules joining together to form a larger molecule. This can happen with malonaldehyde, and the polymeric products are fluorescent (Fig. 4.13). Autoxidizing polyunsaturated fatty acids develop fluorescence even though no compounds with —NH$_2$ groups are present, which may be due to polymers from malonaldehyde and other aldehydes.

4.6.9. The thiobarbituric acid (TBA) test

The TBA test is one of the oldest and most frequently used tests for measuring the peroxidation of fatty acids, membranes, and food products. It is the easiest method to use (the material under test is merely heated with thiobarbituric acid under acidic conditions, and the formation of a pink colour measured at about 532 nm) and it can be applied to crude biological systems. Unfortunately, the simplicity of the test has led many scientists into using it as an index of peroxidation without understanding exactly what it measures.

Small amounts of malonaldehyde are produced during peroxidation (Table 4.2) and can react in the TBA test to generate a coloured product, probably by the reaction below.

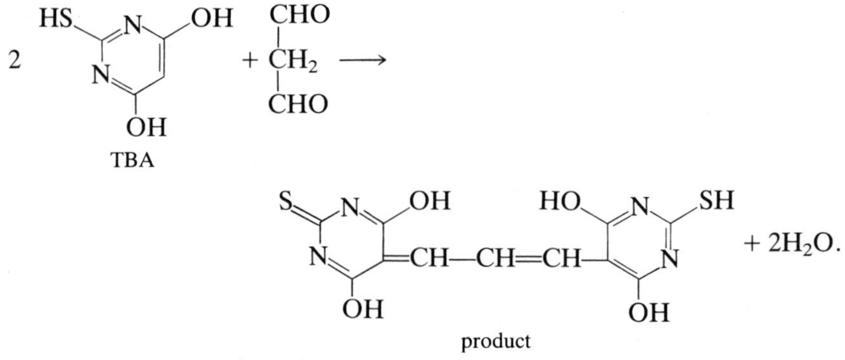

In acid solution the product absorbs light at 532 nm and fluoresces at 553 nm, and it is readily extractable into organic solvents such as butan-1-ol. Construction of a calibration curve for the assay is complicated by the fact that malonaldehyde is unstable and must therefore be prepared immediately before use by hydrolysing its derivatives, 1,1,3,3-tetramethoxypropane or 1,1,3,3-tetraethoxypropane

$$\begin{array}{c} RO \\ \diagdown \\ \diagup \\ RO \end{array} CH-CH_2-CH \begin{array}{c} OR \\ \diagup \\ \diagdown \\ OR \end{array} \qquad \begin{array}{l} R = C_2H_5 \text{ (ethyl)} \\ \text{or } CH_3 \text{ (methyl).} \end{array}$$

In studying the biological properties of malondialdehyde formed in this way it is essential to ensure complete hydrolysis of these derivatives and to bear in mind that the solution will contain four molecules of ethanol (or methanol) per molecule of malonaldehyde. For example, some reports that malonaldehyde is carcinogenic probably originated from the biological activities of partially hydrolysed derivatives of the above compounds.

Because the TBA test is calibrated with malonaldehyde (MDA), the results are often expressed in terms of the amount of MDA produced in a given time. This has sometimes given the impression that the TBA test detects only free MDA, and so measures the amount of free MDA in the peroxidizing lipid system. However, it was shown as long ago as 1958, in studies with peroxidizing fish oil, that 98 per cent of the MDA that reacts in the TBA test was not present in the sample assayed but formed by decomposition of lipid peroxides during the acid heating stage of the TBA assay.

Peroxidizing fatty acids with only one or two $>C=C<$ bonds give little pink colour in the normal TBA assay. Formation of MDA has been suggested by Pryor in the USA to be due to the formation of cyclic peroxides and endoperoxides that undergo fragmentation (Figs. 4.6 and 4.7). Fragmentation requires the presence of iron salts in the reagents used in the test. Indeed, removal of such iron salts prevents TBA-reactivity from developing. This can lead to artefacts in studies of the action of metal chelating agents on lipid peroxidation; they will also affect colour development in the TBA test itself.

Several compounds other than MDA give products that absorb at, or close to, 532 nm on heating with TBA; Table 4.4 lists some of these. Simple measurement at 532 nm after a TBA test could therefore include contributions from these substances, although fluorescence measurements can distinguish the products they form from the 'real' TBA–MDA adduct. Exposure of several carbohydrates and amino acids to hydroxyl radicals produced by ionizing radiation or by chemical systems produces TBA-reactive material. Further, the exact amount of colour developed with these alternative compounds depends on the type and strength of acid used in the TBA test, and on the time of heating.

Table 4.4. Reaction of biological molecules with thiobarbituric acid (TBA)

Compound tested	Absorbance maxima of product (nm)	TBA reactivity	
		Fluorescence of product	
		excitation wavelength (nm)	emission wavelength (nm)
Malonaldehyde	532 (acid)	532	553 (strong)
Ethanal + sucrose	447, 532	532	556 (weak)
Biliverdin	460, 532, 560	522, 560	580 (weak)
Glyoxal	460, 522, 550	524	540 (weak)
Furfuraldehyde	510	Not significant	

Finally, it must be pointed out that any free MDA that is formed *in vivo* will be rapidly metabolized. For example, it seems to be a substrate for mitochondrial aldehyde dehydrogenase activity, being oxidized into malonic acid that is then decarboxylated to carbon dioxide and acetate. Indeed, it is perhaps fortunate that most of the MDA measured in the TBA test is formed from lipid peroxides during the test itself, since if it only measured free MDA the test would be far less useful!

4.6.10. Summary—which is the method of choice?

The simple answer is "none of them". Each assay measures something different. Diene conjugation tells one about the early stages of peroxidation, as does iodometric analysis of lipid hydroperoxides. In the absence of metal ions to decompose lipid hydroperoxides there will be little formation of hydrocarbon gases, carbonyl compounds, or their fluorescent complexes, which does not necessarily mean therefore that nothing is happening. The TBA test can still be used because of decomposition of peroxides during the assay itself. Note that traces of iron salts must be present in the reagents to allow this, however. Analysis of hydrocarbon gases is a measure of a minor side-reaction. Changes in the mechanism of hydroperoxide decomposition might conceivably alter the amounts produced without any change in the overall rate of peroxidation.

Whatever method is chosen, one should think clearly *what* is being measured and *how* it relates to the overall lipid peroxidation process. Whenever possible, two or more different assay methods should be used.

4.7. Protection against lipid peroxidation

4.7.1. Structural organization of the lipids

Unsaturated fatty acids dispersed in organic solvents, or with detergents in aqueous solution, can be peroxidized easily. Similarly, phospholipids in simple micellar structures (Fig. 4.3) are peroxidized more rapidly than they are in lipid bilayers. Ingold in Canada has suggested that this may occur because lipid peroxy radicals are more polar than other hydrocarbon side-chains, and tend to move away from the membrane interior towards the surface of the bilayer, hence reducing the efficiency of initiation.

The presence of cholesterol in cell surface membranes can influence their susceptibility to peroxidation, probably both by intercepting some of the radicals present and by affecting the internal structure of the membrane by interaction of its large hydrophobic ring structure with fatty-acid side-chains. Figure 4.14 shows this effect in phospholipid liposomes. Calcium (Ca^{2+}) ions have been reported to alter the rate of the lipid peroxidation observed when liposomes or erythrocyte membranes are exposed to iron salts. Again, this may be an effect on the organization of lipids in the membrane.

As lipid peroxidation proceeds in any membrane, several of the products produced have a detergent-like activity, especially released fatty acids or phospholipids with one of their fatty-acid side-chains removed.

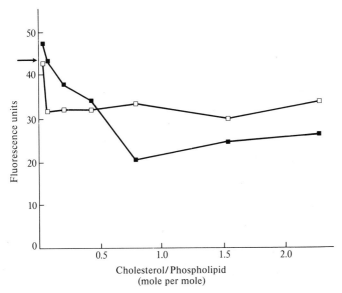

Fig. 4.14. Effect of cholesterol and cholesterol acetate on the rate of peroxidation of ox-brain phospholipid liposomes. Cholesterol (black square-points) or its ester with acetic acid (open squares) was incorporated into liposomes in various amounts during their formation, and peroxidation (induced by iron salt plus ascorbate) was measured by a fluorescence method. Small amounts of cholesterol promote peroxidation, but larger amounts depress it. Arrow indicates peroxidation control with no cholesterol added.

This will contribute to increased membrane disruption and further peroxidation. It is especially true of chloroplast membranes, in which peroxidation activates enzymes that hydrolyse membrane lipids (Chapter 5).

Disruption and damage to cells and tissues therefore increases the rate of lipid peroxidation, which, of course, will make the damage worse. Hence, lipid peroxidation usually accompanies tissue damage, which does not mean that peroxidation initially caused the damage. For example, if a whole organ is removed from a healthy animal, and a tissue sample homogenized and assayed for peroxidation products (e.g. conjugated dienes or TBA-reactivity), few are usually found. However, if the homogenate is allowed to stand on the bench before assay, peroxidation proceeds rapidly. Cell disruption causes the release of intracellular 'pools' of iron salts (Chapter 2) into the medium, and also the release of hydrolytic enzymes from lysosomes, which can degrade metalloproteins and release more metal ions. Microsomes show rapid lipid peroxidation *in vitro* under appropriate conditions but it may be that the membrane rearrangements that occur during homogenization to form this fraction have increased its susceptibility to peroxidation (Chapter 3).

4.7.2 Vitamin E

Vitamin E (otherwise known as *α*-tocopherol), is essential in the diet of animals; lack of it causes a wide variety of symptoms including infertility in female rats (the foetus is resorbed before birth), sterility in male rats, dogs, cocks, rabbits, and monkeys, haemolysis in rats and chicks, muscular degeneration in rabbits, guinea pigs, monkeys, ducks, mice, and minks, 'white-muscle disease' in lambs and calves, and cerebellar degeneration in chicks. By contrast, absence of vitamin E from the human diet does not cause any specific deficiency disease, although there is some evidence that lack of it in premature babies can predispose to haemolytic anaemia, probably due to increased fragility of the erythrocyte membrane (see below). There is no rigorous evidence that human muscular dystrophy or multiple sclerosis respond in any way to vitamin E administration. Other tocopherols exist, but are less likely to be of biological importance (Table 4.5).

Vitamin E is a fat-soluble molecule whose structure is shown in Fig. 4.15. Being hydrophobic, it will tend to concentrate in the interior of membranes. For example, the mitochondrial membranes contain about one molecule of *α*-tocopherol per 2100 molecules of phospholipid but there is much more vitamin E than this in the chloroplast thylakoid membrane and in the outer segment membranes of the retinal rods. Both those membranes must be especially protected against peroxidation *in*

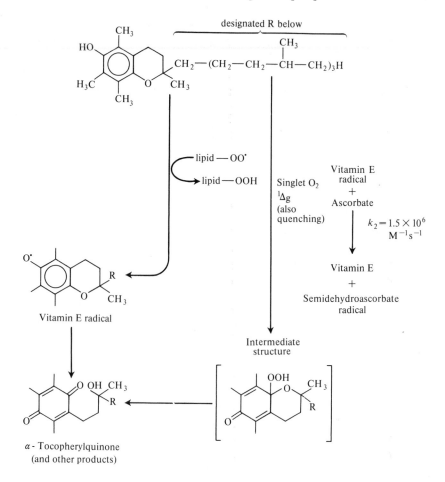

Fig. 4.15. Structure of vitamin E and its reaction with vitamin C. The long, hydrophobic side-chain makes the molecule lipid-soluble whereas the ring structure allows reaction with radicals. During peroxidation of cell membranes *in vitro* some vitamin E becomes converted into the quinone form.

vivo (Chapter 5). Tocopherol is also concentrated in the adrenal glands and in blood lipoproteins. Indeed, it is the major, if not the only, lipid soluble antioxidant in human blood plasma.

Since the damaging effects caused by lack of vitamin E in animals are partially or completely alleviated by feeding synthetic antioxidants (Section 4.7.3.), it seems to function *in vivo* as a protector against lipid peroxidation. How could it do this? Firstly, vitamin E both quenches and reacts with singlet oxygen and could therefore protect the membrane against this species (Chapters 2 and 5). It also reacts with the superoxide radical, (page 62), although this is less important since the reaction is slow, and O_2^- does not initiate lipid peroxidation. Probably of greatest importance in most membranes is the fact that vitamin E can react with lipid peroxy radicals to form vitamin E radicals that are insufficiently reactive to abstract H· from the membrane lipids. It thus interrupts the chain reaction of lipid peroxidation by acting as a *chain terminator*. The vitamin E radicals produced are fairly stable because the unpaired electron on the oxygen atom (Fig. 4.15) can be delocalized into the aromatic ring structure (see Appendix), so increasing stability.

Slater's group in England has shown that the vitamin E radical can be reduced back to vitamin E by ascorbate (Fig. 4.15) and this may be a further mechanism by which high concentrations of vitamin C protect against peroxidation. Reduction would presumably occur close to the membrane surface since it is unlikely that the polar vitamin C molecule would enter the hydrophobic interior. Feeding guinea-pigs on a diet restricted in vitamin C has been reported to decrease the vitamin E content of liver and lung, even though dietary intake of this vitamin was normal, consistent with an interaction between the two vitamins *in vivo*.

Professor J. A. Lucy in England has suggested that vitamin E may also protect against peroxidation by modifying membrane structure (Section 4.7.1.). If phospholipid liposomes are prepared, and peroxidation stimulated by adding iron salts and ascorbic acid, subsequent addition of vitamin E to the reaction mixture has little effect. However, if vitamin E is incorporated into the liposomes during their preparation it has a powerful protective effect as shown in Fig. 4.16. Esterification of the —OH group with acetic acid to give tocopherol acetate, which prevents reaction with lipid—OO· radicals (Fig. 4.15), greatly decreases the protective action at low concentrations which suggests that any 'structural organization' effect makes only a small contribution to protection. The protective effect of vitamin E acetate at higher concentrations (Fig. 4.16) might be attributed to a structural effect, however, and there is evidence from experiments upon cell cultures that the presence of vitamin E can affect the types of fatty acids that become incorporated into membrane lipids.

There is considerable evidence that vitamin E functions to protect against lipid peroxidation *in vivo*, apart from the experiments with synthetic anti-oxidants mentioned above. Feeding unsaturated fats to animals increases their requirement for vitamin E, e.g. for every 1 per cent of corn oil fed to young pigs above 4 per cent of the diet, 100 mg extra vitamin E is required. Chicks fed on lard, a mainly saturated fat, can survive without vitamin E for several weeks. Tissue samples taken from vitamin E-deficient animals show evidence of peroxidation (e.g. as TBA-reactive material), and tissue homogenates or subcellular fractions

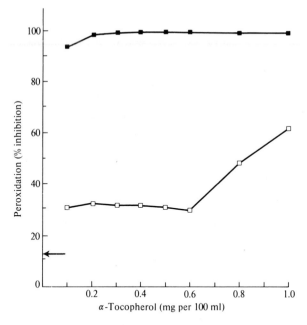

Fig. 4.16. Effect of tocopherol and tocopherol acetate on peroxidation of ox-brain phospholipid liposomes. Liposomes were prepared incorporating vitamin E at the concentrations stated, and the effect on peroxidation stimulated by iron salts and ascorbate was measured. Black square-points, vitamin E; open squares, vitamin E acetate. The arrow shows the inhibition of peroxidation observed when vitamin E at 1.0 mg per 100 ml was added to preformed liposomes. For further details see *Res. Comm. chem. Pathol. Pharmacol.* (1978) **22,** 563–72. Vitamin E added to the outside of preformed liposomes has much smaller inhibitory effects against lipid peroxidation.

from such animals peroxidize much more rapidly than normal when incubated *in vitro*. Vitamin E-deficient mice are more sensitive to the toxic effects of pure oxygen than are controls. Vitamin E-deficient rats exhale more pentane and ethane gas, and accumulate fluorescent 'age-pigments' much more rapidly than normal, especially if they are fed a diet rich in polyunsaturated fatty acids. This increased pigment accumulation is seen especially in the retinal pigment epithelium (Chapter 5).

Although a lack of vitamin E in the diet of adult humans does not produce obvious acute signs of disease, probably because of the presence of other mechanisms of protection against peroxidation (see below), such a lack does increase susceptibility to peroxidation. For example, the amount of pentane in the expired air of some human volunteers during severe exercise could be decreased by feeding them extra vitamin E.

Depletion of body vitamin E stores usually occurs in adult humans only after prolonged intravenous feeding or abnormalities of fat absorption in the gut (since it is a fat-soluble vitamin, it enters the body dissolved in the dietary fats). Erythrocytes from such patients show an increased susceptibility to haemolysis induced *in vitro* by conditions favouring lipid peroxidation. Administration of vitamin E has been reported to decrease the neurological symptoms seen in patients with abnormal fat absorption. Premature babies have low vitamin E levels, and their erythrocytes are more susceptible to haemolysis *in vitro* than normal. Sometimes haemolysis occurs *in vivo*, for reasons unknown, and this *haemolytic syndrome of prematurity* responds to vitamin E therapy. Vitamin E also protects premature babies against retrolental fibroplasia (Chapter 5). The occasional haemolysis occurring in patients with thalassaemia or glucose-6-phosphate dehydrogenase deficiency (Chapters 2 and 3) can be decreased by administration of extra oral vitamin E, and a similar protective effect has been suggested to occur in sickle cell anaemia. In these diseases there is extra 'oxidant stress' or a decrease in other protective mechanisms so that the effects of vitamin E are more readily seen.

Just as there have been dietary 'fads' for the consumption of increased amounts of vitamin C (Chapter 3), vitamin E has also been claimed to be a miracle-worker. Since adult humans can survive for long periods without it unless some other abnormality is present, this is unlikely to be true. If large doses are taken orally, most is not absorbed and is excreted in the faeces. Suggestions have been made that vitamin E may be of use in the treatment of degenerative arterial diseases. It must not be assumed that large doses are harmless, however, since they have been shown in adult human volunteers to decrease the ability of white blood cells to kill invading bacteria and to increase the dietary requirement for vitamin K, necessary for the blood coagulation system (Chapter 7).

Finally, it must not be assumed that the only biological role of vitamin E is to act as an antioxidant. There is evidence that it influences the metabolism of arachidonic acid in platelets and leucocytes (Chapter 7) and that a 'tocopherol oxidase' enzyme is involved in the response of plant tissues to certain hormones.

4.7.3. Other antioxidants

Studies of antioxidants are important because they can be used to protect polymers and foodstuffs against oxidative damage, e.g. during sterilization of food by ionizing radiation, and for possible use in the treatment of patients who have ingested compounds that increase lipid peroxidation *in vivo* (Chapter 6). Table 4.5 summarizes some of the compounds that have been investigated. α-Tocopherol (vitamin E) itself is widely used in many

Table 4.5. Antioxidant inhibitors of peroxidation

Name	Structure	Comments
Tocopherols	α-T = R$_1$, R$_2$, R$_3$, all CH$_3$ β-T = R$_1$, R$_3$, CH$_3$; R$_2$, H γ-T = R$_1$, R$_2$, CH$_3$; R$_3$, H δ-T = R$_1$, CH$_3$; R$_2$, R$_3$, H.	β-, γ-, δ-Tocopherols less good at reacting with peroxy radicals than α-tocopherol; less well absorbed in gut ($\alpha > \gamma > \beta \geq \delta$) (thus less important). (See *J. Am. chem. Soc.* (1981) **103**, 6472–7.)
Butylated hydroxyanisole (BHA)		Very often added to foodstuffs. Acts as an antioxidant by hydrogen donation, which is common to all the phenolic (and amine) antioxidants. For example, addition of BHA to fat (e.g. butter) increases its storage life from a few months to a few years.
Butylated hydroxytoluene (BHT)		Very often added to foodstuffs. Non-toxic, but some evidence for metabolism by liver cytochrome P$_{450}$ system. Very large doses cause liver damage in mice. Detected in the body fat of US citizens!

Table 4.5 (*Continued*)

Propyl gallate

Fairly water-soluble. Good inhibitor of lipid peroxidation. Often added to foodstuffs.

Nordihydroguaiaretic acid (NDGA)

Occurs naturally in resinous exudate of *Larrea divaricata* (American creosote bush) and in some other plants. Often added to foodstuffs and several polymers (e.g. rubber, lubricants).

N,*N'*-Diphenyl-*p*-phenylene diamine (DPPD)

Popular antioxidant *in vitro*. Has been widely used for animal studies of *in vivo* lipid peroxidation.

6 hydroxy-1,4-dimethylcarbazole (HDC)

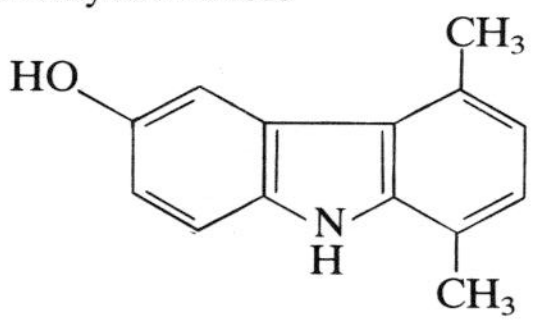

Very powerful inhibitor of lipid peroxidation *in vitro*.

Promethazine

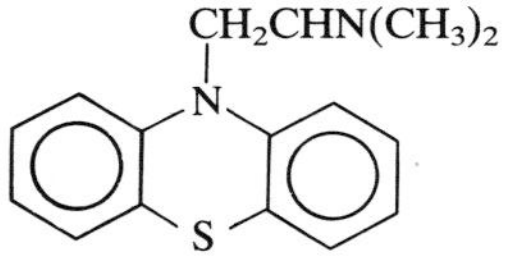

Inhibits lipid peroxidation *in vitro*. Prepared by the drug industry as an anti-histamine and sedative.

Chlorpromazine

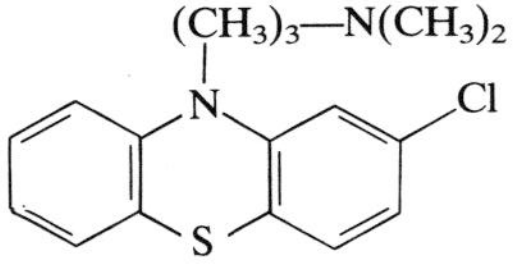

Used in the drug industry as a tranquillizer.

Ethoxyquin (Santoquin)

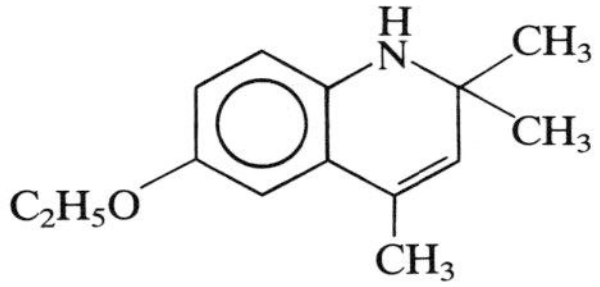

Frequently used in fruit canning. Powerful enzyme inducer *in vivo*. Widely used in longevity experiments in animals (Chapter 8).

foodstuffs although several of the synthetic antioxidants are more efficient in preventing rancidity on a weight-for-weight basis. As we have seen, however, vitamin E is an extremely efficient antioxidant in lipid bilayers. Which antioxidant is selected depends on the particular circumstances. For example, NGDA and BHT protected synthetic diet mixtures containing herring oil against peroxidation induced by γ-irradiation much more effectively than did vitamin E or propyl gallate. The antioxidants in Table 4.5 probably act, as does α-tocopherol, by intercepting lipid—OO$^{\cdot}$ and lipid–O$^{\cdot}$ radicals.

4.7.4. Glutathione peroxidase

We saw in Chapter 3 that glutathione peroxidase, a selenium-containing enzyme found in the cytosol and mitochondria of animal tissues, helps to dispose of hydrogen peroxide by catalysing the reaction

$$2GSH + H_2O_2 \rightarrow 2H_2O + GSSG$$

This enzyme is specific for GSH as a substrate but will act *in vitro* on a range of peroxides in addition to hydrogen peroxide, including cumene hydroperoxide, *tert*-butylhydroperoxide, progesterone 17α-hydroperoxide, several other steroid hydroperoxides (including cholesterol 7β-hydroperoxide at a low rate), and thymine hydroperoxide, as well as linoleic and linolenic acid hydroperoxides. In each case the peroxides (ROOH) are reduced to alcohols (ROH). Indeed, infusion of artificial organic hydroperoxides into animal organs *in situ* causes an increased formation and release of GSSG, much more so if the organs are taken from animals fed on diets deficient in vitamin E. Selenium deficiency in animals produces a variety of diseases that are strikingly similar to those induced by vitamin E deficiency, and to a considerable extent the effects of selenium deficiency can be overcome by giving excess vitamin E and vice versa. Exceptions to this generalization include the observation that selenium cannot protect female rats against foetal reabsorption caused by vitamin E lack, nor can vitamin E protect rats against the damage to the pancreas that occurs on selenium-deficient diets. A combined deficiency of selenium and vitamin E in the diet of animals is usually eventually fatal.

Feeding excess vitamin E to rats has been observed to decrease glutathione peroxidase activity in the tissues. Erythrocytes from selenium-deficient animals are more susceptible to haemolysis *in vitro* under conditions favouring lipid peroxidation than are normal erythrocytes. If rats on E-deficient diets are fed peroxidized methyl linoleate, the glutathione peroxidase activity of the small intestine increases, but not that of the liver. (Presumably peroxides in food lipids are dealt with by the intestinal enzyme.) Injection of peroxidized lipids does increase the activity of the liver enzyme, however. Injection of iron or copper salts into rats fed on diets deficient in both selenium and vitamin E causes a marked increase in the ethane content of the expired breath. The increase is smaller if either vitamin E or selenium is resupplied. Induction of lipid peroxidation by addition of Fe(III) salts chelated with ADP to isolated liver cells caused a decrease in their internal concentration of GSH. Administration of vitamin E to a child with an inborn deficiency of glutathione synthetase in white blood cells was observed to improve the functioning of these cells. Selenium-deficient rats are more susceptible to the toxic effects of elevated oxygen concentrations than normal.

This accumulation of evidence indicates that glutathione peroxidase plays an important role in protection against lipid peroxidation *in vivo*. As is usual, however, things are not so simple! Feeding a selenium-deficient diet to rats causes a drop in glutathione peroxidase activity in liver and erythrocytes, measured with hydrogen peroxide as substrate, to virtually zero within four weeks. However, if cumene hydro-peroxide is used as substrate, substantial activity remains in liver but not in erythrocytes. Further study has revealed that several animal tissues contain a *non-selenium glutathione peroxidase activity* that acts on artificial organic hydroperoxides but not on hydrogen peroxide. This activity appears to be due to some of the glutathione-S-transferase enzymes involved in the conjugation of GSH with 'foreign compounds' (Chapter 3). Not all these transferases will act on hydroperoxides; for example, the transferases found in the bovine eye lens will not, nor will 'transferase c' in guinea-pig liver. The balance between selenium-dependent glutathione peroxidase activity and the non-selenium-dependent activity varies in different body tissues and animal species as summarized in Table 4.6. Perfused livers from rats fed a selenium-deficient diet release no GSSG when hydrogen peroxide is infused, but do so when *tert*-butylhydroperoxide is infused, indicating that these non-selenium enzymes can function in whole organs. Non-selenium peroxidase activity (like 'real' glutathione peroxidase) is mainly located in the cytosol of tissues with small amounts in the mitochondria.

In view of their action upon hydroperoxides *in vitro*, one would expect both types of glutathione peroxidase to protect against lipid peroxidation in membranes by reducing lipid hydroperoxides (lipid—OOH) to stable hydroxyacids (lipid—OH), and thus preventing their decomposition to form alkoxy radicals. McCay *et al.* in the USA showed that peroxidation of rat liver microsomes in the presence of NADPH and ADP—Fe(III) was inhibited by addition of a soluble fraction from rat liver in the presence of GSH. However, the inhibition of peroxidation was not accompanied by formation of hydroxyacids in the membrane. Separation

Table 4.6. Selenium and non-selenium glutathione peroxidase activities in different animals

Animal	Organ studied	Percentage of total glutathione peroxidase activity detected that was due to the non-selenium enzymes
Rat	adrenal	38
	spleen	0
	liver	35
	lung	0
	heart	0
	kidney	31
	testis	91
Hamster	liver	43
Sheep	liver	81
Pig	liver	67
Chicken	liver	70
Human	liver	84
Rabbit	liver	50 (approx)
Mouse	liver	30

Total activity was measured using cumene hydroperoxide as substrate. Results are mostly abstracted from H. Sies *et al.* (1982) *Proc. Third Int. Symp. Oxidases Relat. Redox Systems* (eds. T. E. King *et al.*), Pergamon Press, Oxford, p. 169.

of the proteins present in the 'soluble fraction' showed that this GSH-dependent inhibition of peroxidation was due to a GSH transferase and that purified selenium-containing glutathione peroxidase itself had no inhibitory effect. In agreement with this, Purdy and Tappel in the USA found that glutathione peroxidase has little action on peroxidized fatty-acid side-chains of lecithin incorporated into liposomes, but they suggested that *in vivo* phospholipase enzymes cleave peroxidized fatty acids from the membrane lipids and thus allow glutathione peroxidase to act upon them. Italian scientists have purified, from pig liver and heart, an enzyme that inhibits peroxidation of lecithin liposomes and can reduce hydroperoxides in such liposomes as well as cumene or *tert*-butyl hydroperoxides. However, this enzyme had little activity on linoleic acid hydroperoxide and was not identical with selenium-containing glutathione peroxidase.

Obviously, the situation is confused and much more work needs to be done to investigate the mechanism by which selenium- and non-selenium-enzymes protect against peroxidation *in vivo*, as a great mass of evidence indicates that they do. One should not lose sight of the fact, however, that removal of hydrogen peroxide by the selenium enzyme *in vivo* will decrease the formation of hydroxyl radicals by the Fenton reaction and will thus prevent one route for the initiation of peroxidation. Perhaps the concentration of research effort on organic hydroperoxide metabolism has been misplaced.

Whether or not vitamin E is required in the diet of humans is debatable, as discussed previously, but the situation is clearer for selenium. This follows the discovery in the People's Republic of China of a selenium-responsive degenerative heart-muscle disease known as *Keshan disease*. Careful epidemiological studies showed that the incidence of this disease in different areas was correlated with that of various degenerative diseases in animals and that both it, and the animal diseases, could be prevented by administration of sodium selenite (Na_2SeO_3). In all affected areas, selenium concentrations in foodstuffs and in animals or humans eating them were found to be extremely low. A study of blood selenium concentrations led Chinese scientists to conclude that the minimum adequate value for humans is in the range 0.03–$0.04\,\mu g\,g^{-1}$, corresponding to a minimum daily dietary intake of $30\,\mu g$. Normal daily intakes in advanced countries are in the range 60–200 µg per day. Larger amounts have toxic effects and selenium-accumulating plants such as *Astragalus* species that can poison cattle are a particular nuisance to farmers in several parts of the world. Indeed, injection of large doses of selenite into rats appears to *increase* lipid peroxidation as measured by ethane exhalation! One case of Keshan disease has been reported from Long Island, New York, although the patient also had a zinc deficiency. Selenium deficiency might contribute to the symptoms of severe malnutrition. On the other hand, residents of low-selenium areas in Finland and New Zealand appear to suffer no acute ill-effects although it has been suggested that low blood selenium predisposes to heart disease. It must not, of course, be assumed that Keshan disease is *caused* by lack of active glutathione peroxidase although this enzyme activity is very low in sufferers from the disease. Selenium may well have other biochemical roles, including perhaps an involvement in haem metabolism, and it has been observed to protect against cancerous tumours induced in animals by certain carcinogens (Chapter 8).

H_2O_2-dependent glutathione peroxidase activity is not found in plant tissues, but some GSH transferases are present and might be able to act on hydroperoxides, although this has not been tested. Japanese scientists have identified an enzyme system in the microsomal fraction of pea seeds that catalyses hydroxylation of some organic compounds (e.g. indole or phenol) in the presence of hydroperoxides such as linoleic acid hydroperoxide or even hydrogen peroxide. The overall reaction may be written as follows, where SH is the substrate and ROOH the peroxide:

$$SH + ROOH \rightarrow SOH + ROH$$

The physiological significance of this *peroxygenase* enzyme system has yet to be established, however. Cytochrome P_{450} can bring about a similar reaction (Chapter 3).

4.7.5. Superoxide dismutase and catalase

Since hydroxyl radicals can initiate lipid peroxidation, enzymes that remove hydrogen peroxide and thus prevent formation of these radicals from it may be regarded as protective mechanisms. Hence glutathione peroxidase, catalase, ascorbate peroxidase, and perhaps the 'non-specific peroxidases' (Chapter 3) can help to protect against lipid peroxidation *in vivo*. Superoxide dismutase, which often inhibits formation of OH˙ from hydrogen peroxide in the presence of iron salts, could also be classified as a protective mechanism, especially if the direct reaction of O_2^- with hydroperoxides is important *in vivo*.

4.7.6. Extracellular antioxidants

The extracellular fluids of the human body, such as blood plasma, tissue fluid, sweat, cerebrospinal fluid, synovial fluid, or seminal plasma contain little superoxide dismutase, catalase, or glutathione peroxidase activities, although plasma from rats, mice or rabbits is richer in SOD activity. That small amount of SOD activity that is present in animal or human extracellular fluids has been shown by Marklund in Sweden to be largely due to a copper enzyme with a molecular weight of 135,000; much greater than the molecular weight of normal CuZn SOD (Chapter 3). This copper enzyme is a glycoprotein and is a true catalyst of O_2^- dismutation. It was originally named 'extracellular SOD' (EC-SOD), but small amounts of it are present within body tissues, especially lung. Even in lung, however, EC-SOD accounts for less than 10 per cent of total tissue SOD activity. Its biological significance is unknown.

If extracellular fluids are added to a lipid peroxidation system *in vitro* e.g. phospholipid liposomes in the presence of ascorbate and iron salts, they show a powerful inhibitory effect. This extracellular antioxidant activity has been neglected in recent years because of the interest in such enzymes as superoxide dismutase and glutathione peroxidase, and so our knowledge of it is grossly inadequate. Indeed, it has only been properly investigated in one system: blood serum. The blood lipoproteins contain vitamin E, dissolved in the lipid, which probably helps to protect them against peroxidation. However, fractionation studies by Stocks, Dormandy, and others in England showed that most of the protective activity against peroxidation was associated with two proteins, transferrin and caeruloplasmin (Table 4.7). Transferrin, and the related protein lac-

Table 4.7. Antioxidant activity in human serum

Group	Number of subjects	Antioxidant activity (% inhibition)	Caeruloplasmin (mg l^{-1})	Transferrin (g l^{-1})	Available iron-binding capacity (μ mol l^{-1})
Normal males, age 19–64 years	18	60.2 ± 9.3	270.4 ± 42.8	2.67 ± 0.67	49.6 ± 6.8
Normal females, age 19–64 years	21	73.4 ± 12.2	349.8 ± 110.2	2.45 ± 0.46	46.9 ± 7.2
Normals, all, mean age 32 years	39	66.9 ± 14.1	310.3 ± 128.3	2.56 ± 0.59	48.1 ± 8.4
Females on contraceptives, age 20–35 years	9	86.9 ± 7.3	484 ± 116.0	2.65 ± 0.66	51.8 ± 4.5
Newborn infants, full term	15	14.2 ± 4.8	90.1 ± 56.2	2.49 ± 0.42	44.6 ± 7.3
Thalassaemia (major), mean age 9 years	23	46.3 ± 9.4	297 ± 65.2	2.80 ± 0.42	3.2 ± 1.9
Wilson's disease, all		31.5 ± 11.7	68.2 ± 7.2		
Wilson's disease					
(I) caeruloplasmin > 50		39.6 ± 11.9	154 ± 5.1		
(II) caeruloplasmin < 50		24.1 ± 5.9	22 ± 9.0		
Primary biliary cirrhosis	30	86.1 ± 14.3	502 ± 12.3		
Alcoholic cirrhosis	12	63.9 ± 17.6	374 ± 15.9		
Chronic active hepatitis	15	62.6 ± 17.4	388 ± 10.2		
Liver cirrhosis (unknown cause)	10	62.1 ± 21.0	350 ± 74.2		

Antioxidant activity was measured as the ability of serum samples to inhibit the peroxidation of an ox-brain tissue homogenate.
Data from Gutteridge and Stocks (1981) *CRC Crit. Rev. Clin. Lab. Sci.* **14**, 257–329.

toferrin, can protect against peroxidation simply by binding iron salts (Chapter 2) since they are only partially iron-loaded *in vivo*, except in some disease states such as thalassaemia. Iron-saturated transferrin or lactoferrin lose their protective ability.

The inhibitory action of caeruloplasmin on lipid peroxidation is not fully understood. Like most copper proteins it will react with O_2^- although it does *not* catalyse O_2^- dismutation (Chapter 3). The concentration required to intercept O_2^- at a significant rate is much greater than that present in plasma. Further, caeruloplasmin is a powerful inhibitor of lipid peroxidation stimulated *in vitro* by iron salts and ascorbate even though superoxide dismutase itself has little protective effect in this system. Caeruloplasmin has been observed to protect red blood cells against haemolysis due to lipid peroxidation stimulated by added iron salts.

We saw in Chapter 3 that caeruloplasmin has a *ferroxidase* activity, i.e. it catalyses oxidation of Fe(II) to Fe(III). This might explain part of its action in inhibiting peroxidation stimulated by iron salts, since Fe(III) has a smaller stimulatory effect than does Fe(II). It is not the whole story, however, since caeruloplasmin inhibits peroxidation of liposomes or of erythrocytes stimulated by copper salts as well, perhaps by non-specific binding of copper ions.

Caeruloplasmin is one of the acute-phase proteins: a group of proteins whose synthesis by the liver in animals is increased in response to tissue damage, e.g. due to trauma, or in both acute and chronic inflammation (Table 4.8). It is tempting to speculate that increased caeruloplasmin synthesis is a mechanism to protect the tissues against further damage since, as we have already seen, disrupted tissues are more susceptible to lipid peroxidation.

Table 4.8. Some of the acute-phase plasma proteins in humans

Protein	Typical increase in concentration	Biological function
Caeruloplasmin	50%	See text
C3	50%	Third component of complement reaction with antigens
Antiproteases (e.g. α-antitrypsin)	2- to 4-fold	Inhibit proteolytic enzymes
Haptoglobin	2- to 4-fold	Binds haemoglobin
C-reactive protein (CRP)	Several-hundred-fold (very little present normally)	Function unknown (discovered by its ability to precipitate the 'C-polysaccharide' of *Pneumococci)*
Serum amyloid A protein (SAA)	Several-hundred-fold	See Chapter 8

Unlike serum, human cerebrospinal fluid contains micromolar concentrations of non-protein-bound iron salt (Chapter 2). Protection appears to be achieved by antioxidants in the fluid that are not identical with catalase, glutathione peroxidase, or superoxide dismutase, and their nature requires further investigation. Similarly, the antioxidant activity of seminal plasma appears to be important in protecting spermatozoa against oxidant injury.

Clearly, there is a lot about antioxidant protection *in vivo* that we do not yet know!

Watch out for autoxidation!

Arachidonic Acid Warning

The Department of Health and Social Security has issued a Safety Information Bulletin, *SIB(7)10,* containing a warning about storage of opened vials of arachidonic acid and its salts. This followed an incident in which a 50 mg sample of the substance violently self-ignited on reaching room temperature after being frozen for several months. The DHSS is anxious to draw this incident to the attention of all laboratory workers.

4.8. Further reading

Abrams, B. A., Gutteridge, J. M. C., Stocks, J., Friedman, M., and Dormandy, T. L. (1973). Vitamin E in neonatal hyperbilirubinaemia. *Archs Dis. Child.* **48,** 721.

Barber, A. A., and Bernheim, F. (1967). Lipid peroxidation: its measurement, occurrence and significance in animal tissues. *Adv. Gerontol. Res.* **2,** 355.

Barber, D. J. W., and Thomas, J. K. (1978). Reactions of radicals with lecithin bilayers. *Radiat. Res.* **74,** 51.

Barclay, L. R. C., and Ingold, K. U. (1981). Autoxidation of biological molecules. 2. The autoxidation of a model membrane. A comparison of the autoxidation of egg lecithin phosphatidylcholine in water and in chlorobenzene. *J. Am. Chem. Soc.* **103,** 6478.

Cadenas, E., and Sies, H. (1982). Low-level chemiluminescence of liver microsomal fractions initiated by *tert*-butylhydroperoxide. *Eur. J. Biochem.* **124,** 349.

Diplock, A. T. (1981). Metabolic and functional defects in selenium deficiency. *Philos. Trans. R. Soc.* **B294,** 105.

Dormandy, T. L. (1978). Free-radical oxidation and antioxidants. *Lancet* **i,** 647.

De Duve, C., and Hayaishi, O. (eds) (1978). *Tocopherol, oxygen and biomembranes.* Elsevier/North Holland, Amsterdam.

Daub, M. E. (1982). Cercosporin, a photosensitizing toxin from *Cercospora* species. *Phytopathol.* **72,** 370.

Finean, J. B., Coleman, R., and Michell, R. H. (1978). *Membranes and their cellular functions*, 2nd edn. Blackwells, Oxford.

Gardner, H. W. (1979). Lipid hydroperoxide reactivity with proteins and aminoacids—a review. *J. Agric. Food Chem.* **27**, 220.

Gray, J. J. (1978). Measurement of lipid oxidation: a review. *J. Am. Oil Chem. Soc.* **55**, 539.

Gutteridge, J. M. C. (1978). The membrane effects of vitamin E, cholesterol and their acetates on peroxidative susceptibility. *Res. Commun. chem. Pathol. Pharmacol.* **22**, 563.

Gutteridge, J. M. C. (1982). Fluorescent products of phospholipid peroxidation: formation and inhibition in model systems. In *Ceroid-Lipofuscinoses* (Batten's disease) (eds. D. Armstrong, N. Koppang, and J. A. Rider). Elsevier Biomedical, Amsterdam.

Gutteridge, J. M. C. (1982). Free-radical damage to lipids, amino acids, carbohydrates and nucleic acids determined by thiobarbituric acid reactivity. *Int. J. Biochem.* **14**, 649.

Gutteridge, J. M. C. and Quinlan, G. J. (1983). Malondialdehyde formation from lipid peroxides in the thiobarbituric acid test: the role of lipid radicals, iron salts, and metal chelators. *J. Appl. Biochem.* **5**, 293.

Gutteridge, J. M. C., and Stocks, J. (1981). Caeruloplasmin: physiological and pathological perspectives. *CRC Crit. Rev. Clin. Lab. Sci.* **14**, 257.

Jones, R., Mann, T., and Sherins, R. J. (1978). Adverse effects of peroxidised lipids on human spermatozoa. *Proc. R. Soc. (Lond.)* **B201**, 413.

Kaschnitz, R. M., and Hatefi, Y. (1975). Lipid oxidation in biological membranes. Electron transfer proteins as initiators of lipid autoxidation. *Archs Biochem. Biophys.* **171**, 292.

Kuhn, H., Gotze, R., Schewe, T., and Rapoport, S. M. (1981). Quasi-lipoxygenase activity of haemoglobin. *Eur. J. Biochem.* **120**, 161.

Marklund, S. L. (1982). Human copper-containing superoxide dismutase of high molecular weight. *Proc. Nat. Acad. Sci. US* **79**, 7634.

McBrien, D. C. H., and Slater, T. F. (eds) (1982). *Free radicals, lipid peroxidation and cancer.* Academic Press, London.

Muller, D. P. R., Lloyd, J. K., and Wolff, O. H. (1983). Vitamin E and neurological function. *Lancet* **i**, 225.

O'Brien, P. J. (1969). Intracellular mechanisms for the decomposition of lipid peroxide. Decomposition of a lipid peroxide by metal ions, haem compounds and nucleophiles. *Can. J. Biochem.* **47**, 485.

O'Brien, P. J. (1978). Hydroperoxides and superoxides in microsomal oxidations. *Pharmacol. Therap.* **A2**, 517.

Rosen, B. M. and Rauckman, E. J. (1981). Spin trapping of free radicals during hepatic microsomal lipid peroxidation. *Proc. Nat. Acad. Sci. US* **78**, 7346.

Shimizu, T., Kondo, K., and Hayaishi, O. (1981). Role of prostaglandin endoperoxides in serum thiobarbituric acid reaction. *Archs Biochem. Biophys.* **206**, 271.

Smith, L. L. (1982). *Cholesterol autoxidation.* Plenum Press, New York and London.

Tappel, A. L. (1979). Measurement of and protection from *in vivo* lipid peroxidation. In *Biochemical and clinical aspects of oxygen* (ed. W. S. Caughey). Academic Press, New York.

Tien, M., Morehouse, L. A., Bucher, J. R., and Aust, S. D. (1982). The multiple effects of EDTA in several model lipid peroxidation systems. *Archs Biochem. Biophys.* **218**, 450.

Tien, M., Svingen, B. A., and Aust, S. D. (1981). Superoxide-dependent lipid peroxidation. *Fed. Proc.* **40**, 179.

Wills, E. D. (1966). Mechanisms of lipid peroxide formation in animal tissues. *Biochem. J.* **99,** 667.

Wills, E. D. (1969). Lipid peroxide formation in microsomes. General considerations *Biochem. J.* **113,** 315.

Wills, E. D. (1969). Lipid peroxide formation in microsomes. The role of non-haem iron. *Biochem. J.* **113,** 325.

Wilson, R. B. (1976). Lipid peroxidation and atherosclerosis. *CRC Crit. Rev. Sci. Nutr.* **8,** 325.

Yagi, K (ed) (1982). *Lipid peroxides in biology and medicine.* Academic Press, New York.

5 Protection against radical damage: systems with problems

In previous chapters we have described the problems faced by several different organs and tissues in coping with oxygen, and the ways in which they can deal with oxygen radicals and lipid peroxidation. For example, the lung is exposed to the highest concentration of oxygen of any body tissue and it uses SOD, catalase, and glutathione peroxidase for protection (Chapter 3). Human erythrocytes have to keep their membranes intact for 120 days in the face of a constant flux of O_2^- from haemoglobin, even though they cannot readily replace membrane lipids damaged by peroxidation. Hence they rely on α-tocopherol, SOD, catalase, and glutathione peroxidase (Chapters 3 and 4). The swimbladder of at least one fish has a very high SOD activity to enable it to cope with high oxygen concentrations (Chapter 3). Even the earthworm *Eisenia foetida* increases its respiratory rate with temperature and also its SOD activity, presumably to cope with the increased O_2^- generation at higher temperatures.

The purpose of this chapter, however, is to focus on two systems that have exceptionally difficult problems—the chloroplasts of higher plants, and the mammalian eye. These very different systems have a lot in common.

5.1. The chloroplasts of higher plants

The chloroplasts present in the leaves of higher plants can be seen under the electron microscope to be bounded by an outer *envelope* consisting of two membranes separated by an electron-translucent space of about 10 nm (Fig. 5.1). The envelope encloses the *stroma* of the chloroplast, in which floats a complex internal membrane structure. The stroma is an aqueous solution containing various low-molecular-weight compounds plus a high concentration of proteins, most of which are the enzymes necessary to convert carbon dioxide into carbohydrate by a complex metabolic pathway known as the *Calvin cycle*. For its operation the Calvin cycle requires ATP and NADPH. The first enzyme in the Calvin cycle catalyses reaction of the 5-carbon sugar ribulose, 1,5-bisphosphate with carbon dioxide to form two molecules of phosphoglyceric acid. It is called 'ribulose bisphosphate carboxylase.

The internal membrane structure of the chloroplast, which floats in the stroma, is extremely complex. In the electron micrograph shown in Fig. 5.1, two distinct features may be recognized, i.e. regions of closely stacked membranes known as *grana* that are interconnected by a three-dimensional network of membranes known as the *stroma thylakoids.* The membranes of this complicated internal structure contain the photosynthetic pigments (mainly chlorophylls a and b, green pigments that absorb light in the blue and red regions of the spectrum, together with *carotenoids* and *xanthophylls,* yellow pigments absorbing blue light). They function to produce the NADPH and, by the process of *photophosphorylation*, the ATP needed to drive CO_2-fixation by the Calvin cycle in the stroma. The structure of chlorophylls a and b is described in Chapter 2.

Absorption of light energy by chlorophyll or other pigment molecules causes electrons to move into higher energy states (*excited singlet states*). Absorption of blue light by chlorophyll results in formation of a more excited state (second excited state) than does absorption of red light, but

Fig. 5.1. Electron micrograph of chloroplasts in the leaves of higher plants.

the second excited state loses its excess energy as heat extremely quickly, so that absorption of either red or blue light effectively produces the same first excited state of the chlorophyll molecules. Energy so absorbed can be lost by re-emission of light to give fluorescence; it can be lost as heat or it can be transferred to another molecule, e.g. an adjacent chlorophyll, the first molecule returning to the ground state, while the second becomes excited.

Certain special molecules of chlorophyll a in the thylakoid membrane (*reaction centre chlorophylls*) can, when excited, lose an electron to a neighbouring electron acceptor (A), producing a (chlorophyll$^+$ A$^-$) pair. This initial charge separation is the basic reaction of photosynthesis. Light energy absorbed by other chlorophyll a or chlorophyll b molecules can be transferred quite efficiently in the thylakoid membrane to the reaction centre chlorophylls. Indeed, each reaction centre chlorophyll is associated with a *light-harvesting array* of other pigment molecules that can channel energy to it.

Two different types of reaction centre, known as P$_{700}$ and P$_{680}$, can be detected in the chloroplasts of higher plants. Each contains chlorophyll a and is served by its own light-harvesting pigment system; the complexes containing P$_{700}$ and its harvesting pigments are referred to as photosystems I and those containing P$_{680}$ as photosystems II. Stroma thylakoids are especially rich in photosystems I whereas the grana are enriched in photosystems II. Electrons ejected from the P$_{680}$ reaction centre of a photosystem II are accepted by a molecule of unknown structure, usually referred to as 'Q'. From Q electrons pass 'down' an electron transport chain that contains plastoquinone, plastocyanin, and cytochrome f, and they eventually enter photosystems I to replace the electrons lost on excitation of the P$_{700}$ reaction centres (Fig. 5.2). The molecule which accepts excited electrons from photosystems I has not been conclusively identified, but it contains iron (not in the form of a haem ring) and sulphur. From this primary acceptor, electrons pass to other iron–sulphur proteins bound to the membranes, and then onto a soluble (stromal) iron–sulphur protein, known as *ferredoxin*. Soluble ferredoxin from higher plants contains two iron ions and two atoms of sulphur per molecule, and it acts as a one-electron acceptor. Reduced ferredoxin can then, in the presence of a reductase enzyme, donate electrons to NADP$^+$ to give NADPH (Fig. 5.3).

The electrons ejected from the photosystem II reaction centres are replaced by the splitting of water molecules, accompanied by oxygen evolution. The detailed chemistry of the water-splitting reaction is not understood, but a 'charge-accumulating mechanism' is present in photosystem II. It stores up to four units of positive charge, corresponding to loss of 4 electrons from photosystem II. When fully 'charged', it

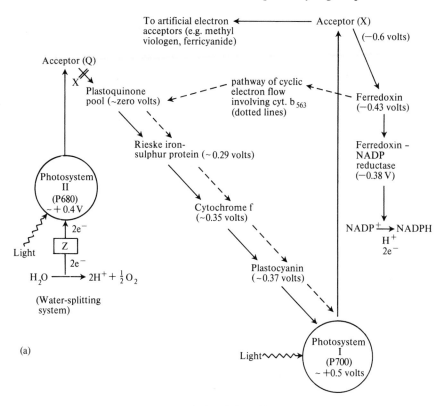

Fig. 5.2. The electron transport chain of the chloroplast. This current scheme of photosynthetic electron transport is often known as the Hill–Bendall scheme or the 'Z'-scheme. The sequence of carriers is presented as an energy diagram using redox potentials (shown in parentheses in volts). The lower the redox potential of a substance, the better is its electron-donating capacity. Hence a component with a negative redox potential is theoretically capable of donating electrons to another component with a less negative, zero, or positive redox potential. The redox potentials of the components are obtained by titrating them with reagents of known redox potential. X represents the site of action of the commonly used photosynthetic inhibitor DCMU. Q is a molecule of unknown structure, plastoquinone is a diphenol that resembles coenzyme Q (Chapter 3), cytochrome f is a typical cytochrome, and plastocyanin is a copper protein. Electrons ejected from photosystem II are replaced by the splitting of water. The acceptor Q passes its electrons into an electron transport chain, in which a plastoquinone pool is involved. Like coenzyme Q in mitochondria, oxidized plastoquinone (a quinone) can accept one electron to form a semiquinone or two to form a diphenol. This 'pool' of plastoquinone molecules may collect electrons from several different photosystems II and pass them onto an iron-sulphur protein (the Rieske iron-sulphur protein) and hence onto cytochrome f and plastocyanin. As electrons flow from Q to plastocyanin, sufficient energy is released to drive the synthesis of ATP from ADP and phosphate (*photophosphorylation*). Plastocyanin donates electrons to PSI to replace those ejected from the reaction centre.

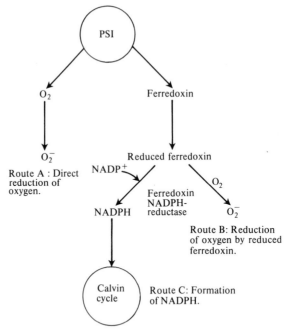

Fig. 5.3. Routes of electron flow from reduced photosystem I. Reduced ferredoxin reduces oxygen to O_2^- as well as passing electrons onto $NADP^+$. The electron acceptor of PSI itself slowly reduces oxygen to O_2^-.

can then take four electrons from two water molecules to generate a molecule of oxygen. This charge accumulator is usually designated as 'S' (like 'Q', we don't know what it is!) so we can write:

$$S \xrightarrow{-1e^-} S^+ \xrightarrow{-1e^-} S^{2+} \xrightarrow{-1e^-} S^{3+} \xrightarrow{-1e^-} S^{4+}$$

$$S^{4+} + 2H_2O \longrightarrow 4H^+ + S + O_2$$

The transition metal manganese is intimately involved in the production of oxygen, but what it does, exactly, is not yet clear.

5.1.1. The problems

Chloroplasts are especially prone to oxygen-toxicity effects, firstly because their internal oxygen concentration in the light will always be greater than that in the surrounding atmosphere, due to oxygen-production in photosystems II. Secondly, the lipids present in the chloroplast envelope, and thylakoids contain a high percentage of polyunsaturated fatty acids, and are thus very susceptible to peroxidation.

Thirdly, we saw in Chapter 2 that illuminated chlorophyll can sensitize the formation of singlet oxygen, which is especially damaging to membrane lipids (Chapter 4). Prolonged illumination of chloroplast thylakoids *in vitro* causes marked lipid peroxidation. Indeed, when polyunsaturated fatty acids are simply mixed with chlorophyll and the mixture illuminated, they are rapidly peroxidized. The deterioration of isolated, illuminated chloroplast thylakoids is an extremely complicated process. In addition to lipid peroxidation and the formation of reactive aldehydes by peroxide decomposition, there is actual hydrolysis of lipids to release fatty acids. Both esterified and released fatty acids undergo peroxidation, and the released fatty acids themselves cause membrane damage and inhibit photosynthesis. Lipid hydrolysis is due to the action of lipase enzymes which normally show little activity in chloroplasts, but seem to be 'unmasked' during membrane deterioration.

Since the trapping of light by chlorophyll causes formation of excited states, then production of singlet oxygen would be expected, since there is plenty of oxygen around. If operation of the electron transport chain were prevented, the reaction-centre chlorophylls could not dispose of their excitation energy, nor could they accept energy from their harvesting pigments and so singlet O_2 formation should be greatly increased. Indeed, about 50 per cent of all known herbicides act by inhibiting the electron transport chain. This stops plant growth because ATP and NADPH can no longer be provided for the Calvin cycle, but many scientists have shown that the damaging effects to the plant are stimulated by light and oxygen. Treating chloroplasts or algal cells with the inhibitors diuron (DCMU, see Fig. 5.2) or monuron (CMU) causes chloroplast damage, lipid peroxidation and bleaching of the chlorophyll molecule—effects that can be decreased by the singlet-oxygen scavenger DABCO (Chapter 2). Similar effects can happen if the amount of carbon dioxide supplied to the plant is restricted and so it is imperative that the chloroplast has some mechanism for disposing of singlet oxygen.

Fourthly, the electron-transport chain of chloroplasts, like that of mitochondria and the endoplasmic reticulum (Chapter 3), can 'leak' electrons onto oxygen. Isolated illuminated chloroplast thylakoids slowly take up oxygen in the absence of added electron acceptors. This was first observed by A. H. Mehler in the USA, and is hence often referred to as the *Mehler reaction*. It appears to result from the reduction of oxygen to O_2^- by the electron acceptors of photosystems I. Addition of the stromal protein ferredoxin increases the amount of oxygen uptake, since it is reduced by photosystems I (PSI) much more quickly than is oxygen and the reduced ferredoxin can then itself reduce oxygen

$$Fd_{red} + O_2 \rightarrow O_2^- + Fd_{ox}.$$

In vivo, however, reduced ferredoxin also passes electrons onto $NADP^+$ via ferredoxin–$NADP^+$ reductase (Fig. 5.2). Thus, electrons from photosystems I can pass through at least three routes, as shown in Fig. 5.3, of which route C is preferred. If the supply of $NADP^+$ were limited, however, the rate of electron flow along pathway C would be expected to be decreased, and more O_2^- should be made by route B and, to a lesser extent, by route A (Fig. 5.3). In some experiments, however, isolated intact chloroplasts have been shown to reduce some oxygen even at fairly low light intensities.

Lastly, oxygen has an effect on the reaction catalysed by ribulose biphosphate carboxylase. One of the chemical intermediates formed during the action of the enzyme on ribulose bisphosphate is capable of reacting with molecular oxygen as shown in Fig. 5.4. As a result of this *oxygenase* activity of the enzyme, ribulose bisphosphate becomes converted into phosphoglyceric acid and phosphoglycollic acid. The phosphoglycollate so produced is further metabolized by a series of reactions known as the *photorespiratory pathway*, a metabolic pathway that eventually converts two molecules of phosphoglycollate into one

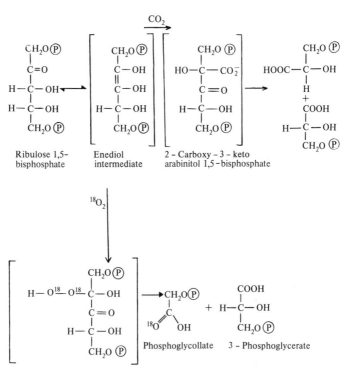

Fig. 5.4. Chemistry of ribulose bisphosphate carboxylase.

molecule each of carbon dioxide and phosphoglycerate, which can re-enter the Calvin cycle. Hence, carbon dioxide previously fixed into the Calvin cycle intermediates is lost and has to be refixed, with expenditure of more energy. Both oxygen and carbon dioxide compete for the enzyme intermediate (Fig. 5.4). The affinity of the carboxylase for carbon dioxide is much greater than that for oxygen, but at elevated oxygen concentrations, CO_2-fixation into the Calvin cycle is decreased and CO_2-loss by photorespiration is increased. At some O_2/CO_2 ratios, a point will be reached at which the leaf is no longer achieving any net CO_2-fixation, and therefore plant growth is completely inhibited. At high O_2/CO_2 ratios there may be actual loss of carbon from the plant as photorespiration exceeds CO_2-fixation into the Calvin cycle: a situation which will result in death of the plant if continued for a long period.

5.1.2. The solutions

The thylakoid membranes are especially rich in α-tocopherol, which interferes with the chain reaction of lipid peroxidation (Chapter 4) and directly scavenges singlet oxygen. The products of reaction of tocopherol with singlet oxygen include tocopherylquinone, which is found in chloroplasts and, as might be expected, increases in amount when they are illuminated. As with animal tissues, however, it must not be assumed that α-tocopherol in plants serves *only* as a protection against membrane damage. Other roles are perhaps suggested by the fact that a tocopherol oxidase enzyme, whose activity is under hormonal control, has been detected in many plant tissues, although this enzyme has so far resisted purification, and its subcellular location is unknown.

Carotenoids, which are important constituents of chloroplast membranes, quench singlet oxygen extremely rapidly and can therefore help to protect chlorophyll and membranes against damage. Carotenoids are quickly destroyed on illumination of thylakoids *in vitro*, presumably as they absorb any singlet oxygen formed. Their protective function *in vivo* is well illustrated by mutant strains of maize which do not synthesize them: illumination of such plants under aerobic conditions causes rapid bleaching of chlorophylls and destruction of chloroplast membranes. Illumination under anaerobic conditions causes much less damage, since singlet oxygen cannot then be generated. Similar destructive effects of illumination in the presence of oxygen are seen in normal plants in which carotenoid biosynthesis has been inhibited by certain herbicides. The carotenoids present in chloroplasts are two main types: the carotenes (e.g. β-carotene, whose structure is shown in Figure 5.5) and xanthophylls, which are oxygen-containing derivatives of carotenes.

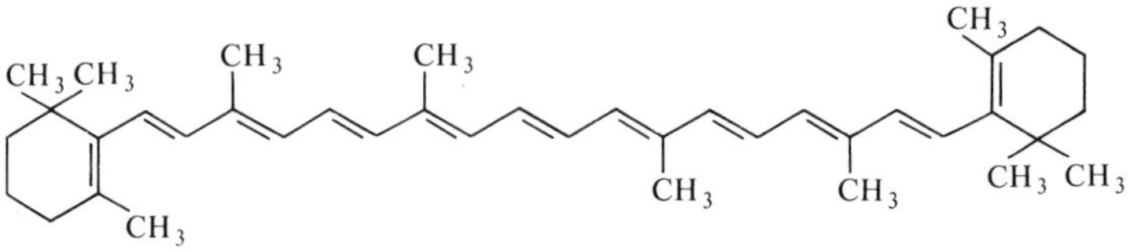

Fig. 5.5. The structure of β-carotene. Carotenes are orange-yellow pigments. The excited states of chlorophyll can transfer energy onto carotene, which dissipates it harmlessly and thus reduces chlorophyll-sensitized formation of singlet oxygen. Carotenes are also good quenchers of any singlet oxygen formed.

Carotenoids are also able to absorb energy from, and so diminish the concentration of, those excited states of chlorophyll that lead to singlet oxygen formation. Hence they have a dual role: decreasing formation of singlet oxygen *in vivo*, and helping to remove any that does happen to be formed. They may also react directly with peroxy and alkoxy radicals, and so interfere with the chain reaction of lipid peroxidation.

Superoxide produced in the chloroplast is dealt with by a superoxide dismutase enzyme. Israeli scientists have shown that the SOD activity of developing tomatoes correlates with the resistance of the fruits to the 'sunscald' damage induced by exposure to heat and high light intensities (a nuisance for growers in hot countries who want to sell their tomatoes!). This observation indicates that SOD plays an important protective role in illuminated pigment systems. In the blue-green alga *Anabaena cylindrica*, nitrogen fixation is carried out in specialized cells called heterocysts which lack the oxygen-evolving photosystems II. Heterocyst SOD activity is much lower than that of the photosynthetic cells. In spinach chloroplasts, a Cu–Zn SOD is present, some of which is bound to the thylakoids, and the rest free in the stroma. No MnSOD or FeSOD has been detected in spinach chloroplasts, although there have been reports of a MnSOD in the chloroplasts of some other plants (possibly due to contamination of the chloroplast fraction assayed with mitochondria), and the FeSOD in *Brassica campestris* leaves (Chapter 3) is found in the chloroplast.

Careful work by Professor Asada's group in Japan has shown that the hydrogen peroxide produced by illuminated chloroplasts is derived from O_2^- arising from photosystem I and ferredoxin. Like all CuZnSODs, the chloroplast enzyme is inactivated on prolonged exposure to hydrogen peroxide. If this were permitted, then O_2^- could react with H_2O_2 to give $OH^\cdot$ in a catalysed Haber–Weiss reaction (Chapter 3). In addition, German scientists have shown that hydrogen peroxide can interact directly with reduced ferredoxin to form $OH^\cdot$. Hydrogen peroxide can also inhibit the Calvin cycle by inactivating the enzymes fructose bisphosphatase and sedoheptulose bisphosphatase. Thus it is imperative that the illuminated chloroplast disposes of hydrogen peroxide, yet these organelles contain no catalase, and glutathione peroxidase is not found in plant tissue. Both ascorbate and glutathione are, however, involved in removing hydrogen peroxide, since chloroplasts contain an ascorbate peroxidase activity, and GSH can reduce dehydroascorbate back to ascorbate. Figure 5.6 summarizes this 'ascorbate–glutathione cycle', a scheme first proposed in the authors' laboratory and greatly strengthened by the discovery of ascorbate peroxidase in chloroplasts by D. Groden and E. Beck in Germany. Ascorbate and GSH also scavenge OH$^{\cdot}$ and singlet oxygen, and ascorbate can react with O_2^- (Chapter 2), so they may help to protect against these species as well. The chloroplast stroma contains GSH at concentrations in the range 1–4 mM and ascorbic acid at concentrations of 10–20 mM and in both cases the ratios of reduced to oxidized form (GSH/GSSG and ascorbate/dehydroascorbate) are kept high under both light- and dark-conditions. Thus direct scavenging of radicals by ascorbate is quite likely in view of the high concentrations present. Addition of hydrogen peroxide to chloroplasts in the light causes transient oxidation of both GSH and of ascorbate, as would be expected (Fig. 5.6). Ascorbate might also reduce tocopheryl radicals at the surface of the thylakoid membrane, so regenerating α-tocopherol for further use in protection against lipid peroxidation (Chapter 4). Ascorbate peroxidase purified from the alga *Euglena* can reduce not only hydrogen peroxide but also some artificial organic hydroperoxides. These enzymes might therefore be capable of acting on lipid peroxides *in vivo*, and further experiments are required on this point.

In summary, superoxide, hydrogen peroxide, singlet oxygen, the hydroxyl radical (OH$^{\cdot}$), and lipid peroxides are species whose formation is damaging to plant tissues and must therefore be carefully controlled. Chloroplasts are especially affected by them because of a high internal oxygen concentration, the presence of molecules which can reduce oxygen to O_2^- (e.g. reduced ferredoxin), and the presence of pigments

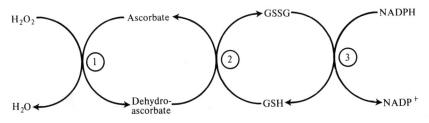

Fig. 5.6. An 'ascorbate–glutathione cycle' in chloroplasts. Enzymes involved: 1, Ascorbate peroxidase; 2, dehydroascorbate reductase (Chapter 2); 3, glutathione reductase. (NADH-dependent mechanisms for reducing dehydroascorbate have also been reported in chloroplasts by Italian and American scientists.)

which can sensitize the formation of singlet oxygen. To allow their continued functioning, chloroplasts have multiple protective mechanisms against these. The main ones are SOD (which removes O_2^- and hence prevents formation of $OH^{\cdot}$ from it), ascorbic acid (which reacts rapidly with O_2^-, $OH^{\cdot}$, and singlet oxygen, and helps remove hydrogen peroxide by the ascorbate peroxidase reaction), reduced glutathione (which reacts rapidly with $OH^{\cdot}$ and singlet oxygen, protects enzyme —SH groups, and helps to regenerate ascorbate from dehydroascorbate), α-tocopherol (which inhibits the chain-reaction of lipid peroxidation and rapidly scavenges singlet oxygen), and carotenoids (which decrease formation of singlet oxygen by absorbing excess excitation energy from chlorophyll by direct transfer, and which also quench singlet oxygen). Not until these systems are overloaded should toxic effects appear.

If spinach chloroplasts or isolated leaf cells are incubated in air from which carbon dioxide has been removed, their ability to fix carbon dioxide supplied to them subsequently is impaired. It may be that in the absence of carbon dioxide, no $NADP^+$ is available to accept electrons from photosystem I so that they are 'shunted off' to oxygen (Fig. 5.3). Any excess excitation of chlorophyll would cause more singlet oxygen to be formed. It has been suggested that the function of the metabolic pathway of photorespiration, which at first sight seems pointless in that it releases carbon dioxide from the plant which then has to be re-fixed into the Calvin cycle with expenditure of more energy, is actually to ensure that some carbon dioxide is always present within the leaf to prevent the damage that would result from excessive formation of O_2^- and singlet oxygen. This suggestion is currently the subject of intense investigation by Australian scientists. Another area of investigation is *chilling injury*. Many plants show depressed photosynthesis after exposure to low temperatures. Chilling injury is greatest in the presence of light and oxygen, and is then accompanied by membrane damage and bleaching of chlorophyll. Perhaps the membrane reorganisations that occur at low temperatures facilitate singlet oxygen formation and associated lipid peroxidation. Strains of the alga *Chlorella* with high SOD activities have been shown to be more resistant to chilling injury than strains with normal SOD activity.

5.2. The eye

Retrolental fibroplasia (sometimes called 'retinopathy of prematurity') is a condition that can lead to blindness; it is a complication of the use of elevated oxygen concentrations in incubators for premature babies (Chapter 1). Although its incidence can be controlled by restricting the oxygen concentrations used, there is evidence that the weakest babies need a high oxygen concentration if they are to survive at all. Fortunately, there is now some evidence that administration of large doses of α-tocopherol (Chapter 4) to such babies lessens the risk of their developing severe retrolental fibroplasia and (possibly) brain haemorrhage. The latter is a common finding in premature babies who die within the first week of life and can lead to handicap in surviving babies.

Cataract, an opacity of the lens, is one of the leading causes of blindness in the world. The lens is surrounded by a capsule, and held in place behind the iris by ligaments (Fig. 5.7). The anterior of it is covered by a single layer of epithelial cells which are metabolically highly active and undergo cell division, elongation, and development to form the *lens fibres*. Newly-formed fibres continuously push the older ones towards the centre of the lens to form the so-called *lens nucleus*. No blood vessels

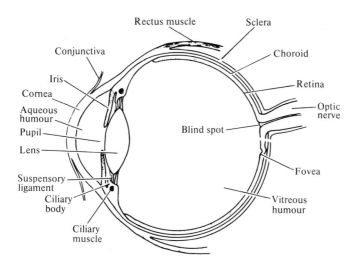

Fig. 5.7. Structure of the human eye. The sclera is a tough fibrous coat, inside which is the choroid layer containing a network of blood vessels that supply food and oxygen to the eye. The choroid is highly pigmented. The aqueous humour is a watery fluid whereas the vitreous humour is more jelly-like. Pressure of these fluids outward on the sclera maintains the shape of the eye. The transparent cornea, lens, and conjunctiva obtain food and oxygen by diffusion from the aqueous humour—they contain no blood vessels. The lens is suspended in position by ligaments. The retina contains two types of light-sensitive cells: rods and cones. Only the cones are sensitive to coloured light, but the rods are more responsive to dim light (hence the perception of colour decreases as light intensity decreases). Any marked variation in the pressure of the humours can alter the blood flow in the eye. For example, *ocular hypertension* (too much pressure) can restrict blood flow and cause ischaemic damage; *glaucoma* is the name of a group of diseases in which this occurs.

enter the lens, and it derives food and oxygen by diffusion through the aqueous and vitreous humours (Fig. 5.7). The epithelium actively transports sodium ions out of the lens, and potassium ions into it, using an ion-transporting ATPase enzyme, which is especially sensitive to damage by O_2^--generating systems.

We saw in Chapter 2 that the lens contains sensitizers of singlet oxygen formation, and singlet oxygen can damage and cross-link lens proteins. Ionizing radiation often induces cataract, perhaps by increased formation of hydroxyl and superoxide radicals. Indeed, a considerable amount of evidence has accumulated implicating prolonged exposure to ultraviolet radiation as a factor in lens ageing and cataract formation. Proteins isolated from human cataractous lenses contain considerable quantities of methionine sulphoxide residues (Chapter 3). UV-induced degradation of tryptophan, a reaction that produces O_2^- (Chapter 3), has been reported in lens proteins.

The vitreous humour (Fig. 5.7) contains hyaluronic acid, which is attacked and depolymerized on exposure to O_2^--generating systems (Chapter 3) causing loss of viscosity. This reaction is due to O_2^--dependent formation of hydroxyl radicals (Chapter 3) in the presence of traces of iron salts. Indeed, introduction of iron salts into the eye causes severe problems; this can happen as a result of penetration by iron objects or by intravitreal haemorrhage. Both can cause severe visual impairment, sometimes leading to blindness. The lipids present in the membranes of retinal rod cells (Fig. 5.7) contain a high percentage of polyunsaturated fatty-acid side-chains, and are thus susceptible to lipid peroxidation. The pigment they contain, rhodopsin, can sensitize the formation of singlet oxygen (Chapter 2); and exposure of frog retina to light of the wavelengths absorbed by rhodopsin *in vitro* has been shown to induce lipid peroxidation. Lipid hydroperoxides are increased in the rabbit retina after exposure to elevated oxygen concentrations or to ionizing radiation; and injection of preformed lipid hydroperoxides into the eyes of rabbits causes severe retinal damage. Dietary deficiencies of selenium and vitamin E in rats cause marked loss of polyunsaturated fatty acids and accumulation of peroxidation products in the retinal pigment epithelium.

Therefore the eye has a lot of problems, and one would expect a corresponding degree of protection. Indeed, the concentration of GSH in the lens of several species is as high as that in the liver, being especially concentrated in the epithelium. GSH may serve to protect the thiol-groups of the lens proteins known as *crystallins*, preventing them from aggregating together to form opaque clusters. The GSH/GSSG ratio is normally kept high by the activity of a lens glutathione reductase, which obtains NADPH by operation of the pentose phosphate pathway in the lens (Chapter 3). Glaucoma (defined in the caption for Fig. 5.7), by decreasing the supply of glucose to the lens, interferes with this pathway and so restricts NADPH supply. Both glutathione peroxidase and catalase activities are present in the lens, as is methionine sulphoxide reductase (Chapter 3). Glutathione-S-transferase enzymes have also been detected in bovine lens but they do not use lipid peroxides as substrates, in contrast to the liver enzymes (Chapter 4). Bovine retina, on the other hand, does contain a glutathione-S-transferase that can act on lipid peroxides. The vitreous humor contains high amounts of an iron-binding protein that resembles transferrin; transferrin-bound iron is much less effective than low-molecular weight iron complexes in stimulating lipid peroxidation or hydroxyl radical formation.

Feeding aminotriazole to rabbits to inhibit catalase raises the concentration of hydrogen peroxide in the lens and induces cataracts; so, at least in this animal, the glutathione peroxidase alone cannot cope with the normal rates of H_2O_2-generation. By contrast, cataract does not develop in mice with an inborn deficiency in catalase activity. The concentration of hydrogen peroxide in the aqueous humour of human patients with cataract is significantly higher than normal, and a common feature of nearly all types of cataract is that lens GSH concentrations fall.

Ascorbic acid is present at high concentrations in the lens, cornea, and aqueous humour of man, monkey, and many other animals (e.g. 1–2 millimoles per litre in human aqueous humour). Its ability to scavenge O_2^-, $OH^\bullet$, and singlet oxygen may be of importance although, on the other hand, it has been suggested that light-induced degradation of ascorbate in the aqueous humour may be a source of hydrogen peroxide that has to be removed by glutathione peroxidase and catalase. Ascorbate has been observed to protect the lens ion-transporting ATPase enzyme against damage by a O_2^--generating system *in vitro*. Consistent with a protective role for ascorbate *in vivo* is the observation that nocturnal animals (e.g. cats) have lower ascorbate concentrations in the eye. Rat lens has virtually no ascorbate.

Superoxide dismutase is present in all parts of the eye, but problems in the assay of eye tissue have made comparison of the SOD activities in the different parts rather difficult. Rat lens has a higher SOD activity than human lens, and so ascorbate may play a greater role in scavenging O_2^- in the latter case. The SOD activities of eye tissues are inhibited by cyanide, and thus can be attributed to the copper–zinc enzymes (Chapter 3). CuZnSODs are susceptible to inactivation by hydrogen peroxide. Indeed, feeding aminotriazole to rabbits (see above) causes a decrease in lens CuZnSOD activity, probably as a result of the accumulated hydrogen peroxide. Finally, rod outer segments are especially rich in vitamin E, and the retina is usually damaged in vitamin-E-deficient animals. Feeding extra vitamin E decreases the severity of the cataracts seen when aminotriazole is administered to rabbits.

Thus the eye, like the chloroplast, has to put up with the problems caused by light, and relies heavily on ascorbate, GSH, and tocopherol as protective mechanisms, whilst methionine sulphoxide reductase may help regenerate methionine in proteins after oxidation by singlet oxygen (Chapter 3). On the other hand, the increased content of methionine sulphoxide residues known to be present in proteins from cataractous lenses is not due to any defect in this enzyme, since a comparison of normal and cataractous human lens showed similar activities.

Carotenoids may be important in protection of light receptors against singlet oxygen in some living organisms. In the compound eye of the housefly a carotenoid is present in large amounts (4–10 molecules per rhodopsin molecule). Carotenoid has also been reported in the lateral eye of the crab *Limulus*. The corneas of puffer fishes are clear in the dark but become yellow in the light, owing to the migration of carotenoid pigment in chromatophore cells.

5.3. Further reading

The effect of deficiency of vitamins E and A on the retina. (1980). *Nutr. Rev.* **38,** 386.

Armstrong, D., Santangelo, G., and Connole, E. (1981). The distribution of peroxide regulating enzymes in the canine eye. *Curr. Eye Res.* **1,** 225.

Bhuyan, K. C., and Bhuyan, D. K., (1979). Mechanisms of cataractogenesis induced by 3-amino 1,2,4 triazole. In *Biochemical and clinical aspects of oxygen* (ed. W. S. Caughey) p. 785. Academic Press.

Dodge, A. D. (1982). Oxygen radicals and herbicide action. *Biochem. Soc. Trans.* **10,** 73.

Elstner, E. F. (1982). Oxygen activation and oxygen toxicity. *Annu. Rev. Plant Physiol.* **33,** 73.

Finer, N. N. *et al.* (1982). Effect of intramuscular vitamin E on frequency and severity of retrolental fibroplasia. A controlled trial. *Lancet* **i,** 1087.

Foyer, C. H. and Hall, D. O. (1980). Oxygen metabolism in the active chloroplast. *Trends in biochem. Sci.* **June,** 188.

Halliwell, B. (1984). *Chloroplast Metabolism: the structure and function of chloroplasts in green leaf cells.* Revised edn. Clarendon Press, Oxford.

Halliwell, B. (1982). Ascorbic acid and the illuminated chloroplast. *Advances in chemistry series* **200,** 263.

Katz, M. L., *et al.* (1982). Effects of antioxidant nutrient deficiency on the retina and retinal pigment epithelium of albino rats: a light and electron microscopic study. *Exp. Eye Res.* **34,** 339.

Kirschfeld, K. (1982). Carotenoid pigments: their possible role in protecting against photo-oxidation in eyes and photoreceptor cells. *Proc. R. Soc. Lond.* **B216,** 71.

Krinsky, N. I., and Deneke, S. M. (1982). Interaction of oxygen and oxy-radicals with carotenoids. *J. Nat. Cancer Inst.* **69,** 205.

Law, M. Y., Charles, S. A., and Halliwell, B., (1983). Glutathione and ascorbic acid in spinach chloroplasts. The effect of hydrogen peroxide and of paraquat. *Biochem. J.* **210**, 899.

Nakano, Y. and Asada, K. (1980). Spinach chloroplasts scavenge H_2O_2 on illumination. *Plant and Cell Physiol.* **21**, 1295.

Spector, A., Scotto, R., Weissbach, H., and Brot, N. (1982). Lens methionine sulphoxide reductase. *Biochem. biophys. Res. Commun.* **108**, 429.

Srivastava, S. K. (ed.) *Red blood cell and lens metabolism.* Elsevier/North Holland Inc.

Varma, S. D., Kumar, S., and Richards, R. D., (1979) Light-induced damage to ocular lens cation pump. Prevention by vitamin C. *Proc. natn. Acad. Sci. U.S.A.* **76**, 3504.

Wah-Kow, Y., Smyth, D. A., and Gibbs, M. (1982). Oxidation of reduced pyridine nucleotide by a system using ascorbate and H_2O_2 from plants and algae. *Plant Physiol.* **69**, 72.

6 Free radicals and toxicology

As a result of the 'Spanish cooking-oil crisis', which is discussed later in this chapter, radicals such as superoxide, and the enzyme superoxide dismutase were mentioned on BBC television and in up-market British newspapers. As usual, the Americans were well ahead of us in that for some time it has been possible to buy superoxide dismutase tablets (sometimes with added catalase) in health food stores to protect oneself against ageing (with total disregard for the action of digestive enzymes on dietary proteins!).

To return to the serious subject matter of this chapter, we have seen that oxygen radicals can damage cells, and that damaged tissues show an increased susceptibility to lipid peroxidation, even if the tissue damage was not initially caused by a radical mechanism. Any agent that could abstract a hydrogen atom from a membrane lipid and initiate peroxidation would obviously be unpleasant, and some toxic compounds that can do this are considered later in this chapter. The increase in the activities of superoxide dismutase and H_2O_2-removing enzymes that occurs when many organisms are exposed to elevated oxygen concentrations (Chapter 3) implies that the amounts of these enzymes present normally in some tissues are sufficient only to cope with the usual rates of O_2^-- and H_2O_2-generation. Hence any compound that increased O_2^-- and H_2O_2-production in such tissues at normal oxygen concentrations would be expected to be toxic.

Let us now illustrate the above points by looking at some examples.

6.1. Bipyridyl herbicides

We saw in Chapter 5 that a number of herbicides act by inhibiting electron transport in chloroplasts or by interfering with carotenoid synthesis.

The bipyridyl herbicides *paraquat* and *diquat* act in a rather different way. *Bipyridyl* means that the structure contains two pyridine rings, aromatic rings in which one carbon atom is replaced by a nitrogen atom. The rings may be joined either by their number-2 carbon atoms, or by their number-4 carbon atoms (the N counting as atom number-one), to give 2,2′-bipyridyl or 4,4′-bipyridyl respectively. The dash is used to indicate an atom in the second ring. In paraquat, a methyl group is

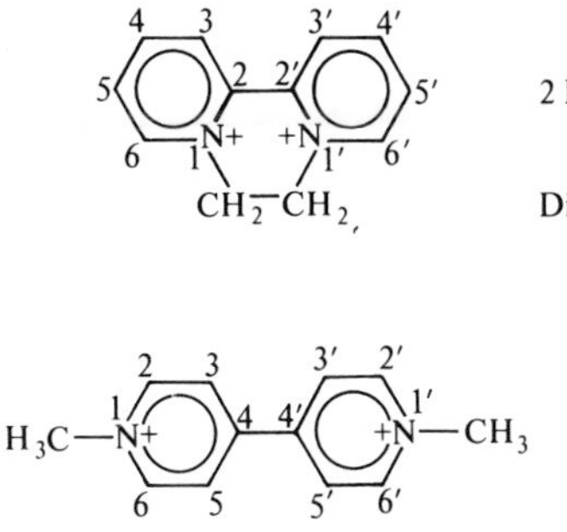

Fig. 6.1. Structures of the herbicides diquat and paraquat.

attached to each nitrogen, giving a full chemical name as 1,1'-*dimethyl-4,4'-dipyridylium ion* (Fig. 6.1). Each N gains a positive charge because it has four bonds and is thus utilizing its lone pair of electrons (see Appendix). In diquat, the two nitrogen atoms are joined by an ethylene group, to give 1,1'-*ethylene-2,2'-dipyridylium ion* (Fig. 6.1). Paraquat is usually manufactured as a salt with chloride (Cl^-) ion, and diquat with bromide (Br^-) ion. Upon chemical reduction, 4,4'-bipyridyl compounds form coloured solutions; and a number of such compounds, under the general name 'viologens', have been used for this reason to study the chemistry of reduction and oxidation reactions. Hence paraquat is sometimes called *methyl viologen*. The colour is produced by addition of one electron to the compounds to form a stable (in the absence of oxygen) radical that has a characteristic ESR spectrum, and that absorbs visible light (λ_{max} in the visible range is at 603 nm for paraquat). The extra electron is delocalized over both ring structures with partial neutralization of the positive charge on each nitrogen-atom.

The bipyridyl herbicides, discovered by Imperial Chemical Industries in England, are extremely valuable because of their broad toxicity to a wide range of plants, and because they are completely inactivated on contact with soil, being tightly bound to clay minerals and destroyed by various microorganisms. Hence a field of weeds can be sprayed with these herbicides and, after the dead plants are cleared, replanting can be done at once. It was soon observed that the killing of green plants by either paraquat or diquat is greatly accelerated by light and is slowed if oxygen is removed from the environment of the plant by flushing with nitrogen gas. Later work showed that both paraquat and diquat cross the chloroplast envelope easily, and can accept electrons from the non-haem-iron proteins associated with the photosystem I and also from the flavin at the active site of ferredoxin–NADP reductase (Chapter 5), in both cases becoming reduced to their radical forms. If these experiments were carried out in the absence of oxygen, the bipyridyl radicals could be identified in herbicide-treated chloroplasts by means of their absorption and ESR spectra. On admission of oxygen (and, of course, chloroplasts normally have a high internal oxygen-concentration because of its production during photosynthesis) the radicals disappear because they react with oxygen extremely rapidly. If BP^{2+} is used to represent the herbicides, the reactions may be written as

$$BP^{2+} \xrightarrow[\text{chain (1 electron)}]{\text{electron-transport}} BP^{\cdot+}$$

$$BP^{\cdot+} + O_2 \rightarrow BP^{2+} + O_2^- \qquad k = 7.7 \times 10^8 \, \text{M}^{-1}\text{s}^{-1} \qquad \text{(for paraquat)}$$

Hence the superoxide radical, O_2^-, is formed. Thus treatment of illuminated chloroplasts *in vitro* with paraquat or diquat leads to a rapid uptake of oxygen as the herbicides are continuously reduced and re-oxidized. The O_2^- is presumably converted into hydrogen peroxide by chloroplast superoxide dismutase (Chapter 5). Since chloroplasts contain no catalase, hydrogen peroxide is dealt with by the ascorbate–glutathione cycle (Chapter 5), but as Table 6.1 shows, GSH and ascorbate are quickly oxidized and inactivation of Calvin cycle enzymes such as fructose bisphosphatase occurs, so that CO_2-fixation stops. An additional reason for inhibition of CO_2-fixation is that diversion of electrons from photosystem I onto the herbicides will decrease the supply of NADPH both for the Calvin cycle, and for glutathione reductase activity. In studies upon whole flax leaves, A. D. Dodge in England observed that inhibition of CO_2-fixation is followed by leakage of ions from the leaves, and accumulation within them of thiobarbituric acid-reactive material, indicative of lipid peroxidation. Analysis of total lipids extracted from the leaves showed destruction of many fatty-acid side-chains. Electron microscopy revealed deterioration and breakdown of a number of cellular membranes, including the chloroplast thylakoids and the tonoplast, the membrane which surrounds the central vacuole of the plant cell and is often very close to the envelopes of chloroplasts. The central vacuole contains a number of hydrolytic enzymes, and often accumulates organic acids to a high concentration; so release of its contents into the rest of the cell will potentiate the damage. Similar studies upon bean-leaves showed that paraquat caused decreased membrane fluidity and increased lipid destruction, accompanied by accumulation of TBA-reactive material. Comparable results have been obtained with several green algae. It does not, of course, follow that lipid peroxidation is the primary cause of paraquat-induced leaf death, but it certainly does not help. Inhibition of CO_2-fixation is a much earlier phenomenon than is detectable lipid peroxidation.

That these damaging effects are related to an increased production of

Table 6.1. Effect of paraquat on illuminated spinach chloroplasts

Time after paraquat addition (minutes)	GSH ($\text{mmol}\,l^{-1}$)	GSSG ($\text{mmol}\,l^{-1}$)	GSH/GSSG ratio	[Ascorbate] in reduced form ($\text{mmol}\,l^{-1}$)	Fructose bisphosphatase activity (enzyme units mg^{-1} chlorophyll)
0	6.5	0.25	26	13	62
2	3	3.2	0.9	8	20
5	3.1	1.5	2.1	6	8
10	3.0	0.8	3.8	4	0

Intact spinach-leaf chloroplasts were isolated and treated with paraquat in the light. The stromal contents of GSH, GSSG, and ascorbate were measured, as was the activity of fructose bisphosphatase, an enzyme essential to the Calvin cycle. Such rapid changes are not seen if paraquat is added to darkened chloroplasts, since the electron-transport chain is not then active to reduce paraquat. Data from *Biochem. J.* **210**, 899–903.

O_2^- and hydrogen peroxide in the presence of paraquat or diquat is shown by a number of experiments. A low-molecular-weight copper chelate that reacts with O_2^- has been shown to decrease the toxicity of paraquat to flax cotyledons. Paraquat induces synthesis of MnSOD in the green alga *Chlorella*. Some strains of ryegrass are relatively resistant to paraquat even though it is still taken up at a normal rate by the plants, and the leaves of the resistant plants have been shown to contain significantly more superoxide dismutase (SOD) and catalase activities than those from sensitive plants. Although leaf catalase is located in peroxisomes rather than in chloroplasts (Chapter 5), hydrogen peroxide is a fairly non-polar molecule and can presumably diffuse across the chloroplast envelope to be dealt with in peroxisomes, which are often seen to be closely associated with chloroplasts in electron micrographs of leaf sections. Paraquat-resistant strains of *Conyza* also have increased SOD activities. Addition of paraquat to old cultures of the blue-green alga *Gloeocapsa* inhibited its ability to fix nitrogen, an effect that could be mimicked by addition of hydrogen peroxide. Paraquat and diquat are also toxic to several non-green plant tissues. For example, they cause increased membrane permeability and accumulation of TBA-reactive material in some fungi such as *Aspergillus niger* and *Mucor hiemalis*. If this is due to increased production of O_2^- and/or of hydrogen peroxide, then reduction of the herbicides must be achieved *in vivo*. A number of flavoprotein enzymes, including glutathione reductase, have been shown to be capable of reducing paraquat and/or diquat; if they can penetrate into the active sites of the enzymes, these herbicides seem able to take electrons from the flavin ring and then, in the presence of oxygen, to generate O_2^-.

Perhaps the most detailed studies of the role of oxygen radicals in paraquat toxicity have been performed with the bacterium *E. coli*, largely by Hassan and Fridovich in the USA. Paraquat added to a culture of *E. coli* is taken up by the cells and rapidly reduced. Under anaerobic conditions, the bipyridyl radical can be detected by observing its absorption spectrum. Extracts of *E. coli* cells have been observed to reduce paraquat if NADPH is added, a reaction that is catalysed by a soluble enzyme, probably a flavoprotein. In the presence of air, as would be expected, the radical disappears, and O_2^- is generated. Addition of paraquat to aerobically-grown *E. coli* induces a rapid synthesis of manganese-SOD activity, the same enzyme induced upon exposure of the bacteria to elevated oxygen concentrations (Chapter 3). Catalase activity increases as well. *E. coli* cells whose SOD activity had been increased by paraquat treatment were more resistant to elevated O_2 than normal cells and, *vice versa*, cells with raised SOD due to previous exposure to increased oxygen concentrations were more resistant to the toxic effects of paraquat. If induction of MnSOD by the bacteria was prevented, either by adding puromycin (an inhibitor of protein synthesis) or by a poor growth medium, then the toxicity of paraquat to the cells was greatly increased. Hassan and Fridovich further observed that addition of SOD to the growth medium offered some protection to *E. coli* against damage by paraquat. This at first sight is surprising, since O_2^- cannot cross the cell membrane and the SOD protein cannot enter the bacterial cells. The scientists were able to show, however, that just as paraquat can easily enter the cells, some of the paraquat radical can leak out of them and react with oxygen in the surrounding medium to give O_2^-, causing extracellular damage. The amount of leakage is inversely proportional to the intracellular oxygen concentration. Paraquat, like oxygen, has been shown to cause mutations in those *Salmonella typhimurium* strains that are used to test for mutagenic ability in the 'Ames test'. *S. typhimurium* cells containing high SOD activities were more resistant to the toxic and mutagenic effects of paraquat than were cells with normal SOD activity. O. R. Brown's laboratory in the USA has shown that addition of paraquat to *E. coli* under aerobic conditions causes inhibition of the dihydroxyacid dehydratase enzyme involved in the biosynthesis of branched-chain amino acids. Providing amino acids to the growth medium partially relieved the inhibitory effect of paraquat on bacterial growth. Addition of nicotinamide, a precursor of NAD^+, gave further relief. These results are very similar to the effect produced by high-pressure oxygen on *E. coli* (Chapter 1) and they suggest that paraquat and elevated O_2 have a common mechanism of toxicity which, from the studies of Fridovich, seems to be an increased generation of O_2^- and hydrogen peroxide *in vivo*.

As mentioned in Chapter 3, several other compounds cause intracellular O_2^- generation and an increase in the synthesis of MnSOD, catalase, and/or peroxidase activities when they are added to *E. coli* under aerobic conditions. These include the antibiotic streptonigrin, juglone, menadione, pyocyanine (a pigment produced by *Pseudomonas aeruginosa*), and methylene blue. Again, toxicity appears to be mediated by increased generation of O_2^- since, for example, strains of *E. coli* with elevated SOD activity are resistant to the effect of streptonigrin.

The main problem in the agricultural use of paraquat and diquat is that they are poisonous to several animal species, including fish, rat, mouse, cat, dog, sheep, cow, and man! Many cases of children drinking herbicides, carelessly put in lemonade or other soft-drink bottles, have been reported, and paraquat is sometimes used in suicide attempts. Oral intake of paraquat first results in local effects—irritation of the mouth, throat, and oesophagus, and sometimes vomiting and diarrhoea. Fortunately, gut absorption of bipyridyl herbicides is fairly slow, and life may often be saved by washing out the stomach and intestines repeatedly with saline solutions. Administration of suspensions of clays that absorb the herbicides (e.g. bentonite or Fullers earth), and dialysis of blood are often carried out as well.

The major lethal effect of paraquat in animals and humans is damage to the lungs. The membranous pneumocyte cells that line the alveoli (sometimes called type I alveolar cells) begin to swell and are eventually destroyed, an effect accompanied by oedema, capillary congestion, and a mild inflammation. Type II cells, or granular pneumocytes, are also damaged. The synthesis of surfactant, which lowers the surface tension of the lung linings and allows expansion, is decreased. As a result of both this and the oedema, gas exchange is hindered. In animals which survive, the damaged lung tissue is replaced by inelastic fibres that cause permanent interference with lung expansion. This major damaging effect on the lung is due to the fact that the granular pneumocytes, and probably several other lung cell types, actively accumulate paraquat, i.e. it is taken up even against a concentration gradient. The diamino compound putrescine ($H_3\overset{+}{N}(CH_2)_4\overset{+}{N}H_3$) has been shown to block this uptake in isolated rat lung slices. Other tissues, including liver and kidney, are damaged more slowly by paraquat. Large doses of diquat also affect the lung but it is not the major target tissue. In most diquat-exposed animals the intestines are especially attacked, becoming distended, and ceasing their normal peristaltic movements.

The special action of paraquat on the lungs means that poisoning can occur not only by oral intake but also by inhaling paraquat droplets from crop spraying. Paraquat has allegedly caused lung damage in the USA to 'pot'-smokers who obtained their marijuana from paraquat-treated Mexican plants. Bipyridyl herbicides can be slowly absorbed from the skin and they produce local irritation upon contact, and interfere with fingernail growth if present at high concentrations. Inhalation of the solid powders can cause nose-bleeds.

Microsomal fractions from several animal tissues, including lung, have been observed to reduce paraquat in the presence of NADPH and the paraquat radical can then react with oxygen to form O_2^-. It seems likely that paraquat can take electrons from the flavoprotein NADPH cytochrome P_{450} reductase, since antibody against this enzyme inhibits paraquat reduction by microsomes *in vitro*. Intravenous injection of paraquat into rats causes a rapid activation of the pentose phosphate pathway in lung. NADPH is required for the biosynthesis of fatty acids, needed to replace damaged membrane lipids, for surfactant biosynthesis, and for the glutathione reductase reaction. Fatty acid synthesis is decreased in the paraquat-treated lung, and there is an accumulation of 'mixed disulphides' formed between protein —SH groups and oxidized glutathione (see Chapter 3). Presumably hydrogen peroxide formed from O_2^- in the superoxide dismutase reaction is acted upon by glutathione peroxidase in the cytosol to form GSSG, which cannot be reduced quickly enough by glutathione reductase, and so combines with proteins. Feeding of nicotinamide has been reported to decrease the toxic effects of ingested paraquat in rats, an effect reminiscent of that in *E. coli* (see above).

The damaging effects of paraquat on lung tissue *in vivo*, and on isolated lung cells, are greatly potentiated at high oxygen concentrations, illustrating the importance of the reaction of the paraquat radical with O_2 *in vivo*. Selenium-deficient rats are more sensitive to paraquat poisoning than are animals fed on a normal diet, which suggests that the selenoprotein glutathione peroxidase has some protective role. There have been a few reports that injection of superoxide dismutase into animals ameliorates the symptoms of paraquat poisoning. Other scientists have not found this effect, however, and even when seen, it is fairly small. Injected superoxide dismutase cannot enter lung cells, and it is rapidly cleared from the body by the kidneys; so a large effect would not be expected. It may be that in lung, as in bacteria (see above), some of the paraquat radical can diffuse out of the cells and generate O_2^- in the surrounding tissue fluid. SOD would, of course, offer protection against this. *Intracellular* SOD may be an important protective mechanism, however, since the toxicity of paraquat to rat kidney cells in culture could be inversely correlated to their content of SOD after paraquat treatment. Further, injection of mice with diethyldithiocarbamate to inhibit CuZn-SOD (Chapter 2) greatly enhances paraquat toxicity. This inhibitor decreases lung SOD activity rapidly, being followed by a slower drop in glutathione peroxidase activity. Treatment of adult rats with bacterial endotoxin increases catalase, superoxide dismutase, glucose-6-phosphate dehydrogenase, and glutathione peroxidase activities in the lung; and these rats are not only more resistant to elevated oxygen concentrations but also to paraquat.

Addition of paraquat to a mouse lung microsomal fraction in the presence of NADPH stimulated the peroxidation of membrane lipids, as observed by Bus *et al.* in the USA. However, attempts by other workers to extend these observations have led to a mixture of results. For example, pretreatment of mice with the antioxidant N,N'-diphenyl-*p*-phenylenediamine (Chapter 4) prevented this stimulatory effect being seen in microsome fractions subsequently isolated from the animals, yet it did not decrease the toxic effects of paraquat to the animals. Other workers have not observed any stimulation of lipid peroxidation by paraquat in microsomal fractions. On the other hand, light emission induced by infusion of *tert*-butylhydroperoxide into the perfused rat lung, and the lung content of TBA-reactive material, are increased by addition of paraquat to the perfusing medium, as is the production of both ethane and GSSG by the perfused rat liver. However, the ethane content of expired air from whole rats is only slightly increased by paraquat administration, and the basal light emission from perfused lung (i.e. in the absence of infused hydroperoxide) is not increased by paraquat infusion. It seems therefore that whereas paraquat can increase the rate of lipid peroxidation *in vivo* to some extent under certain circumstances, as would be expected if production of O_2^- and hydrogen peroxide is increased (Chapter 4), an increased rate of lipid peroxidation is not the primary damaging effect of paraquat. It seems more likely that tissue damage occurs first, and the damaged tissues might then undergo peroxidation more rapidly (Chapter 4). Since fatty acid synthesis is inhibited by paraquat, damaged lipids cannot be replaced. The toxic effects of O_2^--generating systems *in vitro* have often been attributed to formation of the hydroxyl radical (OH$^\cdot$) by an iron-catalysed Haber–Weiss reaction (Chapter 3) i.e.

$$Fe^{3+} + O_2^- \rightarrow O_2 + Fe^{2+}$$
$$Fe^{2+} + H_2O_2 \rightarrow Fe^{3+} + OH^\cdot + OH^-$$

Net reaction: $O_2^- + H_2O_2 \rightarrow OH^\cdot + OH^- + O_2$.

Indeed, the authors have shown that hydroxyl radicals can be formed in the presence of reduced paraquat under aerobic conditions. Some of these radicals appear to attack the paraquat itself, however. Part of the OH$^\cdot$-formation is not prevented by superoxide dismutase or the iron chelator desferrioxamine (Chapter 2), whereas catalase completely suppresses OH$^\cdot$ formation. Hence, as well as the above reactions, a O_2^--independent production of OH$^\cdot$ in systems containing reduced paraquat and oxygen can be demonstrated. It was at first thought that this happens because the paraquat radical can react directly with hydrogen peroxide

$$PQ^{\cdot+} + H_2O_2 \rightarrow OH^\cdot + OH^- + PQ^{2+}$$

The second-order rate constant for this reaction has been estimated by one group at $2.3 \times 10^3 \, \text{M}^{-1} \text{s}^{-1}$, and by another at only $2.0 \, \text{M}^{-1} \text{s}^{-1}$, however; and so the reaction of $PQ^{\cdot+}$ with oxygen ($k_2 = 7.7 \times 10^8 \, \text{M}^{-1} \text{s}^{-1}$) will be the major reaction except at extremely low oxygen concentrations. While the reaction with hydrogen peroxide might occur in animal cells, which often do show low internal oxygen concentrations, it is unlikely to happen in chloroplasts. Elstner in Germany has shown that reduced paraquat and hydrogen peroxide can react under low oxygen concentrations to form a powerfully oxidizing species not identical with OH$\cdot$. They have named it the 'crypto-OH$\cdot$ radical' but elucidation of its nature requires further investigation.

Presumably the toxicity of bipyridyl herbicides to tissues other than the lung is mediated by their reduction and subsequent oxidation. As shown in Table 6.2, NADPH-cytochrome P_{450} reductase is widely distributed in animal tissues. Even in tissues without this enzyme, reduction may be

Table 6.2. Distribution of cytochrome P_{450} and its reductase in microsomal fractions obtained by centrifugation of different rat tissues

Organ examined	NADPH–cytochrome P_{450} reductase	Cytochrome P_{450}
Liver (adult)	46.1 ± 5.0	497 ± 40
Liver (foetal)	3.7 ± 0.4	Not detected
Liver (new born)	18.9 ± 1.4	210
Kidney	21.2 ± 2.5	76 ± 11
Lung†	18.1 ± 2.4	41 ± 14
Brain	8.6 ± 2.0	30 ± 14
Spleen	5.5 ± 0.8	6
Testis	7.9 ± 0.8	46 ± 7
Adrenal	41.5 ± 6.6	604 ± 209
Stomach	11.1 ± 0.9	—
Small intestine	46.9 ± 5.8	Not detected
Mammary gland (late pregnant)	13.0 ± 0.9	—

† Tissues such as the lung are histologically very complex and the cytochrome P_{450} and reductase may be concentrated in only one or two cell types.

Data were abstracted from Benedetto *et al.* (1981) *Biochim. Biophys. Acta* **677**, 363–72. Cytochrome P_{450} reductase units are given in nmol min^{-1} mg^{-1} protein and P_{450} is given in picomoles mg^{-1} protein. The absolute values should not be taken too seriously since microsomal fractions are very heterogeneous.

achieved by other flavoproteins. Feeding diquat to rats causes cataract formation, possibly due to its reduction by lens glutathione reductase. The haemolysis of erythrocytes sometimes seen in paraquat-poisoned humans might be due to a reduction of this compound by erythrocyte glutathione reductase.

6.2. Alloxan and streptozotocin

Injection of alloxan into animals causes degeneration of the β-cells in the islets of Langerhans of the pancreas. Since these cells synthesize the hormone insulin, alloxan is often used to induce diabetes in experimental animals. Its structure is indicated below. Two-electron reduction of it gives *dialuric acid*

alloxan   dialuric acid

   An intermediate radical, formed by one-electron reduction of alloxan, also exists. Dialuric acid is unstable in aqueous solution and undergoes oxidation, eventually to alloxan, that is accompanied by reduction of oxygen to O_2^-. Solutions of dialuric acid have been observed to stimulate peroxidation of membrane lipids, and to inhibit the growth of several bacterial strains *in vitro*. Hence any body tissue that can take up and reduce alloxan will be at risk, because the resulting dialuric acid can then form O_2^- and hence hydrogen peroxide and possibly the hydroxyl radical in the presence of traces of transition-metal ions. When alloxan is injected into rats, it accumulates in the islets of Langerhans and in the liver, but not in other body tissues. Whereas the liver contains high activities of superoxide dismutase, catalase, and glutathione peroxidase, the activities of these enzymes in the islet cells are moderate by comparison. Further, isolated islet cells are capable of reducing alloxan to dialuric acid at a high rate. It is probably this combination of factors that makes these cells so sensitive to alloxan. Addition of alloxan to isolated islet cells causes membrane damage and cell death, effects which can be decreased *in vitro* by addition of superoxide dismutase, catalase, compounds that can react with the hydroxyl radical, including mannitol and dimethylsulphoxide (Chapter 2), or the iron chelator DETAPAC, which inhibits the iron-catalysed Haber–Weiss reaction. Extensive DNA strand-breaks, possibly due to the action of OH˙, have been observed in the islet cells of alloxan-treated rats, but not in the liver cells. How reduction of alloxan in the β-cells is achieved is not known, but both NADH and NADPH may be electron donors.
   That these observations upon isolated islet cells are relevant to the action of alloxan *in vivo* is shown by a number of experiments. Grankvist *et al.* in Sweden found that injection, into mice, of CuZnSOD, attached to a high-molecular-weight polymer to reduce its clearance by the kidneys, protects them against the diabetes normally produced by a subsequent injection of alloxan. Heikkila in the USA has observed a

similar protective effect of injected DETAPAC whereas EDTA, which does not prevent the iron-catalysed Haber–Weiss reaction, did not protect. Dimethylurea, a powerful scavenger of OH', can also protect mice *in vivo,* as can dimethylsulphoxide, thiourea, or the alcohols methanol, ethanol, propanol, or butanol. It must be noted, however, that the effect of butanol is ambiguous since it produces an elevation of blood glucose concentration that can itself offer protection against alloxan. Injection of alloxan into rats deficient in vitamin E increased their exhalation of pentane and ethane, an effect that was decreased by prior injection of the OH' scavenger, mannitol. These effects were not seen in normally fed rats, however. The anti-cancer drug ICRF-159, a derivative of EDTA, has also been observed to protect against alloxan, possibly by chelating iron salts and preventing OH'-production although this remains to be established.

Another drug that has been used to induce diabetes in experimental animals is streptozotocin, a nitrosourea compound produced by *Strepto-myces achromogenes* (Fig. 6.2). Streptozotocin induces DNA strand-breakage in islet cells. Injection of it into rats has been observed to decrease the CuZnSOD activity of retina, erythrocytes, and islet cells, but not that in other tissues. Vitamin E has been reported to decrease the diabetogenicity of streptozotocin in rats. Injected SOD has been claimed to decrease the diabetogenic effects of streptozotocin by some scientists, but others have not confirmed this, and much more work is required on its mechanisms of action. The effect may depend very much on the exact timing of the SOD injection. Streptozotocin may well damage DNA directly. It has antibacterial activity and some activity against cancerous tumours in animals, apparently by inhibiting DNA synthesis.

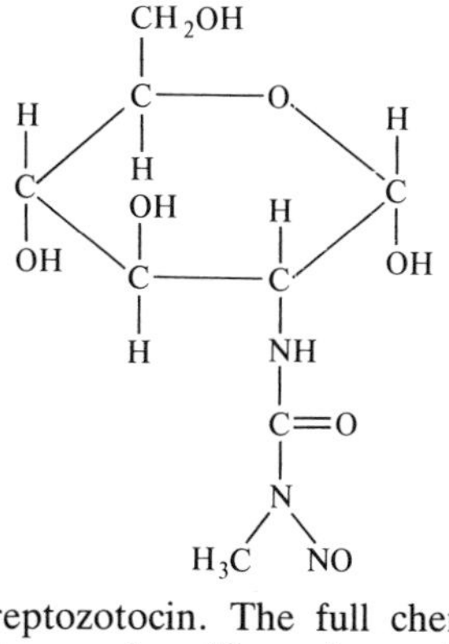

Fig. 6.2. Structures of streptozotocin. The full chemical name of the drug is *N*-methylnitrosocarbamylglucosamine. The α-form of the drug is shown; in the β-form the arrangement of the —H and —OH groups on carbon 1 (next to the oxygen in the ring) is reversed. The drug is normally a mixture of α- and β-forms.

6.3. Substituted dihydroxyphenylalanines

As Fig. 6.3 shows, DOPA, DOPAMINE, adrenalin, and noradrenalin are derivatives of the aromatic amino acid phenylalanine and they are often referred to collectively as *catecholamines*. As described in Chapter 3, the oxidations of adrenalin and noradrenalin can produce O_2^- and hydrogen peroxide in a very complicated series of reactions; DOPA and DOPAMINE do the same. In all cases the rate of oxidation is greatly accelerated by the presence of transition-metal ions, and it produces not only oxygen-derived species but also quinones (e.g. DOPA quinone, Fig. 6.3) and semiquinones. These can combine with various cellular constituents to form covalent bonds, usually with thiol groups.

Treatment of animals or humans with large quantities of manganese salts decreases the amount of DOPAMINE in parts of the brain, and produces symptoms such as muscular tremors, possibly related to increased oxidation *in vivo*. The disorder 'manganese madness' or 'locura manganica' has been observed in the mining villages of northern Chile. In its later stages, this disease has some clinical resemblance to Parkinson's disease.

Whereas the rate of oxidation of adrenalin and other compounds at physiological pH would be expected to be slow at the concentrations of transition metals likely to be present *in vivo* (Chapter 2), the related compounds 6-hydroxydopamine and 6-aminodopamine (Fig. 6.3) oxidize much more rapidly. During the oxidation, semiquinones and quinones are formed (Fig. 6.3 shows the two quinones that are produced by 6-hydroxydopamine) and spin-trapping experiments have shown the production of both superoxide and hydroxyl radicals. Formation of OH$^{\cdot}$ during 6-hydroxydopamine oxidation *in vitro* is decreased by desferrioxamine, suggesting that it is formed by an iron-catalysed Haber–Weiss reaction. The superoxide radical participates in the oxidation of further molecules, and hence addition of superoxide dismutase decreases the observed rate of oxidation of 6-hydroxydopamine *in vitro*. Indeed, this has been made the basis of an assay for SOD activity.

When injected into experimental animals, both 6-hydroxydopamine and 6-aminodopamine cause a rapid and specific damage to catecholamine nerve terminals. Because of this, 6-hydroxydopamine is widely used as a research tool in experiments investigating the physiological roles of such nerve terminals. The selective action seems to be due to the specific uptake of these compounds by catecholamine neurones, and it is tempting to attribute the toxicity to increased formation of O_2^-, hydrogen peroxide, and OH$^{\cdot}$ *in vivo*. The reaction of semiquinones and quinones with —SH groups on proteins might also produce damaging effects. Indeed, D. G. Graham in the USA showed that the toxicity of

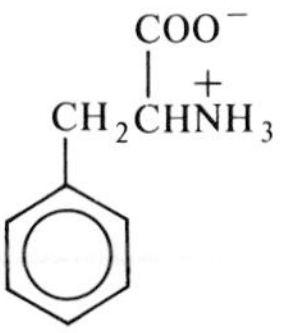

Phenylalanine

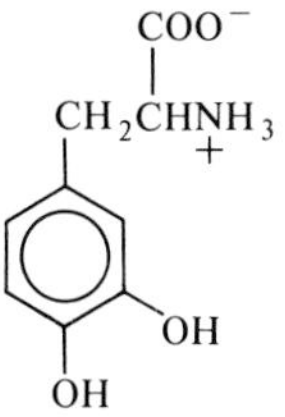

3,4-Dihydroxyphenylalanine (DOPA)

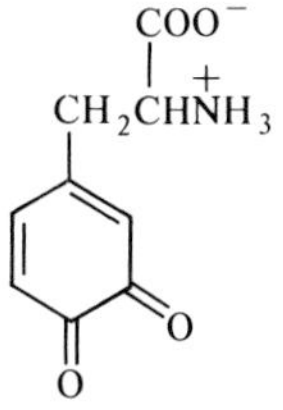

DOPA quinone

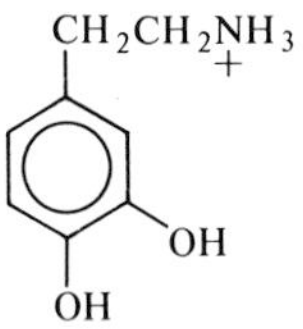

DOPAMINE

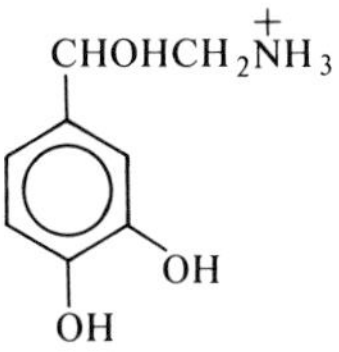

Noradrenalin

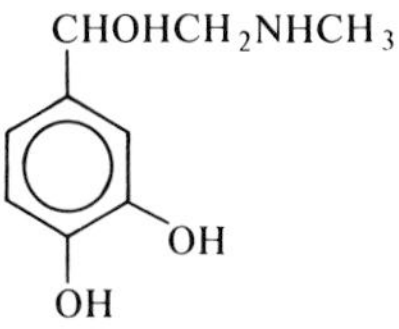

Adrenalin

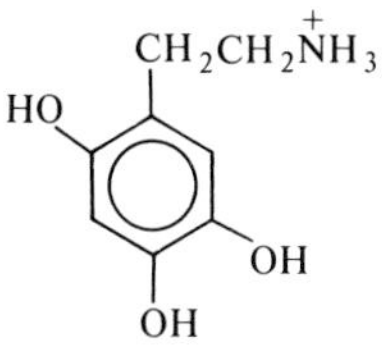

6-Hydroxydopamine

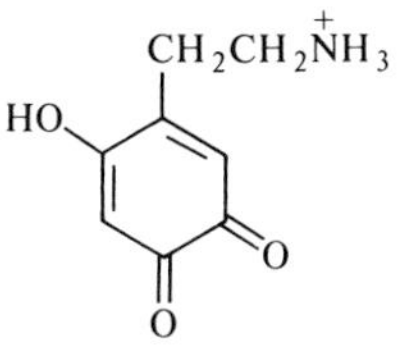

...*ortho*-quinone

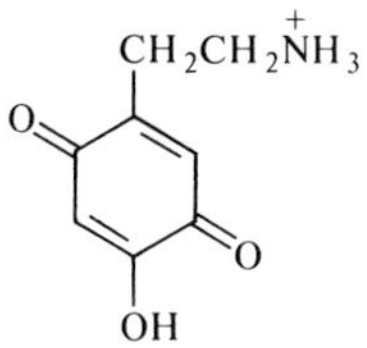

...*para*-quinone

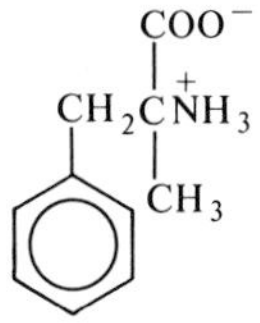

Methyldopa

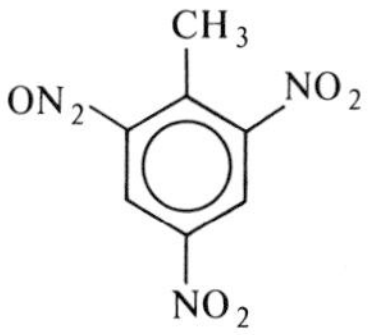

Trinitrotoluene (TNT)

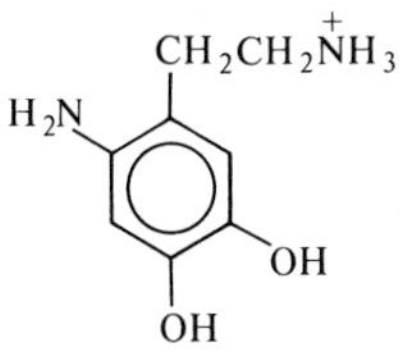

6-Aminodopamine

Fig. 6.3. Structures of dopa derivatives and some other aromatic compounds.

various phenolic compounds to neuroblastoma cells in culture could be correlated with their rate of oxidation.

Figure 6.3 additionally shows the structure of methyldopa, a drug used to lower elevated blood pressure in humans. One of its side effects is an impairment of liver metabolism. Incubation of liver microsomes with methyldopa in the presence of NADPH causes a covalent binding of the drug to the microsomal membranes that can be inhibited either by GSH or by superoxide dismutase. It seems that O_2^- produced by microsomes when NADPH is present (Chapter 3) can accelerate the oxidation of methyldopa, and the quinones and semiquinones formed can attack —SH groups in the microsomal proteins (hence the protection by GSH).

It has recently been suggested that the toxicity of the hydrocarbon benzene to animals and humans is due to its conversion *in vivo* into various phenols, some of which can undergo oxidation to give quinones, semiquinones, and oxygen radicals. Benzene particularly damages the bone marrow. The explosive trinitrotoluene (TNT) has been reported to lead to O_2^- production when incubated with rat liver microsomes plus NADPH. Its structure is shown in Fig. 6.3.

6.4. Cigarette smoke

Soot from urban air, car exhaust fumes, and cigarette smoke all contain considerable numbers of free radicals that can be detected using ESR methods. Both cellulose and glass-fibre filters have been used by Pryor's group in the USA, and others, to trap radicals in the tar present in cigarette smoke. The tar in each puff of smoke contains about 10^{14} radicals. Most of them are highly stable and persist on the filters for hours. Figure 6.4 shows their ESR spectra. Gas-phase cigarette smoke radicals have been studied using the spin-trapping technique (Chapter 2) and Fig. 6.4 shows a typical spectrum obtained with phenyl-*tert*-butyl-nitrone (PBN). The gas phase contains about 10^{15} radicals per puff, both peroxy radicals (ROO˙) and, to a lesser extent, carbon-centred radicals being present. The smoke can be drawn as much as 180 cm down a glass tube without a significant decrease in radical concentration, which has interesting implications for what happens in the lungs! Cigarette smoke contains high concentrations of the gaseous compounds nitric oxide (NO) and nitrogen dioxide (NO_2). The latter molecule is itself a radical and is capable of attacking compounds with $C{=}C$ bonds. Nitric oxide reacts with oxygen to form more nitrogen dioxide. Hydrocarbons containing one or two double bonds are present in cigarette smoke and may react with nitrogen dioxide in a number of ways, e.g.

$$NO \xrightarrow{O_2} NO_2 \xrightarrow[\text{hydrocarbons}]{\text{unsaturated}} R˙ \xrightarrow{O_2} ROO˙ \xrightarrow{NO} RO˙$$

$$\text{slow reaction} \qquad\qquad\qquad\qquad \text{peroxy radicals}$$

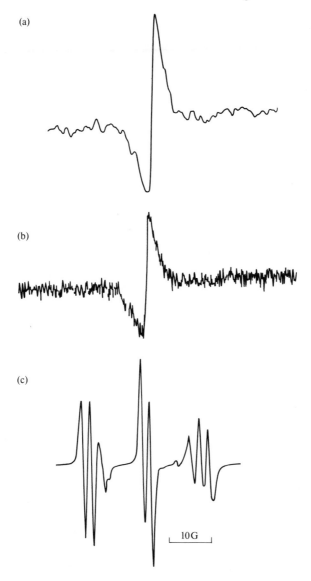

Fig. 6.4. Detection of radicals in cigarette smoke by electron spin resonance. (a) ESR signal from glass-wool filter after smoke from 4 cigarettes has been drawn through it. The g-value is 2.002. (b) ESR signal from a glass-fibre filter after smoke from 4 cigarettes has been drawn through it. The g-value of the centre of the spectrum is 2.002. (c) ESR spectrum of the spin adduct formed by the trapping of radicals in cigarette smoke with PBN in benzene. By courtesy of Prof. W. A. Pryor.

A second reaction is of the type:

$$NO_2 + \text{—HC}{=}\text{CH—CH}_2\text{—} \rightarrow HNO_2 + \text{—CH—CH—CH—}$$

(a radical   nitrous acid   (unpaired electron delocalized
itself)   between 3 carbon atoms)

The various radicals formed can then combine with each other to give non-radical products, so that a steady-state concentration of radicals is established. Nitrogen dioxide can abstract a hydrogen atom from linoleic and linolenic acids, and has been shown to stimulate the peroxidation of membrane lipids both *in vitro* and *in vivo*. For example, exposure of rats to air polluted with nitrogen dioxide increases the amounts of conjugated dienes subsequently detected in lung extracts, and also increases the rats' exhalation of ethane gas (Chapter 4). Clearly, the presence of nitrogen dioxide in cigarette smoke and in the gases produced during combustion of many nitrogen-containing compounds gives considerable potential for damage. Further, both nitric oxide and nitrogen dioxide can react with hydrogen peroxide to produce hydroxyl radicals,

$$NO + H_2O_2 \rightarrow HNO_2 + OH^{\bullet}$$
$$NO_2 + H_2O_2 \rightarrow HNO_3 + OH^{\bullet}.$$

Since cigarette smoke has been reported to stimulate production of hydrogen peroxide by the pulmonary alveolar macrophage cells (see Chapter 7 for a discussion of these), these reactions are feasible *in vivo*. Among the many damaging effects of hydroxyl radical in the lungs is its ability to attack and inactivate the α_1-*antiprotease inhibitor*, a glyco-protein which the body produces in order to inhibit the action of a number of proteases, including the enzyme *elastase*, released by phagocytic cells (Chapter 7). If elastase action is not inhibited, then it will hydrolyse the protein elastin within the lung. Elastin is the major component of elastic fibres, which can stretch to several times their length and then rapidly return to their starting length when tension is released. It is very important in allowing the correct expansion and contraction of the lung. An inborn deficiency of the α_1-antiprotease inhibitor in humans often causes a disease called *emphysema* in which lung capacity is lost. Statistical evidence indicates that, even in humans with normal inhibitor levels, smoking predisposes to the development of emphysema, which may be due to localized inactivation of the protease inhibitor by $OH^{\bullet}$. Lung lavage fluids from chronic smokers sometimes, but by no means always, show sub-normal levels of antiprotease activity. The $OH^{\bullet}$ seems to oxidize a methionine residue essential to the activity of the antiprotease. Further, nitric oxide and nitrogen dioxide can directly oxidize cysteine and GSH to their disulphides, which could lead to inactivation of thiol-dependent enzymes *in vivo*.

Cigarette tar consists mainly of complex aromatic hydrocarbons, and the stable free radicals present are probably unpaired electrons delocalized over large aromatic ring structures. When tar is fractionated, most of the free radicals travel with that part containing polyphenolic material. Among the hydrocarbons present in the tar fraction are some which are carcinogenic, and others which assist the action of the carcinogens. These are considered further in Chapter 8. The burning of cigarette paper also produces radicals.

In summary, we recommend that you give up smoking and, having done that, to keep a considerable distance away from other smokers. This will enable you to avoid all the radicals formed!

6.5. Other air pollutants

Oxides of nitrogen are produced not only in cigarette smoke but also in motor vehicle exhausts and in the combustion of most biological materials.

As we saw in Chapter 4, ozone is also a common air pollutant. If you require additional copies of this book, we strongly recommend that you purchase them rather than using a photocopier. This is merely in the interests of your own health, since photocopiers have been reported to produce significant amounts of ozone which will peroxidize your lung lipids!

Another air pollutant found in urban areas is sulphur dioxide (SO_2), a colourless choking gas formed by the combustion of fuels containing sulphur (e.g. low-grade coals). It contains no unpaired electrons, and thus is not a free radical. Sulphur dioxide dissolves in water to reversibly form the sulphite and bisulphite ions:

$$SO_2 + H_2O \rightleftharpoons H_2SO_3 \rightleftharpoons H^+ + HSO_3^{2-} \rightleftharpoons H^+ + SO_3^{2-}$$
$$\text{sulphurous acid} \qquad \text{bisulphite ion} \qquad \text{sulphite ion}$$

In one of the earliest uses of SOD to probe a radical reaction, McCord and Fridovich in the USA studied the ability of sulphite ions in solution to undergo oxidation. Sulphite ions are unstable and react with oxygen to eventually form sulphate ions, SO_4^{2-}. They observed that the oxidation of SO_3^{2-} in aqueous solution at pH 7.0 was not slowed by addition of SOD. However, the presence of EDTA decreased the rate threefold and changed the mechanism of the oxidation, in that SOD was then a powerful inhibitor of it. It appears that under certain circumstances, O_2^- can participate in sulphite oxidation. Asada in Japan showed that illuminated chloroplasts can accelerate the oxidation of SO_3^{2-}, an effect inhibited by SOD and presumably due to the production of O_2^- by electron-acceptors of photosystem I (Chapter 5). Young poplar-leaves have more SOD activity than old poplar-leaves, and are more resistant to the toxic effects of sulphur dioxide, whereas spraying spinach-leaves with diethyldithiocarbamate to inhibit CuZnSOD increases their sensitivity to sulphur dioxide. These observations suggest that an interaction of SO_3^{2-} and O_2^- is relevant *in vivo*. Indeed, exposure of poplar-leaves to concentrations of sulphur dioxide too low to cause visible injury induces a striking increase in leaf SOD activity. Horseradish peroxidase will catalyse oxidation of sulphite ions, among its many other reactions (Chapter 2), to form sulphite radical $SO_3^{\bullet-}$. Exposure of plants to SO_2-polluted air causes accumulation of hydrogen peroxide within chloroplasts, destruction of chlorophyll and carotenoids, and membrane damage accompanied by the accumulation of TBA-reactive material and increased production of ethane gas. Indeed, SO_3^{2-} has been shown to stimulate the peroxidation of linoleic and linolenic acid emulsions *in vitro*, probably by the interaction of a sulphite radical with preformed lipid hydroperoxides and/or by abstraction of hydrogen atoms (Chapter 4). Radicals such as $SO_3^{\bullet-}$, $SO_4^{\bullet-}$, and $SO_5^{\bullet-}$ are known to be formed during sulphite oxidation. Peroxidation of rat lung extracts, liver microsomes, and mitochondria can be induced by SO_3^{2-} *in vitro* in the presence of transition-metal ions, such as iron salts.

6.6. Haemolytic and anti-malarial drugs

We saw in Chapter 3 that the haemoglobin molecule acts as a source of O_2^- *in vivo*, producing the Fe(III)-form methaemoglobin. Methaemoglobin has been known for some time to catalyse certain peroxidase-like reactions *in vitro*, such as the oxidation of benzidine by hydrogen peroxide. Reactions of this type have often been used to detect blood in faeces, and during forensic examinations. Oxyhaemoglobin can even catalyse a hydroxylation of the aromatic compound aniline ($C_6H_5NH_2$) in the presence of NADPH.

However, a number of drugs are known that can penetrate to the active site of the haemoglobin molecule, and react with it. One of the most studied is phenylhydrazine and its derivative acetylphenylhydrazine (Fig. 6.5). Injection of these compounds into animals causes haemolysis and the bone marrow responds by putting immature erythrocytes into the circulation. Indeed reticulocytes, the precursors of erythrocytes, are often obtained for studies of cellular maturation and differentiation by injecting animals with phenylhydrazine and removing reticulocyte-rich blood several days later.

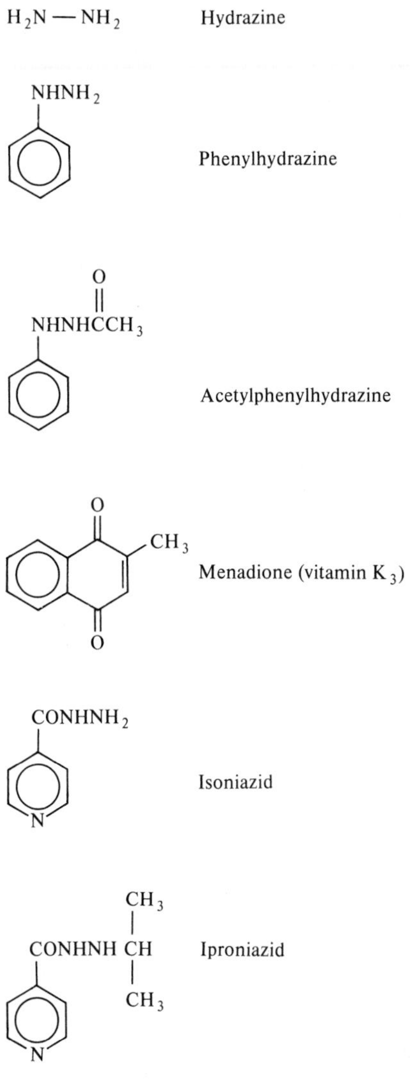

$$H_2N \text{---} NH_2 \qquad \text{Hydrazine}$$

NHNH$_2$

Phenylhydrazine

$$\overset{\displaystyle O}{\overset{\displaystyle \|}{NHNHCCH_3}}$$

Acetylphenylhydrazine

CH$_3$

Menadione (vitamin K$_3$)

CONHNH$_2$

Isoniazid

$$\text{CONHNH}\ \overset{\displaystyle CH_3}{\underset{\displaystyle CH_3}{CH}} \qquad \text{Iproniazid}$$

Fig. 6.5. Structures of some haemolytic drugs.

Phenylhydrazine and its derivatives slowly oxidize in aqueous solution to form O_2^- and hydrogen peroxide, a reaction catalysed by traces of transition-metal ions. The first stages in the oxidation can probably be represented by the equations below, in which M^{n+} represents the metal ions and Ph symbolizes the benzene ring to which the rest of the molecule is attached

$$Ph\text{---}NH\text{---}NH_2 + M^{n+} \rightarrow H^+ + Ph\text{---}NH\text{---}NH^{\boldsymbol{\cdot}} + M^{(n-1)+}$$

$$Ph\text{---}NH\text{---}NH^{\boldsymbol{\cdot}} + O_2 \rightarrow H^+ + O_2^- + PhN\text{=}NH$$

Addition of SOD slows down the oxidation process.

However, Goldberg and Stern in the USA showed that the damage done by phenylhydrazine to erythrocytes is not prevented by SOD. They discovered that methaemoglobin, acting as a peroxidase, oxidizes phenylhydrazine in the presence of hydrogen peroxide. Oxyhaemoglobin also oxidizes phenylhydrazine, but hydrogen peroxide is not required in this reaction. These oxidase and peroxidase reactions of haemoglobin form a phenylhydrazine product that can react with oxygen to give O_2^-, and can also lead to formation of the phenyl radical. The reactions may be written:

$$Ph\text{---}NH\text{---}NH_2 \xrightarrow{\text{haemoglobin}} PhN\text{=}NH$$
$$\text{phenyldiazine}$$

$$Ph\text{---}N\text{=}NH + O_2 \longrightarrow O_2^- + PhN\text{=}N^{\boldsymbol{\cdot}} + H^+$$
$$\text{phenyldiazine radical}$$

$$Ph\text{---}N\text{=}N^{\boldsymbol{\cdot}} \longrightarrow Ph^{\boldsymbol{\cdot}} + N_2$$
$$\text{phenyl radical}$$

$$Ph^{\boldsymbol{\cdot}} \xrightarrow[\text{abstraction}]{H^{\boldsymbol{\cdot}}} Ph\text{---}H$$
$$\text{benzene}$$

Hence the end-products of the oxidation are benzene and nitrogen gas. Of these various species, the most damaging seems to be the phenyldiazine radical, which can denature the haemoglobin molecule, and stimulate peroxidation of membrane lipids, causing eventual haemolysis. The haem group is converted into a green product, both cleavage of the ring, and addition of phenyl groups to it, taking place. Oxidative denaturation of haemoglobin forms intracellular precipitates called *Heinz bodies*. There is evidence that damaged haemoglobin can be acted upon and broken down by a special protease system in erythrocytes that somehow recognizes abnormal proteins. (Proteins damaged by exposure to hydroxyl radicals can also be degraded by this system.) It has been pointed out that reticulocytes obtained from phenylhydrazine-treated animals may themselves have suffered some membrane damage.

Although O_2^- and hydrogen peroxide are not required in the initial reaction of oxyhaemoglobin and phenylhydrazine, they are involved in the subsequent decomposition and precipitation reactions. Winterbourn and Carrell in New Zealand have conducted detailed studies on the interaction of acetylphenylhydrazine with oxyhaemoglobin *in vitro*. Addition of catalase to remove hydrogen peroxide decreased the rate of disappearance of oxyhaemoglobin, but addition of SOD actually increased it. Both ascorbate and GSH decreased the reaction, possibly by directly scavenging intermediate radicals (Chapter 3) such as Ph—N=N$^{\cdot}$, Ph$^{\cdot}$, and Ph—NH—NH$^{\cdot}$. Hence, these compounds, as well as the H_2O_2-removing enzymes catalase and glutathione peroxidase, would offer some protection to the erythrocyte *in vivo*. Consistent with this, the toxic effects of phenylhydrazine to mice are increased if they have been maintained on a selenium-deficient diet to lower glutathione peroxidase activity. Erythrocytes from patients with Fanconi's anaemia, in which SOD activities are decreased but glutathione peroxidase and catalase activities are normal (Chapter 8), do not show an increased susceptibility to haemolysis by acetylphenylhydrazine, consistent with the lack of protection by SOD in *in vitro* experiments.

Hydrazine derivatives are widely used in industry, as rocket fuels, and in medicine. For example, hydralazine is used to treat high blood pressure, but it has several side-effects, and prolonged use of it can produce symptoms of rheumatoid arthritis and of the autoimmune disease lupus erythematosus (Chapter 8). Like other hydrazines, this compound oxidizes *in vitro* in the presence of transition-metal ions to form $O_2{}^-$, hydrogen peroxide, and nitrogen-centred radicals. These may well participate in the side-effects of the drug. The metabolism of the anti-tubercular drug *isoniazid* and of the antidepressant *iproniazid* (Fig. 6.5) has been reported to yield hydrazine derivatives *in vivo*, which again might contribute to their side-effects. It has recently been shown that an antioxidant often used in the rubber industry, *N*-isopropyl-*N*′-phenyl-*p*-phenylenediamine, can cause rapid oxidation and denaturation of both pure haemoglobin and also of this protein in erythrocytes. The reaction was about 40 times faster than that induced by phenylhydrazine. Thus antioxidants are not always antioxidants!

The reaction of oxyhaemoglobin with menadione (Fig. 6.5) also results in oxidation and precipitation of the protein. Menadione is a quinone molecule, as Fig. 6.5 shows, and it has been proposed that it first reacts with oxyhaemoglobin to give a semiquinone (SQ$^{\cdot-}$)

$$Hb(Fe^{2+})O_2 + Q \rightleftharpoons \underset{\text{methaemoglobin}}{Hb(Fe^{3+})} + O_2 + \mathbf{SQ^{\cdot-}}$$

The semiquinone can both re-reduce methaemoglobin and can also convert oxygen to the superoxide radical. In addition, menadione can react directly with —SH groups on the haemoglobin molecule itself. If these are written as protein—SH, then the reaction product is

Treatment of erythrocytes *in vitro* with diethyldithiocarbamate to inhibit CuZnSOD has been shown to accelerate the rate of haemolysis induced by *naphthoquinone-2-sulphonate*, a derivative of menadione in which the —CH_3 group is replaced by a sulphonate group to make the compound more soluble in water and facilitate experimentation. Diethyldithio-carbamate also potentiates the toxicity of menadione to liver cells. This is because a number of enzyme systems in liver, including NADPH-cytochrome P_{450} reductase, catalyse one-electron reduction of quinones into semiquinones that then react with oxygen to give O_2^-. Liver cytosol contains high activities of the enzyme *diaphorase*, which, by contrast, catalyses a two-electron reduction of quinones into stable hydroquinones at the expense of NADH or NADPH. Hochstein and Ernster have suggested that the physiological function of diaphorase is as a 'quinone reductase', decreasing formation of O_2^- *in vivo* by removing quinones and thus preventing their reduction to semiquinones by the other enzyme systems. Consistent with this hypothesis, inhibition of diaphorase in isolated hepatocytes by the anti-coagulant drug *dicoumarol* increases the toxicity of menadione to these cells.

The haemolytic effects of menadione *in vivo* are usually minor unless the patient has an inborn defect in one of the erythrocyte defence mechanisms, the commonest being a deficiency in glucose-6-phosphate dehydrogenase activity (Chapter 3). Most patients with a partial deficiency of this enzyme show little haemolysis normally but it can be induced, often severely, by a wide range of drugs and even by ingestion of certain foods such as the broad bean *Vicia faba*. This latter condition, known as favism, is common in certain Mediterranean countries such as Sardinia and Greece, in the Middle East, and parts of south-east Asia. Its distribution therefore follows that of glucose-6-phosphate dehydrogenase deficiency, but not all patients deficient in this enzyme are sensitive to the bean, for an unknown reason. The erythrocytes of patients with favism show a rapid fall in NADPH/$NADP^+$ ratios and GSH concentrations very soon after ingesting the bean. The chemicals that cause this effect are the pyrimidine derivatives, vicine and convicine, which are present at about 0.5 per cent of the total weight of the bean. Both compounds react rapidly with oxygen to form hydrogen peroxide and an oxidized form that can be re-reduced by GSH, thus leading to more H_2O_2-generation.

Drugs that trigger haemolysis in glucose-6-phosphate dehydrogenase deficient patients include the antimalarials primaquine and pamaquine, as summarized in Table 6.3. Mixtures of primaquine with NADPH and with oxyhaemoglobin have been shown to produce hydrogen peroxide *in vitro*, an observation perhaps relevant to its haemolytic effects. Inborn defects in other erythrocyte enzymes can also cause problems, although they occur much more rarely. The oxidation and precipitation of several abnormal haemoglobins, which produces O_2^-, is accelerated by certain drugs (Table 6.3). Infections frequently initiate haemolytic crises in carriers of unstable haemoglobins, which may sometimes be due to a faster denaturation of them during the elevated body temperatures that occur in fever. Some of the denaturation products of haemoglobin might be able to catalyse an interaction of O_2^- and hydrogen peroxide to form hydroxyl radicals.

The prevalence of glucose-6-phosphate dehydrogenase deficiency and sickle-cell anaemia has been related to their ability to confer some protection against the malarial parasite (Chapter 3). Malarial parasites are damaged by systems generating O_2^- and hydrogen peroxide. Clark in Australia has shown that injection of alloxan or *tert*-butylhydroperoxide (Chapter 4) into malaria-infected mice kills a large number of the parasites. The effect of alloxan can be overcome by pretreatment of the mice with desferrioxamine, an inhibitor of the iron-catalysed Haber–Weiss reaction. Thus any defect in protection of erythrocytes against oxidative damage, or an increased rate of radical production *in vivo*, should favour the eradication of malarial parasites within such erythrocytes. Several other protozoan parasites seem to be equally sensitive to free-radicals, which raises prospects for the design of new drugs. *Trypanosoma brucei*, for example, is unable to synthesize haem and lacks catalase-activity. It can be killed by increasing its rate of intracellular H_2O_2-generation, e.g. by adding menadione, and killing is accelerated by the addition of free haem.

6.7. Ethanol

Most people drink solutions of ethanol (ethyl alcohol) as alcoholic drinks, and even in teetotallers it is formed in small amounts by gut bacteria. Ethanol is very soluble both in water and in organic solvents, and can cross cell membranes readily. It penetrates the blood–brain barrier and affects the central nervous system. One beneficial effect that it has, apart from the social ones, is to partially protect experimental animals against alloxan-induced diabetes (Section 6.2). It probably does this by

Table 6.3. Inborn deficiencies of erythrocyte enzymes in relation to drug-induced haemolysis

Abnormality	Prevalence	Usual clinical feature	Drugs inducing haemolysis	Normal clinical use of drug
Glucose-6-phosphate dehydrogenase deficiency (X-linked, recessive)	Very common in patients in tropical or Mediterranean areas or their descendants	Some RBC damage often detected in laboratory tests but severe haemolysis *in vivo* very rare	Fava beans Furazolidone Nitrofurantoin Nitrofurazone Pamaquine Primaquine Sulphonamides	— (See Section 6.10) Antimalarials Antibacterial
Glutathione peroxidase deficiency (recessive)	Very rare	Often none, sometimes severe haemolysis	Sulphonamides Nitrofurantoin	Antibacterial (Section 6.10)
Glutathione reductase deficiency (recessive)	Very rare, but lack of riboflavin in the diet can also decrease the activity (enzyme has FAD at the active site). The drug BCNU, used in cancer chemotherapy, is a powerful inhibitor of glutathione reductase and can cause erythrocyte damage	Often none, sometimes severe haemolysis	Sulphonamides	Antibacterial
Abnormal haemoglobins (e.g. Hb Torino, Hb Shepherd's Bush, Hb Peterborough, Hb Zurich)	Rare	Often some haemolysis seen related to protein instability. Drug administration can cause a severe haemolytic crisis	Sulphonamides	

Data are largely taken from the article by G. F. Gaetani and L. Luzzatto (1980) in *Pseudo-allergic reactions. Involvement of drugs and chemicals*, Volume 2, (eds P. Dukor *et al.*), S. Karger, Basle, Switzerland.

scavenging the highly reactive hydroxyl radical to form a much less reactive hydroxyethyl radical

$$CH_3CH_2OH + OH^{\cdot} \rightarrow CH_3^{\cdot}CHOH + H_2O$$

Indeed, this reaction is made use of in spin-trapping experiments with DMPO (Chapter 2).

Ethanol absorbed into the body is mainly metabolized in the liver by an alcohol dehydrogenase enzyme to form the aldehyde ethanal (acetaldehyde)

$$CH_3CH_2OH + NAD^+ \rightarrow NADH + H^+ + CH_3CHO$$

Smaller quantities are oxidized by the peroxidatic action of catalase in peroxisomes (Chapter 3) and the MEOS system, in which hydroxyl radicals are involved (Chapter 2). Excessive doses of ethanol severely damage the liver.

In England, it is a criminal offence to drive a motor vehicle with a blood ethanol concentration greater than 80 mg per 100 ml, corresponding to a concentration of 17.4 millimoles per litre. Some 'heavy' drinkers get themselves up to 300 mg per 100 ml of blood, or 65.2 mmol l^{-1}. Such high concentrations have been shown in experimental animals to affect antioxidant protective systems. Injection of large doses of ethanol into rats has been reported to decrease the SOD activity measurable in homogenates of their brains by about twenty-five per cent, and an even greater decrease is produced by repeated injection. Similarly, exposure of cultures of neuronal or glial cells from rats, mice, hamsters, and chicks to 100 mM ethanol *in vitro* decreased their SOD activity. Whether these small changes are sufficient to cause biological damage has yet to be determined. By contrast, other scientists have reported increases in SOD activity in animals treated with ethanol for prolonged periods. Both large doses of ethanol, and smaller doses given repeatedly, have been shown to increase lipid peroxidation in the livers of rats and baboons, as followed by accumulation of conjugated dienes or ethane production. Large doses of ethanol have also been observed to cause a significant decrease in the GSH concentrations of liver and kidney cells of rats, but not in other tissues. Comparable falls in GSH are seen in mice and baboons. Such a drop in GSH might either cause more lipid peroxidation or merely be a consequence of it in view of the reactions catalysed by glutathione peroxidase (Chapter 4). Neither does it follow that an induction of lipid peroxidation is the mechanism by which ethanol damages the liver, since increased peroxidation often accompanies damage caused by other means (Chapter 4). There is some evidence that the damage is done by ethanal rather than ethanol itself. Ethanal is metabolized mainly by an aldehyde dehydrogenase enzyme that converts it into ethanoic acid (acetic acid,

CH_3COOH), but is possible that some of it could be acted upon by the enzyme aldehyde oxidase in the liver, which is known to produce O_2^- (Chapter 3). In the liver of alcoholics, mitochondria, which contain the alcohol dehydrogenase enzyme, are particularly damaged. Ethanal may also react chemically with GSH and so decrease its concentration.

The potential relevance of free radicals to chronic ethanol effects in man is shown by the observation of American scientists that red blood cells from alcoholics have increased SOD content. Curiously, this effect is only seen with black alcoholics and not white ones, for an unknown reason.

6.8. Paracetamol and phenacetin

Paracetamol (sometimes called acetaminophen) is a mild pain-killer that has found increasing use in recent years as a substitute for aspirin. Unlike aspirin, it does not irritate the stomach lining and it appears fairly safe in the recommended dosage. At high doses, however, it is acutely toxic to both the liver and kidneys, and poisoning by paracetamol overdosage in suicide attempts is becoming increasingly common. The paracetamol derivative, phenacetin, is also a mild pain-killer, but it is no longer widely used because of the risk of kidney damage. Some protection against paracetamol toxicity in animals is afforded by giving α-tocopherol or thiol compounds, such as GSH or *N*-acetylcysteine.

Paracetamol is a substrate for the cytochrome P_{450} system (Chapter 3), and phenacetin is converted into paracetamol *in vivo* by removal of the ethyl group. One suggestion is that the action of the P_{450} system produces an *N*-hydroxylation followed by release of water to yield the highly reactive quinoneimine (Fig. 6.6), which can attack various cellular membranes, presumably by combining with protein —SH groups. De Vries in Holland has suggested an alternative pathway in which cytochrome P_{450} converts paracetamol into its semiquinone, which can then either react with —SH groups or reduce oxygen to O_2^-. Further experiments are required to investigate this possibility. Some evidence for paracetamol-induced lipid peroxidation in the livers of animals has been reported, but only if the liver GSH concentration is decreased, e.g. by starvation.

6.9. Halogenated hydrocarbons

One of the first reactions encountered by any student of organic chemistry is that of the hydrocarbon gas methane (CH_4) with chlorine or bromine vapour. The reaction only proceeds in the presence of ultraviolet light, which provides sufficient energy to cause homolytic fission of the covalent

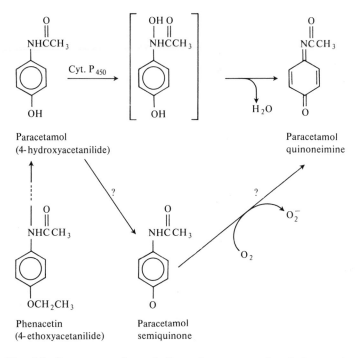

Fig. 6.6. Structure and metabolism of paracetamol and phenacetin.

bond in the halogen molecule. The reaction then proceeds as a typical radical chain reaction, i.e.

Initiation: $$Cl_2 \xrightarrow{uv} Cl^{\cdot} + Cl^{\cdot}$$

Propagation reactions: $$Cl^{\cdot} + CH_4 \longrightarrow CH_3^{\cdot} + HCl$$
$$CH_3^{\cdot} + Cl_2 \longrightarrow CH_3Cl + Cl^{\cdot}$$
$$Cl^{\cdot} + CH_3Cl \longrightarrow HCl + {\cdot}CH_2Cl$$
$${\cdot}CH_2Cl + Cl_2 \longrightarrow CH_2Cl_2 + Cl^{\cdot}, \text{ etc.} \ldots$$

Termination reactions: $$CH_3^{\cdot} + Cl^{\cdot} \longrightarrow CH_3Cl$$
$$CH_3^{\cdot} + CH_3^{\cdot} \longrightarrow C_2H_6.$$

Similar reactions occur with bromine, but much more slowly. Hence the methane is successively converted into chloromethane (CH_3Cl), dichloromethane (CH_2Cl_2), trichloromethane ($CHCl_3$, often called 'chloroform'), and tetrachloromethane (CCl_4, often called 'carbon tetrachloride'). Both $CHCl_3$ and CCl_4 are liquids at room temperature. Chloroform was the first anaesthetic to be employed in surgery, by the Scottish doctor Sir James Young Simpson in 1847, but its liver-damaging properties were quickly discovered. Carbon tetrachloride is a constituent of many dry-cleaning fluids because of its ability to easily dissolve greasy stains. Its lipid solubility allows it to cross cell membranes rapidly, and any tetrachloromethane taken into the body is quickly distributed to all organs. However, its main toxic effects are shown on the liver (and it is much more toxic than chloroform) although there is some injury to other tissues. Even a single small dose of tetrachloromethane to rats produces fat accumulation in the liver due to a blockage in synthesis of the lipoproteins that carry triglyceride away from this organ (Chapter 4). The normal structure of the liver cell endoplasmic reticulum as seen under the electron microscope is distorted, hepatic protein synthesis slows down, and the activity of enzymes located in the endoplasmic reticulum, such as glucose-6-phosphatase and the P_{450} system, rapidly declines, as does the ability of the reticulum to bind Ca^{2+} ions. The nuclear membrane is attacked more slowly. Eventually there is necrosis of liver cells in the central areas of the organ. Yet incubation of microsomal fractions with tetrachloromethane *in vivo* does not directly inactivate these enzymes. As a result of detailed work by the groups of Recknagel in the USA, Slater in England, and Comporti in Italy, it has been concluded that the effects produced by tetrachloromethane are due to the fact that it is a substrate for the cytochrome P_{450} system which is, of course, especially concentrated in the liver (Table 6.2). If microsomes from rat liver are exposed to tetrachloromethane *in vitro*, nothing happens until NADPH is provided to allow NADPH–cytochrome P_{450} reductase to operate. There is then a rapid peroxidation of microsomal lipids accompanied by the inactivation of the above enzymes, and destruction of cytochrome P_{450} itself. Microsomal fractions isolated from CCl_4-treated rats also show increased peroxidation. Fatty-acid side-chains attached to phosphatidylserine (Chapter 4) are especially attacked, perhaps because this lipid is adjacent to P_{450} *in vivo*. Phenobarbital increases the activity of the P_{450} system in liver (Chapter 3). Phenobarbital-treated rats or sheep are more susceptible to tetrachloromethane and show more peroxidation in liver microsomal fractions isolated from them. Administration of antioxidants such as vitamin E, promethazine, propyl gallate, and GSH, or of the compound SKF-525A, which inhibits microsomal drug metabolism, decreases CCl_4-toxicity in animals.

As a result of these observations, it was suggested that tetrachloromethane is metabolized by the P_{450} system to give the trichloromethyl radical ($CCl_3^{\cdot}$)

$$CCl_4 \xrightarrow[P_{450} \text{ system}]{1 \text{ electron}} {}^{\cdot}CCl_3 + Cl^-$$

Spin-trapping experiments upon isolated rat liver microsomes in the presence of NADPH and tetrachloromethane have been carried out by several groups. Addition of the spin-trap, phenyl-*tert*-butylnitrone (PBN, see Chapter 2) to the system gives the ESR signal expected from reaction of the trap with $CCl_3^{\cdot}$

Stable nitroxyl radical

Conversion of tetrachloromethane into $CCl_3^{\cdot}$ appears to be brought about by the cytochrome P_{450} molecule itself (after reduction by the NADPH–cytochrome P_{450} reductase) although it is possible that the reductase might also interact with tetrachloromethane. The trichloromethyl radical is then thought to abstract a hydrogen atom from the membrane lipid, setting off the chain reaction of lipid peroxidation (Chapter 4). Products of peroxidation are known to inhibit protein synthesis and the activity of certain enzymes (Chapter 4). Indeed, liver microsomal fractions from CCl_4-treated rats show an increased amount of protein-bound carbonyl compounds and CCl_4-treated rats exhale more pentane gas than normal, indicative of increased lipid peroxidation *in vivo*. The rats also exhale chloroform vapour, and trichloromethane would be expected to be produced by the combination of a trichloromethyl radical and a hydrogen atom abstracted from a membrane lipid. These effects are more marked if the rats have been fed on diets deficient in vitamin E. Increased ethane production has also been observed in isolated liver cells treated with tetrachloromethane, together with increased light emission, and accumulation of TBA-reactive material. Peroxidation occurred rapidly and was seen before the onset of loss of cell viability. The antioxidant promethazine decreased peroxidation in the liver cells and prevented loss of glucose-6-phosphatase activity, but not the loss of P_{450} itself. Probably cytochrome P_{450} can be directly attacked by $CCl_3^{\cdot}$ or other radical species formed, whereas the inactivation of glucose-6-phosphatase is brought about by products of lipid peroxidation. Thus CCl_4-metabolism is a 'self-limiting' event in that once the P_{450} is destroyed, no more $CCl_3^{\cdot}$ can be formed.

Despite the mass of evidence reviewed above, some questions have been raised about peroxidation of membrane lipids induced by $CCl_3^{\cdot}$ as an explanation of CCl_4-toxicity. Pulse radiolysis studies of the reactivity of the trichloromethyl radical have been carried out by R. L. Willson's group in England, whereupon it was found that its most rapid reaction is with molecular oxygen to form the *trichloromethylperoxy radical*

$$ \cdot CCl_3 + O_2 \rightarrow CCl_3O_2 \cdot $$

$CCl_3O_2 \cdot$ reacts much more rapidly with arachidonic acid ($k = 6 \times 10^6$ at pH 7), promethazine, ascorbate, thiol compounds, and the tyrosine and tryptophan residues of proteins than does the trichloromethyl radical, and so $CCl_3O_2 \cdot$ would seem a more likely candidate for a damaging species. Spin-trapping experiments have so far failed to reveal its presence in CCl_4-treated microsomes, however, although its high reactivity will make it correspondingly more difficult to detect. Formation of the trichloro-methylperoxy radical would explain why small amounts of phosgene gas ($COCl_2$) are produced by CCl_4-treated microsomes. It could simply arise by the reactions below:

$$ CCl_3O_2 \cdot + lipid{-}H \rightarrow CCl_3O_2H + lipid \cdot $$

$$ CCl_3O_2H \rightarrow COCl_2 + HOCl $$

It is possible that the local oxygen concentration in different parts of the liver influences whether $CCl_3 \cdot$ itself covalently binds to biological molecules or combines with O_2 to form $CCl_3O_2 \cdot$, which should be a much better initiator of lipid peroxidation.

Chloroform ($CHCl_3$—trichloromethane) is no longer used as an anaesthetic, but it is still widely employed in industry as a solvent, and small amounts are sometimes added to cough mixtures and mouth-washes, although this is no longer encouraged. Trichloromethane is much less damaging to the liver than is tetrachloromethane, and induces lipid peroxidation in isolated liver microsomes at a much lower rate. This may be because the energy required to cause homolytic fission of trichloro-methane to produce the trichloromethyl radical is greater than with tetrachloromethane. Consistent with this argument, compounds in which homolytic fission is easier such as bromotrichloromethane ($BrCCl_3$) induce peroxidation even more rapidly than does CCl_4 (Table 6.4).

Table 6.4. Lipid peroxidation induced by halogenated hydrocarbons

Reaction	Energy needed for reaction (kcal mole^{-1})	Relative rate of lipid peroxidation
$CCl_4 \rightarrow Cl \cdot + \cdot CCl_3$	68	100
$CHCl_3 \rightarrow H \cdot + \cdot CCl_3$	90	7
$BrCCl_3 \rightarrow Br \cdot + \cdot CCl_3$	49	3650

The rate of peroxidation induced in rat-liver microsomes in the presence of NADPH was measured by the thiobarbituric acid method. Results are expressed relative to the stimulatory effect of CCl_4. Data were abstracted from T. F. Slater and B. C. Sawyer (1971) *Biochem. J.* **123**, 805–14.

Nonetheless, liver damage is still a significant problem in humans exposed to chloroform vapour. The toxicity does appear to be due to metabolism by the P_{450} system, in that inducers of this system potentiate $CHCl_3$-toxicity in animals, and there is covalent binding of the molecule to cell membranes. Phosgene is produced during $CHCl_3$-metabolism.

A frequently used inhalation anaesthetic is 'halothane', which has the structure

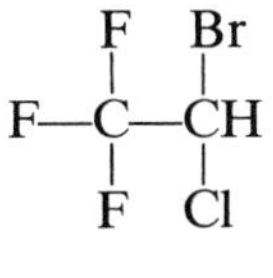

$$
\begin{array}{c}
\quad\; \text{F} \quad \text{Br} \\
\quad\; | \qquad | \\
\text{F—C—CH} \\
\quad\; | \qquad | \\
\quad\; \text{F} \quad\; \text{Cl}
\end{array}
$$

It normally has no significant side-effects but it has been observed to produce liver damage in a few patients, for unknown reasons. Incubation of halothane with rat liver microsomes in the presence of NADPH and the spin-trap PBN (see above) resulted in the formation of an ESR signal. The same signal was observed if the spin trap was fed in an oil emulsion to rats which were allowed to inhale halothane, killed, and the liver lipids extracted and placed in the ESR spectrometer. Anaerobic incubation of halothane with liver microsomes has been observed to produce reactive metabolites that can bind to the membranes, and their formation is associated with increased lipid peroxidation. American and German scientists have suggested that the action of cytochrome P_{450} on halothane causes formation of the radical $F_3C\dot{C}HCl$. Consistent with this, the expired air of rabbits and of human patients exposed to halothane has been shown to contain chlorotrifluoroethane and chlorodifluoroethene, which could be derived from such a radical

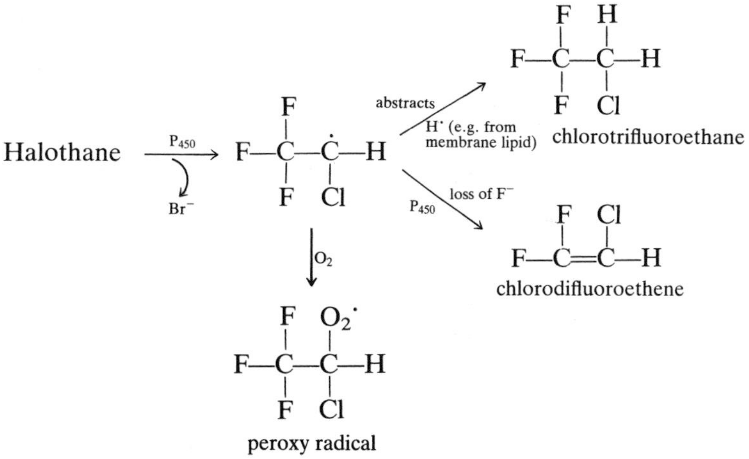

peroxy radical

Formation of chlorodifluoroethene from the radical may be catalysed by cytochrome P_{450} donating an electron to remove fluorine as F^-. The $F_3C\dot{C}HCl$ radical might also combine with oxygen to form a reactive peroxy radical. It therefore seems that radical reactions are important in halothane toxicity, but the reasons for the variability in observed effects in human patients have yet to be established.

6.10. Hypoxic cell sensitizers and related compounds

We saw in Chapter 2 that absence of oxygen decreases the sensitivity of cells to ionizing radiation. Such an observation is particularly relevant in the treatment of large cancerous tumours by radiotherapy. Often, as such tumours grow, areas within them no longer receive an adequate blood supply and thus becomes short of oxygen (hypoxic). Whilst radiation treatment may destroy most of the tumour, the hypoxic cells are more resistant, and can serve as 'nuclei' for subsequent regrowth. There has therefore been some interest in various therapies combining increased oxygen exposure with radiation (Chapter 1), and a greater interest in various drugs that, like oxygen, make hypoxic cells more sensitive to ionizing radiation. Such drugs are collectively known as *hypoxic cell sensitizers*.

In 1973, R. L. Willson in England suggested that metronidazole (Fig. 6.7) might be useful as a hypoxic cell sensitizer. Metronidazole, under the trade name *Flagyl®*, was originally introduced as an effective treatment for diseases caused by anaerobic protozoa such as *Trichomonas vaginalis* and *Giardia lamblia*. Subsequent work showed that it is also effective against *Trypanosoma cruzi*, which causes *Chagas' disease*, an infection particularly common in South America that produces recurrent fever, with damage to the heart and gastro-intestinal system. Metronidazole was later found to be effective against a wide range of anaerobic bacteria. Exposure of *E. coli*, plant tissues, and animal cells in culture to metronidazole was observed to increase their susceptibility to ionizing radiation under anaerobic conditions, and experiments on tumour-bearing animals confirmed this effect *in vivo*. A number of other imidazole compounds have been introduced for chemotherapy of bacterial and protozoal infections, although Flagyl® is still very popular. By contrast, although metronidazole has been used experimentally to supplement radiation treatment in some human cancer patients, several other sensitizers are more promising, especially *misonidazole* (Fig. 6.7). Unfortunately, a side-effect of this drug—damage to peripheral nerves—limits the amounts that can safely be given to patients. There is therefore considerable research to find less toxic sensitizers and/or drugs which minimize the neurotoxicity.

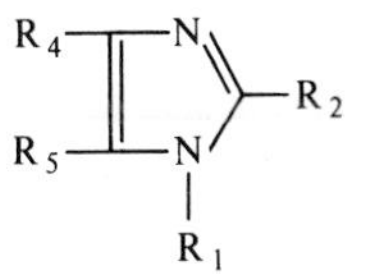

Misonidazole $R_1: CH_2CHCH_2OCH_3$ (OH on CH)
$R_2: NO_2$
$R_4, R_5: H$

Metronidazole $R_1: CH_2CH_2OH$
$R_2: CH_3$
$R_4: H$
$R_5: NO_2$

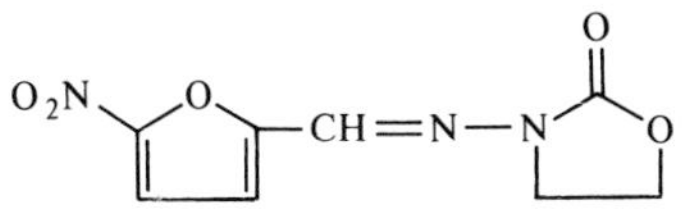

Furazolidone

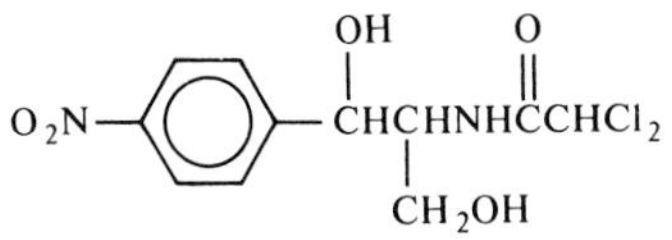

Chloramphenicol

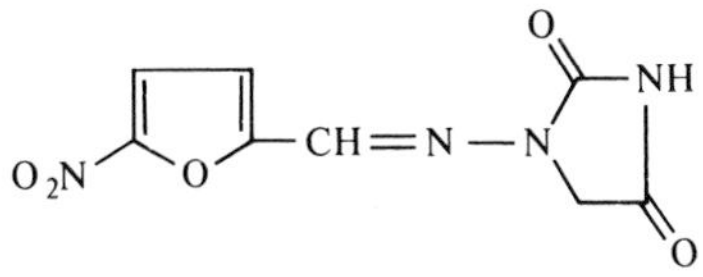

Nitrofurantoin

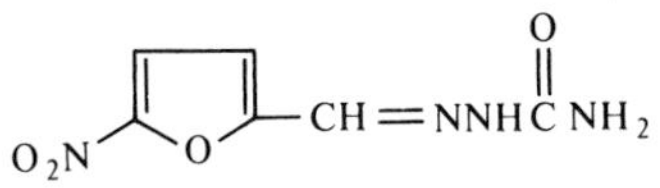

Nitrofurazone

Fig. 6.7. Structure of hypoxic cell-sensitizers and some other nitro-compounds.

As discussed in Chapter 3, the principal reason for the effect of oxygen in potentiating radiation damage seems to be prevention of the repair of organic radicals i.e.

$$RH \xrightarrow{\text{radiation}} R^{\cdot} \quad \text{(radical production from cell components)}$$

$$R^{\cdot} \xrightarrow[\text{ascorbate}]{\text{GSH or}} RH \quad \text{(repair)}$$

$$R^{\cdot} + O_2 \longrightarrow RO_2^{\cdot} \quad \text{(damage fixation)}.$$

Metronidazole and other hypoxic cell sensitizers might act in a similar way to oxygen, forming either a radical adduct (A) or a radical anion (B) that cannot easily be repaired

$$R^\bullet + \text{O}_2\text{N}\text{—[imidazole ring, N–CH}_2\text{CH}_2\text{OH, CH}_3]} \longrightarrow$$

A: $\text{O}=\overset{\bullet}{\text{N}}$, $\overset{|}{\text{OR}}$ attached to imidazole ring with N–CH$_2$CH$_2$OH and CH$_3$

B: $\text{O}=\overset{\bullet}{\text{N}}$, O^- attached to imidazole ring with N–CH$_2$CH$_2$OH and CH$_3$ $+\ R^+$

In addition, it has been suggested that cell sensitizers might interfere with normal recombination reactions between radiation-produced radicals.

Metronidazole, misonidazole, and related compounds are reduced *in vivo* to form radical species, and this seems to be the basis of their toxicity to anaerobic bacteria. It may also explain their toxicity to anaerobic animal cells, an effect seen in addition to radiation sensitization. Reduction of these drugs has been demonstrated *in vitro* using a number of enzyme systems such as xanthine oxidase, the hydrogenase complex of *T. vaginalis* (in which reduced ferredoxin is probably the electron-donating component), and with microsomal fractions from several animal tissues, apparently by the action of NADPH–cytochrome P$_{450}$ reductase. Exposure of DNA to reduced metronidazole causes strand-breakage in the molecule, especially at thymidine residues. The reduction products also bind to microsomal membranes under anaerobic conditions and react with GSH.

Probably the first stage in reduction is formation of a radical anion $RNO_2^{\bullet-}$ (compound B above). The ability of this to react with GSH could potentiate radiation damage by interfering with an important protective mechanism:

$$RNO_2^{\bullet-} + GSH \rightarrow RNO + GS^\bullet + OH^-.$$

Further reduction of $RNO_2^{\bullet-}$ to nitroso compounds (RNO) and to hydroxylamines (RNHOH) has been observed *in vivo*, although which of these species actually causes the DNA damage has not been established. The nitro-drug furazolidone (Fig. 6.7), used in veterinary medicine, is also reduced *in vivo*.

The covalent binding of metronidazole-derived products to microsomes in the presence of NADPH is not seen if oxygen is present, since $RNO_2^{\cdot-}$ reacts quickly to form the superoxide radical:

$$RNO_2^{\cdot-} + O_2 \rightarrow RNO_2 + O_2^-.$$

It is possible that increased O_2^--formation by this reaction might account for some of the observed side-effects of hypoxic cell sensitizers on normal aerobic tissues, although there is no direct evidence for this.

Figure 6.7 shows the structure of nitrofurantoin, an antibacterial drug that produces lung damage as a side effect. It has been suggested that this may be due to reduction of the drug by lung endoplasmic reticulum

$$X\!-\!NO_2 \xrightarrow[\text{electron}]{\text{one}} X\!-\!NO_2^{\cdot-}$$

followed by re-oxidation of the nitro-radical to produce O_2^- and hence hydrogen peroxide. Consistent with this, feeding chicks on selenium-deficient diets to reduce glutathione peroxidase activity greatly potentiates the toxicity of nitrofurantoin. Similar reactions have been suggested to account for the toxicity of nitrofurazone (Fig. 6.7) and its derivatives to *Trypanosoma cruzi*. In the absence of oxygen the nitro-radicals seem to undergo disproportionation to form highly-reactive nitroso compounds

$$2XNO_2^{\cdot-} + 2H^+ \rightarrow X\!-\!N\!\!=\!\!O + XNO_2 + H_2O.$$

XNO may then be further reduced to a hydroxylamine, XNHOH.

Another nitro-compound used to treat humans is chloramphenicol (Fig. 6.7). This antibiotic is active against a wide range of bacteria, but its use is severely restricted since it depresses the action of the bone-marrow and sometimes produces severe and irreversible anaemia. Its action has often been attributed to its ability to inhibit protein synthesis, but this may not be the whole story since it also induces DNA damage *in vivo*. Chloramphenicol in which the nitro group has been reduced *in vitro* will damage isolated DNA. Further, microsomal fractions from rats pre-treated with phenobarbital convert chloramphenicol into a number of products, at least one of which can combine covalently with microsomal protein. However, the microsome-catalysed activation involves the terminal $-CHCl_2$ group rather than the nitro-group (Fig. 6.8). The exact relation of these effects to the observed bone-marrow damage has yet to be determined.

6.11. The 'Spanish cooking-oil' syndrome

In September 1981, Dr J. M. Tabuenca reported on a new syndrome that has caused at least 350 deaths and affected over 20 000 other people in

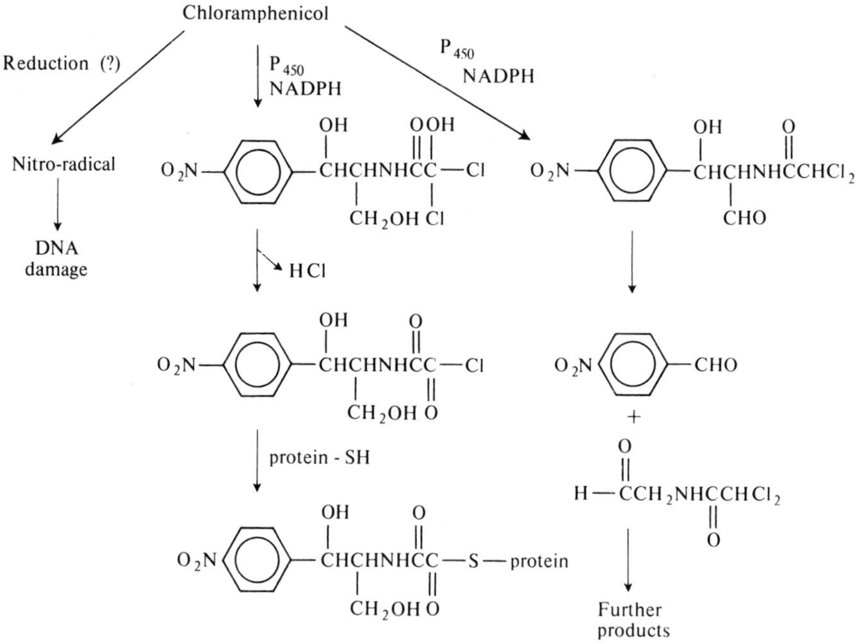

Fig. 6.8. Metabolism of chloramphenicol *in vivo*. Although the reduction of the nitro-group of chloramphenicol can be achieved *in vitro* and the product so obtained damages DNA, it has not been established how it could be reduced *in vivo*.

Spain up to the end of 1982. The symptoms of the disease in its acute stage are respiratory distress, fever, headache, itching, nausea, and sometimes muscular pains and neurological disorders. Lung damage has been responsible for most of the early deaths. When followed up five months later, about ten per cent of patients had developed muscular wasting and weakness, and another forty-nine per cent had suffered some muscular impairment. All sufferers from the disease were found to have consumed 'olive oil' sold by door-to-door salesmen. Analysis showed that the oil in question had never seen an olive tree, and was basically oil obtained from seeds of the rape plant *Brassica rapus*. Rapeseed oil is normally used as an industrial lubricant. Such oil imported into Spain is treated with aniline (phenylamine, $C_6H_5NH_2$) to deter its consumption; and it appears that attempts had been made to refine the aniline out of the rapeseed oil. During this attempted purification, a number of chemical changes took place, resulting in the formation of various toxins. The disease differs from that of simple poisoning by aniline itself.

It has been suggested that part of the pathology of the disease may be related to peroxidation of membrane lipids induced by one or more of the toxins. This does not, of course, mean that peroxidation causes the damage, since increased lipid peroxidation often accompanies cell damage caused in other ways (Chapter 4). The single good thing to come out of this affair is the setting-up, in England, of a society for free-radical research to combine the expertise of chemists, physicists, biochemists, biologists, and clinicians in studying free-radical reactions. (Further information about this international society may be obtained from Professor R. L. Willson, Department of Biochemistry, Brunel University, Uxbridge, Middlesex, England.)

6.12. Further reading

Adams, G. E. (1981). Hypoxia-mediated drugs for radiation and chemotherapy. *Cancer* **48**, 696.

Ahr, H. J., King, L. J., Nastainczyk, W., and Ullrich, V. (1982). The mechanism of reductive dehalogenation of halomethane by liver cytochrome P_{450}. *Biochem. Pharmac.* **31**, 383.

Albano, E., *et al.* (1982). Spin-trapping studies on the free-radical products formed by metabolic activation of CCl_4 in rat liver microsomal fractions, isolated hepatocytes and *in vivo* in the rat. *Biochem. J.* **204**, 593.

Autor, A. P. (ed.) (1982). *Pathology of oxygen.* Academic Press, New York.

Calderbank, A. (1968). The bipyridylium herbicides. *Adv. Pest Control Res.* **8**, 127.

Chevion, M., Navok, T., Glaser, G., and Mager, J. (1982). The chemistry of favism-inducing compounds. The properties of isouramil and divicine and their reaction with glutathione. *Eur. J. Biochem.* **129**, 405.

Cohen, G. (1978). The generation of hydroxyl radicals in biological systems. Toxicological aspects. *Photochem. and Photobiol.* **28**, 669.

Cox, F. E. G. (1983). Oxidant killing of the malaria parasites. *Nature* **302**, 19.

Del Villano, B. C., Miller, S. J., Schacter, L. P., and Tischfield, J. A. (1980). Elevated superoxide dismutase in black alcoholics. *Science* **207**, 911.

De Vries, J. (1981). Hepatotoxic metabolic activation of paracetamol and its derivatives phenacetin and benorilate: oxygenation or electron transfer? *Biochem. Pharmac.* **30**, 399.

Do Campo, R. and Moreno, S. N. J. (1984). Free radical metabolites in the mode of action of chemotherapeutic agents and phagocytic cells on *Trypanosoma cruzi, Rev. Infect. Dis.* **6**, 223.

Donaldson, J., McGregor, D., and Labella, F. (1982). Manganese neurotoxicity: a model for free radical mediated neurodegeneration? *Can. J. Physiol. Pharmac.* **60**, 1398.

Farrington, J. A., Ebert, M., Land, E. J., and Fletcher, K. (1973). Bipyridylium quaternary salts and related compounds. Pulse radiolysis studies of the reaction of paraquat radical with oxygen. Implications for the mode of action of bipyridyl herbicides. *Biochim. Biophys. Acta* **314**, 372.

Fernandez-Pol, J. A., Hamilton, P. D., and Klos, D. J. (1982). Correlation between the loss of the transformed phenotype and an increase in SOD activity in a revertant subclone of sarcoma virus-infected mammalian cells. *Cancer Res.* **42**, 609.

Forni, L. G., *et al.* (1983). Reactions of the trichloromethyl and halothane-derived peroxy radicals with unsaturated fatty acids: a pulse radiolysis study. *Chem.–Biol. Interac.* **45**, 171.

Frank, L. (1981). Prolonged survival after paraquat. Role of the lung antioxidant enzyme systems. *Biochem. Pharmac.* **30**, 2319.

Gandy, S. E., Buse, M. G., and Crouch, R. K. (1982). Protective role of SOD against diabetogenic drugs. *J. clin. Invest.* **70**, 650.

Gilsanz, V. (1982). Late features of toxic syndrome due to denatured rapeseed oil. *Lancet* **i**, 335.

Goldberg, B., Stern, A., and Peisach, J. (1976). The mechanism of superoxide anion generation by the interaction of phenylhydrazine with haemoglobin. *J. biol. Chem.* **251**, 3045.

Grankvist, K., Marklund, S., and Taljedal, I. B. (1981). Superoxide dismutase is a prophylactic against alloxan diabetes. *Nature* **294**, 158.

Harper, D. B. and Harvey, B. M. R. (1978). Mechanisms of paraquat tolerance in perennial ryegrass. Role of superoxide dismutase, catalase and peroxidase. *Plant Cell and Environ.* **1**, 211.

Hassan, H. M. and Fridovich, I. (1979). Paraquat and *E. coli*. Mechanism of production of extracellular superoxide radical. *J. biol. Chem.* **254**, 10846.

Heikkila, R. E. and Cabbat, F. S. (1982). The prevention of alloxan-induced diabetes in mice by the iron chelator detapac. Suggestion of a role for iron in the cytotoxic process. *Experientia* **38**, 378.

Hill, H. A. O. and Thornalley, P. J. (1983). The effect of spin-traps on phenylhydrazine-induced haemolysis. *Biochim. Biophys. Acta* **762**, 42.

Keeling, P. L., Smith, L. L., and Aldridge, W. N. (1982). The formation of mixed disulphides in rat lung following paraquat administration. Correlation with changes in intermediary metabolism. *Biochim. Biophys. Acta* **716**, 249.

Kornbrust, D. J. and Mavis, R. D. (1980). Microsomal lipid peroxidation. Stimulation by carbon tetrachloride. *Mol. Pharmac.* **17**, 408.

Levey, G., Rieger, A. L., and Edwards, J. O. (1981). Rates and mechanisms for oxidation of paraquat and diquat radical cations by several peroxides. *J. Org. Chem.* **46**, 1255.

McBrien, D. C. H. and Slater, T. F. (eds) (1982). *Free radicals, lipid peroxidation and cancer.* Academic Press, London.

McMurray, C. H. and Rice, D. A. (1981). Toxic edible oils. *Nature* **293**, 332.

Mason, R. P. (1982). Free-radical intermediates in the metabolism of toxic chemicals. In *Free radicals in biology,* Vol. V (ed. W. A. Pryor) p. 161. Academic Press, New York.

Misra, H. P. and Fridovich, I. (1976). The oxidation of phenylhydrazine. Superoxide and mechanism. *Biochem.* **15**, 681.

Morris, P. L. *et al.* (1982). A new pathway for the oxidative metabolism of chloramphenicol by rat liver microsomes. *Drug Metab. Disposition* **10**, 439.

Peiser, G. D., Lizada, M. C. C., and Yang, S. F. (1982). Sulphite-induced lipid peroxidation in chloroplasts as determined by ethane production. *Plant Physiol.* **70**, 994.

Pestana, A. and Munoz, E. (1982). Anilides and the Spanish toxic oil syndrome. *Nature* **298**, 608.

Pryor, W. A. (1980). Methods of detecting free radicals and free radical-mediated pathology in environmental toxicology. In *Molecular basis of environmental toxicity* (ed. R. S. Bhatnagar) p. 3. Ann Arbor Science Publishers, USA.

Pryor, W. A., Dooley, M. M., and Church, D. F. (1983). Mechanisms for the reaction of ozone with biological molecules: the source of the toxic effects of ozone. In *Int. Symp. biomed. effects of ozone and related photochemical oxidants* (eds S. E. Lee *et al.*) p. 7. Princeton Scientific Publishers, USA.

Richmond, R. and Halliwell, B. (1982). The formation of hydroxyl radicals from the paraquat radical cation, demonstrated by a highly-specific gas chromatographic technique. The role of superoxide radical anion, hydrogen peroxide and glutathione reductase. *J. inorg. Biochem.* **17**, 95.

Shimazaki, K., Sakaki, T., Kondo, N., and Sugahara, K. (1980). Active oxygen participation in chlorophyll destruction and lipid peroxidation in SO_2-fumigated leaves of spinach. *Plant and Cell Physiol.* **21**, 1193.

Stockley, R. A. (1983). Proteolytic enzymes, their inhibitors and lung diseases. *Clin. Sci.* **64**, 119.

Tabuenca, J. M. (1981). Toxic-allergic syndrome caused by injection of rapeseed oil denatured with aniline. *Lancet* **ii,** 567.

Thor, H. *et al.* (1982). The metabolism of menadione (2-methyl-1-4-naphthoquinone) by isolated hepatocytes. A study of the implications of oxidant stress in intact cells. *J. biol. Chem.* **257**, 12419.

Videla, L. A. and Valenzuela, A. (1982). Alcohol ingestion, liver glutathione and lipoperoxidation: metabolic inter-relations and pathological implications. *Life Sci.* **31**, 2395.

Wardman, P. *et al.* (1982). Reduction of nitroimidazoles in model chemical and biological systems. *Int. J. Radiat. Oncology biol. Phys.* **8**, 777; and other papers in this issue of the journal.

Willson, R. L. (1977). Metronidazole (Flagyl®) in cancer radiotherapy: a historical introduction. In *Metronidazole: Proceedings,* Montreal May 26–28th 1976. Exerpta Medica.

Winterbourn, C. C., French, J. K., and Claridge, R. F. C. (1979). The reaction of menadione with haemoglobin. Mechanism and effect of superoxide dismutase. *Biochem. J.* **179**, 665.

7 Free radicals as useful species

We have already seen a number of cases in which free-radical reactions are employed in living systems for useful purposes, such as the role of highly reactive ferryl species at the active sites of cytochrome P_{450} and peroxidase enzymes, and the part played by hydroxyl radicals in the small amount of ethanol oxidation *in vivo* that is achieved by the microsomal ethanol-oxidizing system (Chapter 2). In the present chapter we will examine some other examples in detail.

7.1. Reduction of ribonucleosides

Deoxyribonucleotides, the precursors of DNA, are formed *in vivo* by the reduction of ribonucleoside diphosphates in the presence of an enzyme called *ribonucleoside diphosphate reductase*. This enzyme catalyses replacement of the 2' —OH group on the ribose sugar ring by a hydrogen atom (Fig. 7.1) and the overall reaction may be written

$$\text{Ribonucleoside diphosphate} + R\begin{array}{c}\diagup\text{SH}\\ \diagdown\text{SH}\end{array} \longrightarrow$$

$$\text{H}_2\text{O} + \text{deoxyribonucleoside diphosphate} + R\begin{array}{c}\diagup\text{S}\\ \hspace{2pt}|\\ \diagdown\text{S}\end{array}$$

where $R(SH)_2$ is a dithiol compound. In most cases $R(SH)_2$ is *thioredoxin*, a small protein with two cysteine —SH groups in close proximity. The oxidized thioredoxin (RS_2) is re-reduced by a reductase enzyme at the expense of NADPH. However, in *E. coli* and in calf thymus tissue, not only thioredoxin but also a related protein called *glutaredoxin* can donate electrons to ribonucleoside diphosphate reductase. Oxidized glutaredoxin is reduced at the expense of GSH, the GSSG so produced being itself reduced by glutathione reductase (Fig. 7.1).

In *E. coli* and in higher cells the ribonucleoside diphosphate reductase enzyme contains two different subunits. One of these binds the substrates and controls the enzyme activity. It contains —SH groups which react directly with the substrate during catalysis, and are then re-reduced by

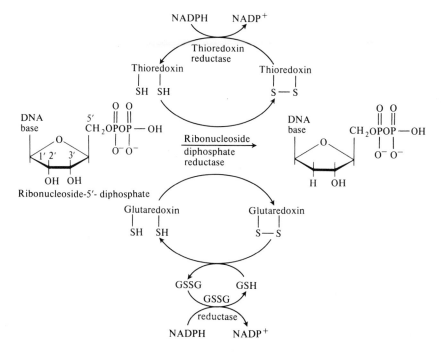

Fig. 7.1. Reduction of ribonucleoside diphosphates. The thioredoxin system operates in all cells containing the ribonucleoside diphosphate reductase activity. In addition, glutaredoxin is present in *E. coli* and calf thymus, and perhaps in other tissues. The relative importance of these two reducing systems *in vivo* is not clear.

thioredoxin or glutaredoxin. The other subunit contains iron in the Fe(III) state and gives a marked ESR signal, which has been identified as coming from a tyrosine residue. The presence of the tyrosine radical is closely linked to the presence of iron; the radical is lost on removal of iron and re-formed an incubation of the enzyme with an iron salt and a thiol compound in the presence of oxygen.

The function of the iron as it binds to the protein is both to generate the tyrosine radical, probably by an iron-catalysed one-electron oxidation, and then to stabilize the radical by some continued interaction with it. The positive charge is delocalized over the whole of the benzene ring structure of the tyrosine. This tyrosine radical is known to participate in the catalytic mechanism of the enzyme, but details have yet to be established. *Hydroxyurea* ($H_2NCONHOH$), a widely used inhibitor of ribonucleoside diphosphate reductase and, hence, of DNA synthesis in cells, appears to act by reacting with the tyrosine radical.

A number of bacteria and algae contain a different ribonucleoside

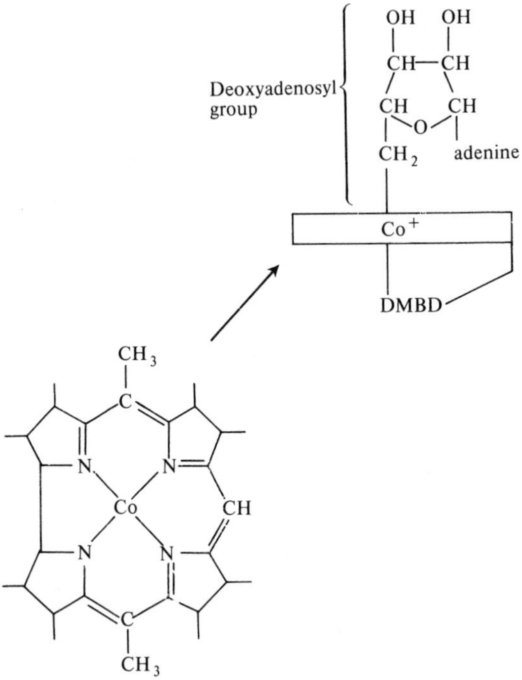

Fig. 7.2. Vitamin B_{12} and its derivatives. Co is cobalt, a transition metal (see Appendix). The corrin ring is represented as a flat plate, four N-atoms being co-ordinated to the cobalt. The fifth coordination position is used to attach a derivative of dimethylbenzimidazole (DMBD, structure not shown) which is also attached to a side chain of the corrin ring. The sixth co-ordination position of the cobalt can be occupied by a number of groups, such as cyanide (CN^-) ion to give *cyanocobalamin*. The cyanide is introduced during the isolation procedure and is not present *in vivo*. The cobalt atom in cobalamin can have a +1, +2, or +3 oxidation state. In hydroxycobalamin, OH^- occupies the sixth co-ordination position and the cobalt is in the Co^{3+} state. This form, called B_{12a}, is reduced by a flavoprotein enzyme to $B_{12r}(Co^{2+})$ which is reduced by a second flavoprotein to give $B_{12s}(Co^+)$. B_{12s} is the substrate for a reaction with ATP that yields $5'$-deoxyadenosyl cobalamin, whose structure is shown. Impaired absorption of cobalamin from the human diet results in pernicious anaemia.

diphosphate reductase in which the reaction involves a deoxyadenosyl cobalamin radical instead of a tyrosine radical. *Cobalamin,* otherwise known as vitamin B_{12}, consists of a *corrin* ring structure with a central cobalt ion. Like the haem ring, the corrin ring is based on four pyrrole units (Fig. 7.2). The radical involved in the reductase reaction seems to be formed by a homolytic cleavage of the bond between the cobalt and the —CH_2 group in deoxyadenosylcobalamin (Fig. 7.2).

7.2. Oxidation, carboxylation, and hydroxylation reactions

As we saw in Chapter 3, the enzyme indoleamine dioxygenase, which catalyses cleavage of the indole rings of tryptophan, tryptamine, and serotonin, has been conclusively shown to involve O_2^- in its reaction mechanism; and it is inhibited by superoxide dismutase. Indeed, decreasing the SOD activity of rabbit intestinal cells by treatment with diethyldithiocarbamate increases the rate of action of the dioxygenase *in vivo*. Similarly, the fungal enzymes, nitropropane dioxygenase and galactose oxidase, are prevented from functioning by high concentrations of SOD. If these enzymes indeed use O_2^- *in vivo*, then any SOD present in the cells containing them cannot be completely effective in removing O_2^- or else the enzymes would not be able to function. Perhaps they are located in parts of the cell that have low SOD activities.

There is no clear evidence for the participation of *free* superoxide and hydroxyl radicals in the mechanism of action of any hydroxylase enzyme, although iron-oxygen radical complexes such as ferryl species are obviously important at the active sites of cytochrome P_{450} and peroxidase. Similarly, the lysyl and prolyl hydroxylases involved in the synthesis of collagen (Chapter 2) require three things for their action: an Fe(II) ion at the active site, a reducing agent (ascorbic acid), and the compound 2-oxoglutarate. The first stage of the hydroxylation catalysed by these enzymes may be the attack of an Fe^{2+}—O_2 complex at the active site upon the carbonyl group of 2-oxoglutarate to give an intermediate which then hydroxylates the substrate. A complex such as Fe^{2+}—O_2 would have considerable resonance contribution from Fe^{3+}—O_2^-, and so the initial step might well be regarded as nucleophilic attack by a form of 'bound-superoxide'. Hydroxylation is not inhibited by SOD because *free* O_2^- is not involved in the catalytic mechanism, but it can be inhibited by low-molecular-weight copper complexes with SOD activity, which can presumably penetrate to the active site and react with 'bound-superoxide'. Great caution should be exercised in the use of such complexes however, since they often act as unspecific enzyme inhibitors.

The synthesis of prothrombin and factors VII and IX, essential constituents of the blood coagulation system, requires the attachment of carboxyl (—COOH) groups to a number of glutamic acid residues to give *γ-carboxyl-glutamic acid* residues in the polypeptide chains.

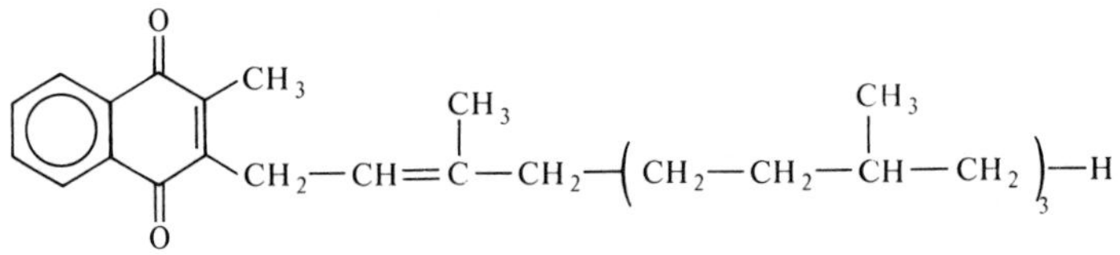

Fig. 7.3. Vitamin K_1

Carboxylation is achieved by an enzyme located in the endoplasmic reticulum of the liver cells, and it requires the fat-soluble vitamin K_1 for its action. As Fig. 7.3 shows, this vitamin is a quinone and it can be enzymically reduced by liver microsomal fractions to a semiquinone form at the expense of NADPH. This semiquinone appears to be required for the carboxylation reaction and, like most molecules of this type (Chapter 2), it can reduce oxygen to O_2^-. Esnouf and Hill in England have observed that the carboxylation reaction in isolated liver microsomes is partially inhibited by large amounts of SOD and by copper-chelates that can scavenge O_2^-. However, inhibitions by O_2^--scavengers should be interpreted with caution in systems containing quinones (Chapter 3).

7.3. Phagocytosis

In the late 1800s, the Russian scientist Metchnikoff observed the engulfment of bacteria by cells from the bloodstream of animals. This process is called *phagocytosis*—the cell 'flows around' the foreign particle

Table 7.1. Distribution of leukocytes ('white blood cells') in human blood

Type of cell	Normal number per μl (10^{-6} l) of blood	Function
Polymorphonuclear leukocytes ('polymorphs')		
neutrophils	2500–7500	Phagocytosis (see text).
eosinophils	40–400	Allergic and hypersensitivity responses (see text).
basophils	0–100	Hypersensitivity reactions (see text).
Monocytes	200–800	Precursors of macrophages.
Lymphocytes	1500–3500	Immune response.

Polymorphs, monocytes, and lymphocytes differ in their precursor cells, morphology, and function. An abnormally high content of white cells in the blood (above 11 000 per μl) is called *leukocytosis*, whilst a decrease below 4000 per μl is termed *leukopenia* or, since neutrophils constitute such a high proportion of white cells, *neutropenia*. Eosinophils differ from neutrophils (Fig. 7.4) in having larger cytoplasmic granules and often a nucleus with only two lobes. Monocytes are the precursors of macrophages in sites of inflammation. Most eosinophils reside in the tissues rather than circulating in the blood.

and encloses it in a plasma membrane vesicle that becomes internalized into the cytoplasm of the phagocytic cell. Most of the phagocytic cells in the human bloodstream are *neutrophils* (Table 7.1), which have a multilobed nucleus (hence they are called *polymorphonuclear* cells) and a large number of cytoplasmic granules, which are of several different types (Fig. 7.4). The *primary* or *azurophil* granules contain the enzymes myeloperoxidase and lysozyme, several proteases, and a number of 'granular cationic proteins'. The *specific* or *secondary granules* contain lysozyme, a protein that binds vitamin B_{12} (*cobalophilin*) and may also contain lactoferrin, an iron-binding protein similar to transferrin. Like transferrin itself, lactoferrin can bind two moles of Fe^{3+} ion per mole of protein (Chapter 2). A whole series of lysosomal enzymes is probably housed within the so-called *tertiary granules*.

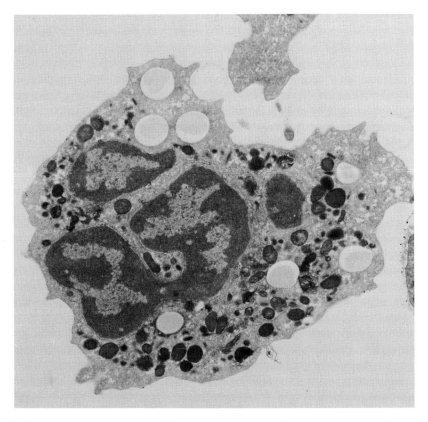

Fig. 7.4. Structure of a neutrophil as seen under the electron microscope. The cell is phagocytosing opsonized latex beads (the white circles). Some are being taken up by the cell flowing around them and others are already present within the cell in vacuoles (×18,000) Photograph by courtesy of Dr A. W. Segal.

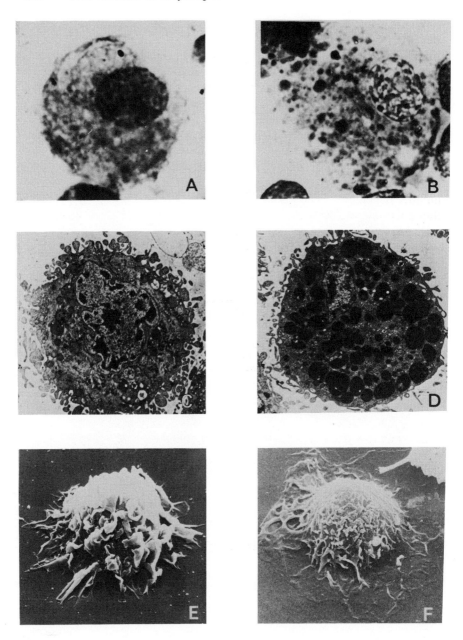

Fig. 7.5. Human pulmonary macrophages. The cells were obtained from the lungs of healthy volunteers. A, C, and E are from non-smokers; B, D, and F from smokers. A and B are light-micrographs. The cells vary from 15–50 μm in diameter. C and D are electron micrographs ($\times$6175); and E and F are scanning electron

When human tissue is injured, an *acute inflammatory response* develops, characterized by swelling, warmth, pain, reddening, and partial immobilization. The arterioles in and around the injured area relax, so that the capillary network becomes engorged with blood (hence the heat and redness). The permeability of the blood vessel walls increases so that more fluid leaks out, causing oedema. This fluid is rich in protein. As they enter the inflamed area, neutrophils often stop on the endothelial cells lining the blood vessels, a phenomenon known as *pavementing* or *margination*. They then push out cytoplasmic pseudopodia and squeeze through the gaps between endothelial cells, crossing the vessel wall and entering the inflamed tissue. The migration appears to be produced by a number of compounds which are formed in the inflamed area and attract the neutrophils (*chemotactic factors*). Such factors are numerous and include products of complement activation.

At a later stage of inflammation, monocytes (Table 7.1) leave the circulation and enter the inflamed area. These cells are less actively mobile and phagocytic than neutrophils, but once in the inflamed area they undergo development and change into *macrophages* (Figs 7.5 and 7.6), which involves increase in their content of lysosomal enzymes, metabolic activity, motility, and phagocytic and microbicidal capacity. Macrophages are larger than neutrophils and do not usually have a lobed nucleus. Apart from their formation during inflammation, they can be found in the lymphatic system, spleen (Fig. 7.6), and as scattered cells in connective tissue and in the lungs. Indeed, the *pulmonary* (or *alveolar*) macrophages lie on the alveolar walls, and are a major defence of the lung against inhaled bacteria and other particles (Fig. 7.5). The Kupffer cells, which form part of the lining of the liver sinusoids, are also macrophages. Macrophages can phagocytose bacteria and dead cells, e.g. degenerated erythrocytes. They can also ingest large amounts of insoluble material and retain it for months or even years (Fig. 7.5). The *lymphokines* released when primed T-lymphocytes react with antigens include factors that are chemotactic for macrophages, and stimulate their activity.

The fluid leaking into the inflamed area contains various antibodies which can bind to bacteria. Both neutrophils and macrophages have

---

micrographs which show the surface of the cells. Large numbers of vesicles are present. Cigarette smoke causes marked changes in the cells, which contain characteristic 'smokers inclusions', these include 'needle-like' or 'fibre-like' structures. These needle-like structures consist of kaolinite, an aluminium silicate present in cigarette smoke that the macrophages can engulf. The cell surface membrane of macrophages is highly ruffled. Photographs by courtesy of Prof. W. G. Hocking.

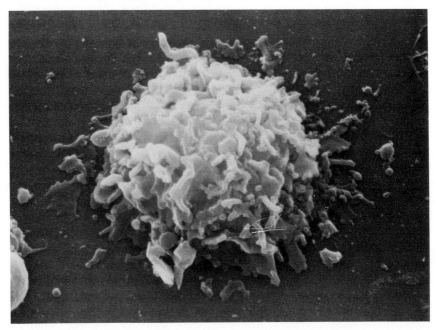

Fig. 7.6. A spleen macrophage. The surface of the cell is shown by scanning electron microscopy ($\times 15\,000$). Note the intensely ruffled surface. Photograph by courtesy of Dr W. Dawson.

surface receptors that can recognize immunoglobulin G antibodies and the C3b component of reacted complement. Coating of bacteria with such host-derived proteins, known collectively as *opsonins,* enables the neutrophils and macrophages to recognize them, although a few bacterial strains have coats that can be recognized directly. Once the bacteria are engulfed, various cytoplasmic granules fuse with, and hence mix their contents with, the vacuole containing the engulfed particle (*phagocytic vacuole*). The engulfed particles are then killed and, if possible, digested within the phagocytic vacuole, the digestion products eventually being expelled from the cell.

After the bacteria causing the lesion have been destroyed (or, if injury was caused by another mechanism such as heat or chemicals, after the insult ceases) there is usually reversal of the inflammatory changes. The vessel walls regain their normal permeability. Most of the emigrated neutrophils probably die, and the fragments are phagocytosed by macrophages. During phagocytic activity, neutrophils and macrophages release lysosomal enzymes into the surrounding fluid, where they contribute to the digestion of inflammatory debris.

If the bacteria are not completely eliminated, however, or the tissue injury continues, chronic inflammation results. There is formation of vascular granulation tissue, which matures into fibrous tissue. Bacteria likely to cause chronic inflammation include those responsible for syphilis and tuberculosis; and it is often induced by silica dust (quartz) or silicate fibres (asbestos) inhaled into the lungs. The fibrous tissue produced during chronic inflammation causes loss of function. Chronic inflammatory lesions are especially rich in macrophages, several of which may fuse together to give *giant cells* with multiple nuclei. The various hydrolytic enzymes secreted by macrophages play an important role in tissue damage during chronic inflammation.

7.3.1. The bacterial killing mechanism

Most studies on the biochemistry of phagocytosis have been carried out upon neutrophils and macrophages, especially pulmonary macrophages since they are the only macrophages that can readily be obtained from humans. Resting neutrophils consume little oxygen since they rely mainly on glycolysis for ATP production and are rich in stored glycogen. By contrast, macrophages possess mitochondria and so utilize oxidative phosphorylation and show more oxygen consumption. At the onset of phagocytosis, however, both cell types show a marked increase in oxygen uptake that is not prevented by cyanide, and so is unrelated to mitochondrial electron transport. This *respiratory burst* can be ten or twenty times the resting respiratory rate in neutrophils, but is less marked in macrophages. At the same time, there is an increased consumption of glucose by the cells, which is fed into the pentose phosphate pathway as shown by the increased release of $^{14}CO_2$ from [1-^{14}C]-glucose. Triggering of the respiratory burst depends on some kind of membrane perturbation, possibly a decrease in membrane potential, since not only opsonized bacteria but also small latex beads (Fig. 7.4), opsonized zymosan (a preparation of yeast cell walls), and certain chemicals can induce it. These chemicals include *phorbol myristate acetate,* a co-carcinogen (see Chapter 8), high concentrations of fluoride ion, certain small peptides such as *N*-formylmethionylleucylphenylalanine (fmet-leu-phe), and concanavalin A, a lectin; (lectins are plant proteins that bind with high affinity to carbohydrate-containing substances, such as the glycoproteins found in cell surface membranes). Hence the act of phagocytosis itself is not required for the respiratory burst to occur.

The exact size of the respiratory burst shown by macrophages depends on how, and from which tissue site, they were obtained. *Resident macrophages,* tissue macrophages that have not yet met foreign materials, are poorly active. More active macrophages can be obtained by infecting animals with micro-organisms such as bacillus Calmette–Guerin (BCG) or by injecting irritating materials such as thioglycollate. The antibacterial behavior of the macrophages obtained by these different procedures is very variable and papers concerning, for example, the relative importance of oxygen radicals in killing by macrophages should be read with this in mind.

If the respiratory burst in neutrophils is prevented by placing them under an atmosphere of nitrogen gas, the killing of some bacterial strains, such as *Bacillus fragilis* and *Clostridium perfringens,* is not impaired. Many of these organisms are anaerobes, and the neutrophil would not be expected to rely upon an oxygen-dependent killing mechanism for species that it would be likely to encounter under conditions of low oxygen concentration. The killing is presumably achieved by the contents of the various vacuoles as they empty into the phagocytic vacuole. The enzyme lysozyme can attack and digest the cell walls of various bacteria, especially gram-positive strains. Lysosomal enzymes might be able to attack some strains and a number of the 'granular cationic proteins' have bactericidial activity. For example, a protein of this type from rabbit and human neutrophils causes breakdown of the cell membranes of *E. coli* and *Salmonella typhimurium.* Lactoferrin can bind iron essential for the growth of certain bacteria, and it has been reported to kill a few bacterial strains directly.

However, the killing of many other bacterial strains by neutrophils is greatly decreased under anaerobic conditions. Indeed, patients suffering from *chronic granulomatous disease (CGD),* an inherited condition in which phagocytosis is normal but the respiratory burst is absent, show persistent and multiple infections, especially in skin, lungs, liver, and bones, by those bacterial strains whose killing by neutrophils requires oxygen.

The oxygen uptake in both neutrophils and macrophages is due to the activation of an enzyme complex associated with the plasma membrane. The system oxidizes NADPH provided by the pentose phosphate pathway in the cytosol into $NADP^+$, the electrons being used to reduce oxygen to the superoxide radical, O_2^-, as was first shown by B. M. Babior's group in the USA. The overall equation can be written

$$NADPH + 2O_2 \rightarrow NADP^+ + H^+ + 2O_2^-.$$

Indeed, very severe inborn deficiencies of glucose-6-phosphate dehydrogenase, the first enzyme in the pentose phosphate pathway, also cause increased susceptibility to infections. The superoxide radical can be detected outside stimulated neutrophils and macrophages by its ability to reduce cytochrome c or nitro-blue tetrazolium, both reactions being inhibited by added SOD. Alternatively, O_2^- can be identified by spin-trapping experiments (Chapter 2). The NADPH oxidase complex is a flavoprotein (FAD is present), and is defective in chronic granuloma-tous disease. There may be different forms of CGD with defects in different parts of the oxidase system. It has also been suggested that a cytochrome b and a quinone molecule are involved in O_2^--generation, reminiscent of the mitochondrial electron-transport chain. The Km for oxygen of the activated oxidase complex in rat neutrophils is within the range of the normal oxygen concentration in body fluids, and so it is possible that the amount of O_2^- produced *in vivo* will depend on O_2 supply.

Since the bacteria or other engulfed particles are 'wrapped up' in a plasma membrane vesicle in the phagocyte cytoplasm, they are exposed to a high flux of O_2^- (Fig. 7.7). In view of its low reactivity in aqueous solution (Chapter 3), it is unlikely that O_2^- itself is responsible for bacterial killing. Some strains of bacteria are quickly killed by hydrogen peroxide, which will be formed from O_2^- by the dismutation reaction

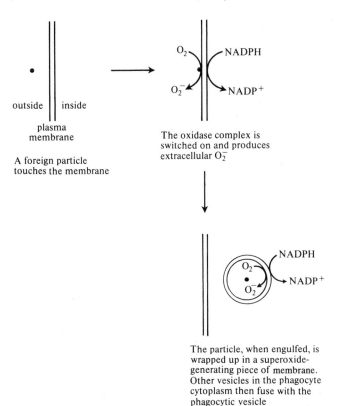

outside | inside

plasma
membrane

A foreign particle
touches the membrane

The oxidase complex is
switched on and produces
extracellular O_2^-

The particle, when engulfed, is
wrapped up in a superoxide-
generating piece of membrane.
Other vesicles in the phagocyte
cytoplasm then fuse with the
phagocytic vesicle

Fig. 7.7. A schematic representation of the respiratory burst. The engulfed particles are exposed to a flux of O_2^- inside the phagocytic vacuole.

(Chapter 3). The hydroxyl radical, $OH^{\cdot}$, formed from O_2^- and hydrogen peroxide by an iron-catalysed Haber–Weiss reaction may sometimes participate in killing, since dimethylsulphoxide, a scavenger of $OH^{\cdot}$ that can penetrate into cells (Chapter 3), can partially protect *Staphylococcus aureus* against killing by human neutrophils. During these experiments, the formation of methane, a product of the reaction between dimethylsulphoxide and $OH^{\cdot}$ (Chapter 2), was observed. Hydroxyl radical has been detected in suspensions of activated neutrophils by its ability to oxidize methional into ethene, to decarboxylate benzoic acid, and by spin-trapping experiments with DMPO (Chapter 2). The killing of *E. coli, S. aureus,* and *Streptococcus viridans* by human neutrophils was decreased by addition of SOD or of catalase attached to latex particles to facilitate their uptake into the phagocytic vesicles, consistent with the operation of a Haber–Weiss reaction to generate $OH^{\cdot}$.

The catalyst required for $OH^{\cdot}$-formation might be 'free' iron salts. In addition it has been shown that *in vitro* the protein lactoferrin, if saturated with iron (2 moles Fe^{3+} per mole of protein), is also a catalyst although not as effective as non-protein-bound iron. Lactoferrin in body fluids usually has only a low degree of saturation *in vivo,* although it might pick up more iron when it enters the phagocytic vacuole by fusion of the vesicles containing it with the vacuolar membranes. Lactoferrin-deficient phagocytes can still generate $OH^{\cdot}$ radicals, however, and so it is not the only available catalyst.

Once the phagocytic vacuole is formed, fusion of it with other granules in the neutrophil cytoplasm empties, among other things, the enzyme myeloperoxidase over the engulfed particle. (Myeloperoxidase is not normally present in macrophages.) Myeloperoxidase is a green haem-containing enzyme that shows 'non-specific' peroxidase activity (Chapter 3) but it catalyses other reactions as well. In the presence of hydrogen peroxide, and chloride or iodide ions, myeloperoxidase can kill a number of bacteria and fungi *in vitro*. Since much more Cl^- than I^- ion is present in the phagocyte cytoplasm, it has been suggested that myeloperoxidase can participate in the killing mechanisms by oxidizing Cl^- into hypochlorous acid, $HOCl$. Hypochlorous acid is highly reactive, being able to oxidize many biological molecules (especially —SH groups), and can itself kill a number of bacteria. It is known to react with hydrogen peroxide at alkaline pH values to form singlet oxygen $^1\Delta g$ (Chapter 2)

$$HOCl + H_2O_2 \rightarrow O_2^* + Cl^- + H_2O + H^+.$$

A further suggestion is that hydroxyl radicals may originate not only from the Haber–Weiss reaction but also from a reaction between O_2^- and

HOCl:

$$HOCl + O_2^- \rightarrow O_2 + OH^{\bullet} + Cl^-$$

Activated phagocytes produce a weak light-emission that can be greatly increased by addition of compounds such as luminol or lucigenin. Indeed, addition of luminol to any O_2^--generating system causes light production (Chapter 3). Light is also produced when luminol is added to a mixture of myeloperoxidase and hydrogen peroxide. The exact origin of the weak basal and of the luminol- or lucigenin-dependent light emission by neutrophils is not clear. Some scientists have suggested an involvement of singlet oxygen, but attempts to use various singlet oxygen scavengers (Chapter 2) in experiments have been frustrated by the fact that HOCl can oxidize a number of them directly, including DABCO, 2,5-diphenylfuran, histidine, and β-carotene. Formation of singlet oxygen by neutrophils *in vivo* cannot yet be regarded as rigorously proved.

Patients with an inborn deficiency of myeloperoxidase in their neutrophils show only minor decreases in resistance to infection, and so this enzyme must be of less importance in bacterial killing than is the respiratory burst. However, neutrophils from these patients do kill bacteria more slowly than normal *in vitro* and myeloperoxidase may be of special importance in protection against fungi such as *Candida albicans*. In neutrophils from patients with chronic granulomatous disease myeloperoxidase is present normally, but the absence of the respiratory burst means that it is not provided with the necessary hydrogen peroxide. Thus both antibacterial mechanisms fail to operate, accounting for the severe symptoms. Indeed, neutrophils from patients with CGD can destroy those strains of bacteria that themselves release hydrogen peroxide, thus providing the myeloperoxidase with the means of operation.

Formation of O_2^- in a respiratory burst has also been observed in the Kupffer cells of the liver, in monocytes, basophils, mast cells, and eosinophils. Indeed, human monocytes show a more marked respiratory burst than do macrophages, although it is not as large as in neutrophils. Monocytes contain a myeloperoxidase-like enzyme which disappears as they differentiate into macrophages. Eosinophils and basophils are rare in the circulation (Table 7.1) but the former cells increase in number in allergic conditions, such as bronchial asthma and hay fever, and during infestations by some parasites, such as schistosomes. Eosinophils phagocytose bacteria very sluggishly, but they do seem to be important in defence against parasitic worms where, of course, the worm is much bigger than the eosinophil. Damage to the parasites can be achieved by oxygen radicals, by an eosinophil peroxidase activity, and by a 'highly basic protein' that attacks cell membranes. All of these are released when the eosinophils contact the parasite. Human eosinophils are especially active in O_2^- production.

Basophils and mast cells, which are present in connective tissue, are rich in histamine, and both may be activated by allergens which combine with immunoglobulin E antibody bound to their surface. Such activation not only produces O_2^- but also pharmacologically active products, including histamine, 5-hydroxytryptamine, and SRS (Section 7.6). These products are responsible for the symptoms produced when patients with allergies are exposed to the allergen that their bound IgE recognizes, such as grass pollen in the case of hay fever, or the house dust mite in many cases of bronchial asthma. Histamine has been reported to stimulate O_2^--production by human eosinophils.

However, activated neutrophils, macrophages, and other cells release O_2^- not only into the phagocyte vacuole but also into their surrounding tissue fluid (Fig. 7.7) where it can give rise to hydrogen peroxide and OH˙. Hydrogen peroxide vapour, possibly originating from alveolar macrophages (although production by mouth bacteria is also possible), has been detected in human breath. Indeed, activated phagocytes are mutagenic in the Ames test. Myeloperoxidase, lysozyme, elastase (Chapter 6), other proteolytic enzymes, and lactoferrin are released as well. Release of these products can help phagocytes to attack opsonized foreign cells, such as cancer cells (this is discussed further in Chapter 8), and parasitic worms, but the extracellular fluids of the human body contain little of the protective enzymes, superoxide dismutase, catalase, and glutathione peroxidase. Released oxygen radicals can kill the phagocytes themselves, probably by inducing damage to the plasma membrane. Neutrophils contain a copper–zinc SOD enzyme in the cytosol, and a manganese SOD in the mitochondria, but these cannot offer protection against externally generated O_2^-. Neutrophils also contain catalase and glutathione peroxidase activities, together with GSH at millimolar concentrations, which help to protect them against hydrogen peroxide. Methionine sulphoxide reductase and large amounts of ascorbic acid are also present. Indeed, neutrophils deficient in glutathione reductase activity are much more rapidly inactivated during phagocytosis than normal, presumably because hydrogen peroxide can easily diffuse out of the phagocytic vacuole into the cell cytoplasm, and GSH cannot be regenerated for glutathione peroxidase activity. The turnover of GSH in mouse peritoneal macrophages has been observed to increase during phagocytosis. Catalase activity seems to be less important in protection against hydrogen peroxide than the glutathione system, and indeed it varies considerably between species. For example, human or guinea-pig neutrophils contain much more catalase than cells from rats and mice. DNA strand breakage has been observed in human neutrophils during the respiratory burst, perhaps reflecting formation of OH˙ by a Haber–Weiss reaction.

Superoxide has been claimed to react with a component of normal human plasma to form a chemotactic factor that attracts more neutrophils into the area, although confirmation of this is awaited. The action of proteases released from phagocytes is normally held in check by 'antiproteases' in the body fluids which inhibit their action, such as serum α_1 antiprotease (sometimes called α_1-antitrypsin). Exposure of antiproteases to OH˙ or to the myeloperoxidase–H_2O_2–Cl^- system prevents their inhibitory action (Chapter 6). Fortunately, some kind of equilibrium will eventually be achieved, since chemotactic factors and released proteases are themselves inactivated by myeloperoxidase and oxygen radicals, and an increased blood flow in an inflamed area will replenish the supply of antiproteases. Binding of iron to lactoferrin released from neutrophils should inhibit iron-dependent lipid peroxidation, and the extracellular protein caeruloplasmin (Chapter 4) also helps to prevent this process as well as scavenging some superoxide radicals. Released lactoferrin appears to promote aggregation of neutrophils.

If controlled in this way, the release of oxygen radicals and enzymes by phagocytic cells probably does no permanent damage. Oxygen radicals probably have a role in mediating vascular permeability in some circumstances. Del Maestro *et al.* in Sweden have shown that perfusion of a hamster cheek pouch with O_2^--generating systems causes increased leakage of material from the vascular network and increased adherence of neutrophils to the vessel walls. Leakage is decreased by including SOD, catalase, or the OH˙-scavenger, dimethylsulphoxide, in the perfusing medium. Vascular endothelial cells are very sensitive to damage both by released proteolytic enzymes and by oxygen-derived species, especially hydrogen peroxide, which can penetrate easily, perhaps to form hydroxyl radicals internally. Neutrophils affect vascular permeability by other mechanisms as well.

It follows, however, that anything causing abnormal activation of phagocytes has the potential to provoke a devastating response. For example, a factor in the serum of patients with the skin disease psoriasis has been reported to increase oxygen-radical production by phagocytes, which are known to accumulate in the epidermal lesions. Macrophages are present in the lesions of myelinated nerves seen in multiple sclerosis. Initial damage to the lungs by exposing animals to elevated oxygen concentrations has been reported to provoke an influx of neutrophils, which become activated and make the damage worse as O_2^- and hydrogen peroxide attack the alveolar cells. Pulmonary macrophages increase their SOD activity on exposure of animals to elevated O_2 (Chapter 3). Cigarette smoke has been reported to increase the production of hydrogen peroxide and O_2^- by pulmonary macrophages, an effect that may be involved in the development of smoker's emphysema (Chapter 6). Silica and asbestos particles can be phagocytosed by pulmonary macrophages. They then appear to cause rupture of the phagocytic vesicles, leading to release of proteolytic enzymes into the macrophage cytoplasm. If the macrophage is killed, the particles will be released to be taken up by other cells with the same effect. The proteases will thus be released into the surrounding lung tissue. Asbestos fibres contain bound metal ions that can catalyse OH$^•$ production.

In 1968, it was discovered that within the first few minutes of blood dialysis, leukopaenia (Table 7.1) occurs, which was traced to the fact that contact of plasma with cellophane in the dialyser causes activation of the complement system, leading to accumulation, aggregation, and activation of neutrophils and monocytes in the lung. This can cause damage by interfering with blood flow, and perhaps by producing oxygen radicals. Indeed, the adult respiratory distress syndrome (ARDS)—acute respiratory failure due to pulmonary oedema—is a frequent result of severe shock, tissue damage due to burns or accidents, or massive infections. The accumulation and activation of neutrophils in the lung plays an important part in the pathology of ARDS. Oxygen radicals are known to produce lung oedema *in vivo* (Chapter 3). Perhaps the most striking consequence of abnormal phagocyte action are seen in the autoimmune diseases, which are considered in detail in Chapter 8.

7.4. Peroxidase and NADH oxidase enzymes

The action of peroxidases in the presence of hydrogen peroxide is important in several areas of metabolism in addition to the leucocyte killing mechanism. For example, addition of iodine to the benzene ring of tyrosine during the synthesis of thyroid hormone is achieved by a peroxidase located in the endoplasmic reticulum of thyroid cells, which uses hydrogen peroxide to oxidize iodide ions into an 'active iodine' that attacks the ring. It has been suggested that xanthine oxidase activity in the thyroid gland provides the hydrogen peroxide necessary for the action of the peroxidase. Salivary peroxidase and lactoperoxidase may have some anti-bacterial action (Chapter 3). The ability of peroxidase to oxidize phenols into quinones in the presence of hydrogen peroxide is made use of by the bombardier beetle, which attacks its enemies by spraying them with a hot quinone-containing fluid. A sac within the insect contains a 25 per cent aqueous solution of hydrogen peroxide plus 10 per cent hydroquinone, and the spray is generated when the contents of the sac are pushed into a reaction chamber containing catalase and peroxidase.

Some wood-destroying fungi release both hydrogen peroxide and peroxidase enzymes extracellularly to aid in the degradation of their substrates. Hydroxyl radicals (but not singlet oxygen) are involved in the mechanism of lignin (wood) degradation by the white-rot fungus *Phanerochaete chrysosporium.* The lignin-destroying activity of the fungus appears after primary growth stops and in response to nitrogen starvation. Both hydrogen peroxide and a peroxide-requiring ligninolytic enzyme are released extracellularly, and evidence for hydroxyl radical formation has been obtained by spin trapping, aromatic hydroxylation, and the formation of ethene (Chapter 2). Scavengers of OH$^{\bullet}$ inhibit lignin degradation.

Lignin itself is a complex polymer derived by the oxidation and polymerization of phenols synthesized from the amino acids phenylalanine and tyrosine (Fig. 7.8). Oxidation of phenols to the phenoxy radicals which polymerize seems to be achieved by one or more peroxidase enzymes bound to the plant cell wall. The source of the hydrogen peroxide required for peroxidase action is not entirely clear, but a likely possibility is that the cell wall peroxidases simultaneously oxidize NADH. This reaction does not require the addition of hydrogen peroxide (Chapter 3), and the NADH oxidation actually generates both O_2^- and hydrogen peroxide, the latter being used by the peroxidase to oxidize phenols. Work by Elstner and others in Germany, and in the authors' laboratory, has shown the reaction mechanism to be complicated.

Changing to a completely different system, the fertilization of sea-urchin eggs induces a rapid uptake of oxygen and a fertilization membrane containing cross-linked tyrosine residues is produced to stop further spermatozoa from entering. Cross-linking is catalysed by a peroxidase (*ovoperoxidase*) present in granules in the egg, and most of the extra oxygen uptake can be accounted for by synthesis of the hydrogen peroxide that it requires (a 'respiratory burst' with a novel purpose). Administration of oestrogen to female rats induces the synthesis of a uterine peroxidase activity that has been suggested to play a role in destruction of this hormone, and might additionally have an antibacterial effect in the uterine fluid. Whether this peroxidase is synthesized by the uterus itself, or is released by uterine eosinophils (it strongly resembles eosinophil peroxidase), is not entirely clear.

A peroxidase enzyme plays a role in the light emission from the boring mollusc *Pholas dactylus.* In most bioluminescent systems, light is produced when an enzyme (called a *luciferase*) acts on a low-molecular-weight substrate (*a luciferin*) to generate an excited state which then emits light as it decays. Detailed work by A. M. Michelson's group in France has shown, however, that *Pholas dactylus* luciferin is itself a protein, and can be induced to emit light when exposed to several systems generating

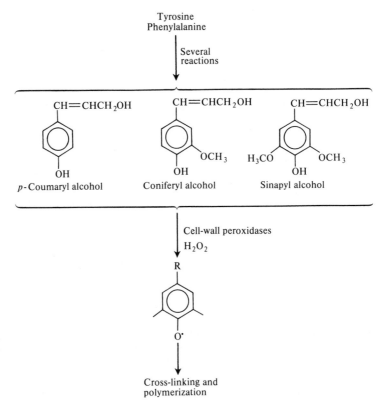

Fig. 7.8. Outline of the structure of lignin found in plant cell walls. Lignin is derived by the oxidation of various phenols into phenoxy radicals which then polymerize. The phenols are derived by adding —OH and —OCH₃ groups onto the benzene ring of the amino acids tyrosine or phenylalanine. Oxidation of phenols to phenoxy radicals is probably achieved by one or more peroxidase enzymes bound to the plant cell walls.

oxygen radicals, such as Fe^{2+} ions in the presence of oxygen (Chapter 3), or a mixture of hypoxanthine and xanthine oxidase. The *Pholas* luciferase is a glycoprotein with peroxidase activity, although it contains copper rather than a haem ring. Luciferase from the earthworm *Diplocardia longa* (a luminescent earthworm several inches long found in Georgia, USA) has also been discovered to be a copper-containing protein with peroxidase activity. All the above systems must be provided with hydrogen peroxide *in vivo* and so the enzymes that normally dispose of hydrogen peroxide cannot be completely effective in their vicinity.

The plasma membranes of some animal cells other than phagocytes contain enzymes that can oxidize NADH and NADPH, including adipose tissue cells, erythrocytes, and renal brush border membranes. The electrons from NAD(P)H oxidation can be accepted *in vitro* by a variety of reagents such as ferricyanide or cytochrome c, but the electron acceptors *in vivo* have not been identified, and there is no clear evidence that the action of NADH or NADPH oxidases in these other tissues can lead to O_2^- generation. However, Mukherjee *et al.* in the USA have suggested that the action of insulin, in increasing glucose transport into adipocytes and decreasing breakdown of stored triglycerides, may be mediated by a stimulation of membrane NAD(P)H oxidase by the hormone, resulting in increased intracellular generation of hydrogen peroxide. Several laboratories have obtained evidence consistent with this proposal, which means that hydrogen peroxide would be acting as an intracellular second messenger of insulin action. However, it is probably premature to accept the proposal fully at this stage. By contrast, the NADH oxidase in renal brush border membranes seems to be involved in transport processes, and a similar role has been suggested for an NADH oxidase system detected in the plasma membrane of corn root cells. Our knowledge of electron-transport systems in plasma membranes is meagre, largely because of the difficulty in obtaining membrane fractions uncontaminated with mitochondria or endoplasmic reticulum. However, any such systems present might well contribute to oxygen radical production by cells, if only by minor 'leakage' of electrons to oxygen to form O_2^- in the way already described for mitochondrial and microsomal electron-transport chains (Chapter 3).

7.5. Fruit ripening and the 'wound response' of plant tissues

The ripening and senescence of fruits seems to be a controlled oxidative process. Studies on pears have shown that, as ripening proceeds, the concentration of free —SH groups in the fruit decreases, and there is an accumulation of hydrogen peroxide and of lipid peroxides. Ripening of pears, or the senescence of rice plant leaves, can be speeded up by treatments that stimulate formation of hydrogen peroxide within the tissue. As pears and bananas ripen, fluorescent products of lipid peroxidation (Chapter 4) accumulate within them. A role for leaf peroxidase activity has been suggested in the breakdown of cell walls, and in the degradation of chlorophyll to yellow and brown pigments during leaf senescence.

Many non-green plant tissues, such as tubers, fruits, and seeds, contain the enzyme lipoxygenase, an iron-containing protein which catalyses a direct reaction of polyunsaturated fatty acids with oxygen to give 13- and 9-hydroperoxides (Fig. 7.9). Low activities of lipoxygenase are sometimes present in green leaves. The first such enzyme purified was from soyabeans, and it is now known as *soybean lipoxygenase I*, since another

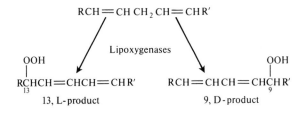

Fig. 7.9. Reactions catalysed by lipoxygenases. Purified lipoxygenases from plant tissue catalyse the peroxidation of fatty acids with a *cis,cis*-1,4-pentadiene structure. For example, linoleic acid may be converted into 13-L or 9-D hydroperoxy derivatives or both, depending on the enzyme. The reactions are highly stereospecific, L- and D- referring to the exact arrangements of groups in space around the carbon atom bearing the —OOH group.

lipoxygenase has since been discovered in this plant. Lipoxygenase I has a pH optimum of 9, and acts on linoleic acid to produce the 13, L-hydroperoxide plus a small amount of 9, D-hydroperoxide (Fig. 7.9). Soybean lipoxygenase I contains one atom of iron per molecule. The enzyme begins its action by abstracting a hydrogen atom from the substrate in a highly stereospecific manner. Bond rearrangement and oxygen insertion follow. Other lipoxygenases act similarly, but they produce different ratios of products, and have different pH optima. The hydroperoxide products interact with the enzyme in a complex way to change it into an active state. The action of lipoxygenases on their substrates can produce a 'co-oxidation' of other added materials, such as thiols, carotenoids, and chlorophylls. Indeed, the enzyme is employed commercially to bleach wheat flour carotenoids during the bread-making process: as the carotenoids are oxidized they lose their yellow-orange colours. Chlorophyll is also bleached in the presence of lipoxygenase and a fatty-acid substrate. It seems that the action of lipoxygenase produces a hydroperoxy radical that is normally reduced to a hydroperoxide but can also interact with an oxidizable co-substrate to cause damage. If L—H is the fatty-acid substrate and XH the co-oxidizable molecule the mechanism may be written

$$\text{L—H} \xrightarrow{\text{enzyme}} \text{L}^{\cdot} \xrightarrow{O_2} \text{LOO}^{\cdot} \xrightarrow{\text{enzyme}} \text{LOOH}$$
Substrate   product

$$\text{LOO}^{\cdot} + \text{XH} \longrightarrow \text{LOOH} + \text{X}^{\cdot}$$

$$\text{X}^{\cdot} + O_2 \longrightarrow \text{co-oxidation products.}$$

Hence lipoxygenase action may cause damage to surrounding tissues.

The subcellular location of lipoxygenase activity in plant tissues is variable, and it is usually found in several subcellular fractions. Indeed, the contamination of plant mitochondrial fractions by lipoxygenase activity has greatly confused studies of their oxygen uptake.

Many fruits, tubers, and seeds (and some leaves) respond to wounding or other tissue damage by initiating a series of complex biochemical reactions (Fig. 7.10). Firstly, a set of hydrolytic enzymes attacks membrane lipids and releases fatty acids, many of which are polyunsaturated. For example, linoleic and linolenic acids represent about seventy-five per cent of the total fatty-acid side-chains in potato tuber lipids. The released fatty acids are then acted upon by lipoxygenases, and the resulting hydroperoxides cleaved, both non-enzymically in the presence of metal complexes (Chapter 4) and by the action of 'cleavage enzymes'. This results in a wide variety of products, including volatile aldehydes and hydrocarbon gases, such as ethane and pentane. Some of the aldehydes so produced have characteristic smells which are responsible for the aroma of damaged plant tissues, such as sliced tomatoes and cucumbers (see legend to Fig. 7.10). The odour of crushed green leaves is sometimes caused by processes similar to these in Fig. 7.10. The reaction sequence shown in Fig. 7.10 can be extremely swift: over thirty per cent of the lipids in potato tuber slices are hydrolysed in less than fifteen minutes even at 3 °C. Slicing a potato greatly increases its uptake of oxygen, both in the lipoxygenase reaction and in the subsequent metabolism of fatty acid products. Indeed, such enzyme oxidation of disrupted plant material can give rise to problems of rancidity and 'off flavour' during processing and storage, that have to be controlled by the use of antioxidants to scavenge lipid—OO˙ radicals (Chapter 4). Plant lipoxygenase activity can be inhibited by the antioxidant propyl gallate.

Formation of lipid peroxides and aldehydes when plant tissues are damaged may play an important role in killing fungi and bacteria

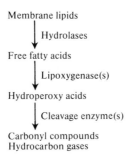

Fig. 7.10. Pathway for the breakdown of membrane lipids induced on wounding plant tissues. The volatile products from the 9-hydroperoxides of 18:2 and 18:3 fatty acids are *cis*-3-nonenal and *cis*-3-, *cis*-6-nonadienal respectively. These are the two main components of the odour of sliced cucumbers. The 13-hydroperoxides give hexanal and cis-3-hexenal.

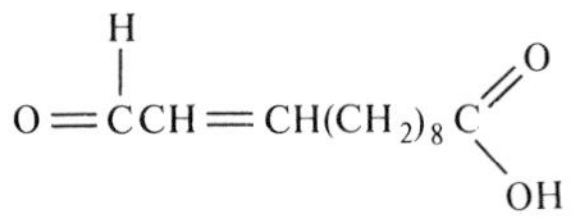

$$O = \overset{\overset{\text{H}}{|}}{\text{C}}\text{CH} = \text{CH(CH}_2)_8\,\text{C}\overset{\nearrow O}{\underset{\searrow OH}{}}$$

Fig. 7.11. Structure of traumatic acid. The correct chemical name is 12-oxo-*trans*-10-dodecenoic acid. Both the aldehyde form (shown) and the dicarboxylic acid form produced by oxidation of the —CHO group are biologically active.

attempting to enter the wound, since several of these products have been shown to be toxic. Damage to certain plant tissues results in formation of *traumatic acid* or *wound hormone* (Fig. 7.11) a compound which induces proliferation of new cells. Traumatic acid is formed by oxidative degradation of polyunsaturated fatty acids by a process similar to that in Fig. 7.10. It seems that the 'wound response' of plant tissues is an example of controlled lipid peroxidation put to a useful purpose.

A lipoxygenase enzyme is found in reticulocytes and it has been suggested to be responsible for the degradation of mitochondria that occurs as these cells mature to form erythrocytes.

7.6. Prostaglandins and cyclic nucleotides

Among the most exciting events in the pharmacological world in recent years has been the elucidation of the chemical structures and biological roles of the prostaglandins and their close relatives, the thromboxanes and prostacyclins. The term prostaglandin was coined in the 1930s to describe compounds present in human semen and sheep vesicular gland that could affect blood pressure in animals and cause contraction of smooth muscle. The first prostaglandins to be isolated were the stable species PGE_2 and $PGF_{2\alpha}$ (Fig. 7.12) but better analytical techniques have demonstrated a series of unstable compounds such as PGG_2, PGH_2, TXA_2 (a thromboxane), and PGI_2 (prostacyclin).

Prostaglandins are synthesized whenever cells are injured and they play an important part in the process of inflammation. For example, injection of E-type prostaglandins into human skin causes dilation of blood vessels and reddening, as well as increased vascular permeability and oedema.

Fig. 7.12. Structure of prostanoic acid, and of thromboxanes and prostaglandins derived from arachidonic acid. Enzymes involved: 1, prostaglandin *endo*-peroxide synthetase (cyclooxygenase); 2, glutathione-S-transferase; 3, prostaglandin *endo*-peroxide E isomerase; 4, prostaglandin *endo*-peroxide reductase; 5, prostaglandin *endo*-peroxide-1-isomerase (6,9, oxocyclase); 6 and 7, prostaglandin *endo*-peroxide: thromboxane A isomerase. Reactions 2 and 3 require glutathione to proceed. Prostaglandins E_3 and E_1 come from other fatty acids.

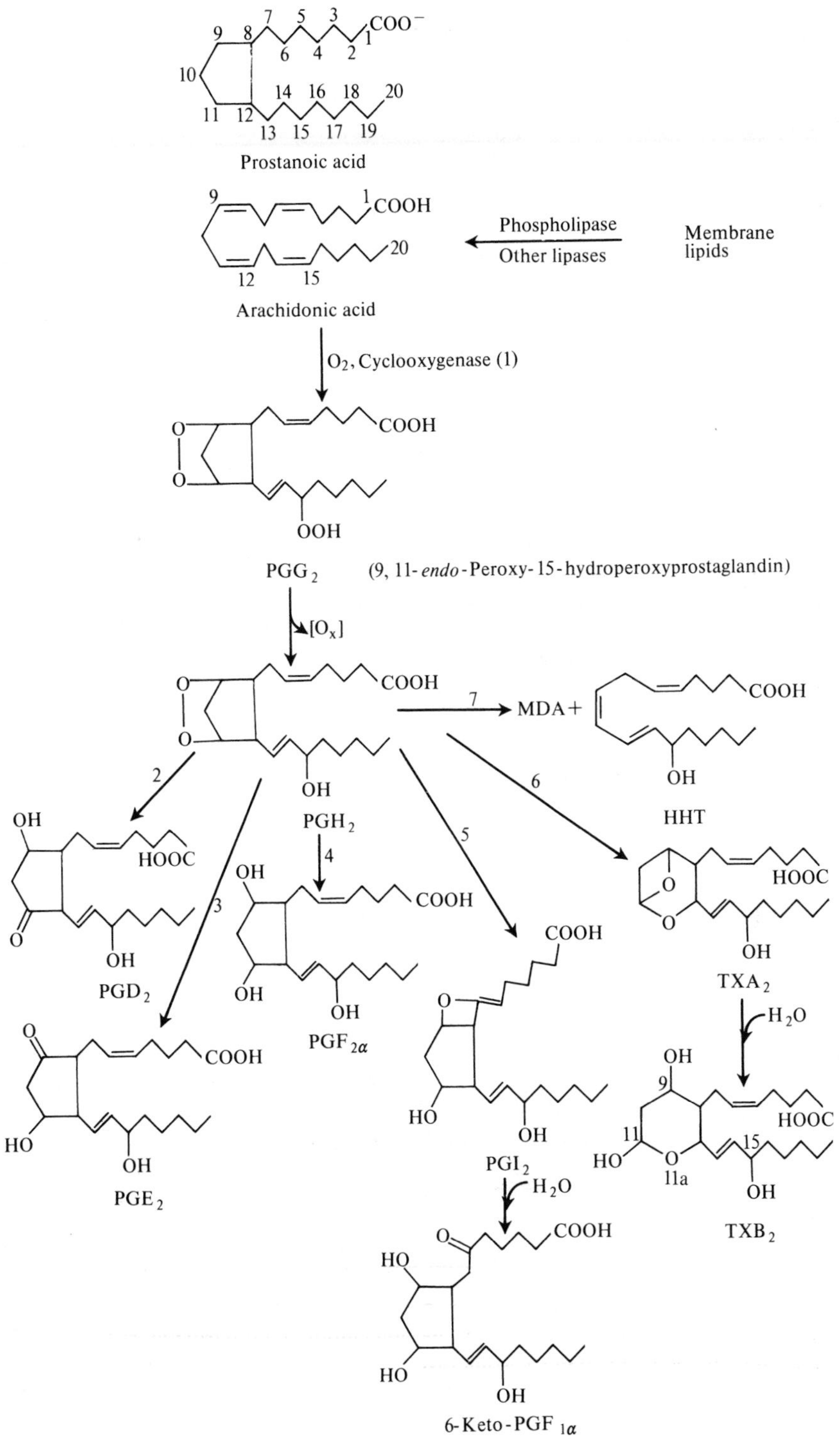

Prostanoic acid

Arachidonic acid

Phospholipase
Other lipases

Membrane
lipids

O_2, Cyclooxygenase (1)

PGG$_2$ (9, 11-*endo*-Peroxy-15-hydroperoxyprostaglandin)

$[O_x]$

MDA+

HHT

PGH$_2$

PGD$_2$

PGF$_{2\alpha}$

PGE$_2$

PGI$_2$

TXA$_2$

H_2O

TXB$_2$

H_2O

6-Keto-PGF$_{1\alpha}$

Activated macrophages are known to produce prostaglandins, prostacy-clin, and thomboxanes. Prostaglandins are intimately involved in the regulation of acid secretion in the stomach, the control of kidney function by hormones, and in platelet metabolism. If the endothelial cells lining blood vessels are damaged, platelets (small cells circulating in the blood in vast numbers) adhere to the injured tissue, most likely to collagen fibres beneath the damaged endothelium. Contact with collagen provokes complex biochemical and shape changes in the platelets, including a release of ADP which causes further aggregation of platelets to form a clump that seals the damaged area. Simultaneously, vascular injury initiates the process of blood coagulation in which a soluble blood protein (fibrinogen) is converted to a tangle of insoluble threads (fibrin) by an enzyme (thrombin). Thrombin also causes release of ADP, 5-hydroxytryptamine, and other components from the platelets, leading to further aggregation. Platelets have been observed to release $O_2^{.-}$ and hydrogen peroxide into the surrounding medium, although the physiological significance of this is unclear.

Chemically, prostaglandins may be regarded as derivatives of the hypothetical *prostanoic acid*, a C_{20} acid containing a 5-membered ring structure (a *cyclopentane* ring) Thromboxanes A_2 and B_2 do not possess a prostanoic acid structure and are thus not prostaglandins, strictly speaking (Fig. 7.12). There are several types of prostaglandins distinguished by the chemical nature and geometry of the groups attached to the ring (e.g. E, F, D), and by the number of double bonds in the side chains (e.g. E_1, E_2, F_1, F_2, etc). Prostaglandins are synthesized *in vivo* from fatty acids that contain *cis* double bonds at positions 8, 11, and 14. The most important fatty acid of this type in humans is arachidonic acid ($C_{20:4}$, double bonds at 5, 8, 11, 14—see Chapter 4), but others include 8,11, 14-eicosatrienoic acid ($C_{20:3}$) and 5,8,11,14,17-eicosapentaenoic acid ($C_{20:5}$). This last fatty acid is very common in the blood lipids of Eskimos, no doubt due to their unusual diet. We shall confine further discussion to products derived from arachidonic acid.

The first stage in prostaglandin formation is to provide a fatty-acid substrate, by the action of a lipase enzyme. Lipases are activated by hormones or by tissue injury, depending on the circumstances. Phospholipase A_2 seems to be particularly important, but other enzymes are involved as well. For example, luteinizing hormone may stimulate a cholesterol esterase enzyme in ovarian tissues, releasing fatty acids attached to the —OH group of cholesterol (Chapter 4). The anti-inflammatory action of a number of steroids is probably due to suppression of fatty-acid formation.

Once a fatty-acid substrate is available, the enzyme *cyclooxygenase* forms an *endo*-peroxide derivative from it. Cyclooxygenase is present in almost all mammalian tissues, utilizes molecular oxygen and requires haem for its action. Treatment of crude or purified cyclooxygenase preparations with GSH and glutathione peroxidase causes a lag period to appear in their subsequent reaction with arachidonic acid. This lag period may be overcome by addition of PGG_2 or several other hydroperoxides, such as those formed by the action of lipoxygenase (Sections 7.5 and 7.7) or during the air-oxidation (autoxidation) of arachidonic acid (Chapter 4). Cumene hydroperoxide and hydrogen peroxide are also effective, although at higher concentrations. Indeed, addition of 5 mM H_2O_2 to their seawater causes spawning in both male and female abalones (large herviborous marine snails valued as a food source in certain parts of the world), apparently by activation of cyclooxygenase and subsequent prostaglandin synthesis. Hence there is an intimate relationship between lipid peroxidation and prostaglandin metabolism, in that very efficient antioxidant protection will slow down prostaglandin synthesis, at least until sufficient PGG_2 is formed to maximally activate cyclooxygenase. It seems that a trace of hydroperoxide is required to interact with iron(III) haem at the enzyme active site to form a peroxy radical, which can then abstract a hydrogen atom from arachidonic acid and start off the reaction.

PGG_2 is then converted into PGH_2 (Fig. 7.12) by a peroxidase activity that co-purifies with cyclooxygenase and thus seems to be part of the same protein. During this reduction process, a highly-reactive radical intermediate, which can be detected by ESR studies at low temperatures (Chapter 2), is released. This reactive radical is capable of oxidizing methional into ethene, converting phenol into polymeric products, attacking propyl gallate, oxidizing adrenalin, paracetamol (Chapter 6), or diphenylisobenzofuran (yet another example of the lack of specificity of this compound as a 'singlet oxygen scavenger'), converting the carcinogen benzpyrene into a quinone, and causing the emission of light from luminol. It thus has many of the properties of hydroxyl radical, although it is known not to be this species, and it is usually written as $[O_x]$. If $[O_x]$ is not scavenged, e.g. by phenol, it can attack the cyclooxygenase itself or subsequent enzymes in the prostaglandin pathway, causing inactivation. Hence addition of antioxidants can sometimes stimulate prostaglandin synthesis by preventing such inactivations. Too much of certain antioxidants can inhibit by removing the radicals leading to the formation of the traces of hydroperoxide that the cyclooxygenase needs to get going, however. Certainly, the vitamin E status of animals is known to affect the behaviour of their platelets. The peroxidase that converts PGG_2 into PGH_2 can act upon several other peroxides *in vitro* to form $[O_x]$, including hydroperoxides of arachidonic acid. A number of 'non-steroidal anti-inflammatory' drugs, especially aspirin and indomethacin (Fig. 7.13),

Indomethacin

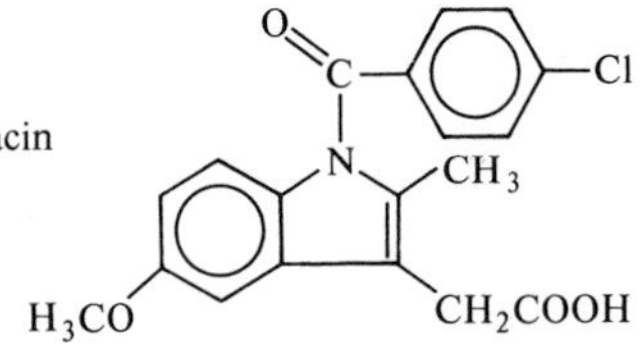

Aspirin
(acetylsalicylic acid)

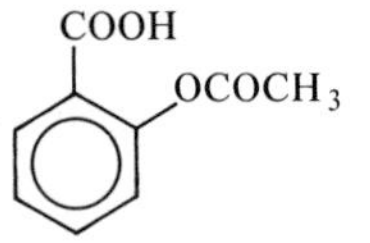

Sulindac

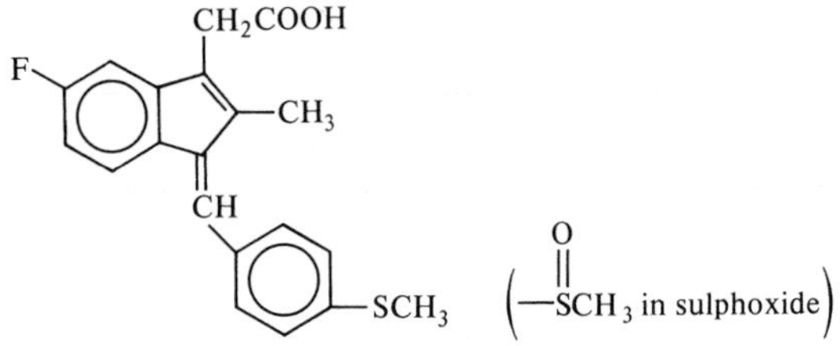

$\left(-\overset{\overset{\displaystyle O}{\|}}{S}CH_3 \text{ in sulphoxide}\right)$

Phenylbutazone

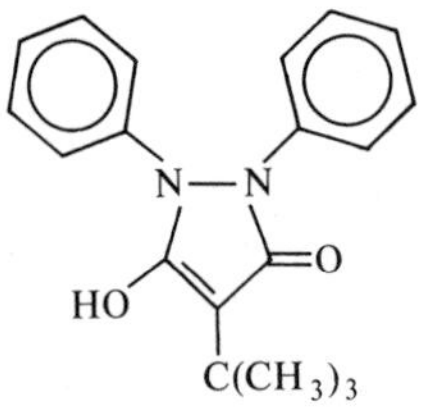

Fig. 7.13. Some non-steroidal anti-inflammatory drugs (NSAIDs). Sulindac (Clinoril®) has been used in the treatment of inflammatory joint disease such as rheumatoid arthritis. It is converted *in vivo* into its sulphide form. Its exact mode of action is unclear although it is known to scavenge the $[O_x]$ species. Reduction of the sulphoxide to sulphide *in vivo* appears to involve the protein thioredoxin (Fig. 7.1).

act by inhibiting cyclooxygenase activity and hence the whole cascade of prostaglandin biosynthesis. Aspirin increases blood coagulation time by decreasing platelet aggregation. Sulindac sulphide (Fig. 7.13) is converted into the sulphoxide by $[O_x]$, which further illustrates the reactivity of this species since neither O_2^- nor hydrogen peroxide can achieve this conversion. $[O_x]$ may be some kind of ferryl radical (Chapter 4).

Both PGG$_2$ and PGH$_2$ are unstable, having a half-life of only minutes under physiological conditions, but they exert profound biological effects, including contraction of smooth muscle and stimulation of platelet aggregation. Their subsequent fate depends on the tissue under study.

They are precursors of the primary prostaglandins PGD_2, PGE_2, and $PGF_{2\alpha}$ (Fig. 7.12), but platelets also convert them into two other compounds, 12-hydroxy-5,8,10-heptadecatrienoic acid (HHT for short) and thromboxane B_2 (TXB_2), which is formed from an unstable precursor thromboxane A_2. The function of HHT is not clear, but it has also been found in lung tissue and seminal vesicles. Its formation involves loss of 3 carbon atoms, probably as malondialdehyde (MDA, Chapter 4). MDA reacts with thiobarbituric acid to give a pink chromogen. Prostaglandin endoperoxides decompose under the acid-heating conditions of the TBA-test (Chapter 4) to form MDA; and so application of the TBA reaction to tissues synthesizing prostaglandins would overestimate the amount of free MDA present. The amount of 'TBA-reactive' material found in rabbit serum decreases if the animals are pretreated with aspirin to inhibit cyclooxygenase, suggesting that prostaglandins are a significant contributor to the observed TBA-reactivity. This should be borne in mind in attempts to use TBA-reactivity of body fluids as an index of lipid peroxidation *in vivo* in various disease states. Indeed, the generation of TBA-reactivity by lipid hydroperoxides may be due to the formation and decomposition of cyclic endoperoxides during the assay (see Chapter 4).

Thromboxane A_2 stimulates platelet aggregation and smooth muscle contraction, and it has also been detected in phagocytosing neutrophils. Blood vessel endothelium synthesizes prostacyclin (PGI_2), which dilates blood vessels, inhibits platelet aggregation, and depresses neutrophil chemotaxis. Prostacyclin is unstable and decomposes to give 6-keto-$PGF_{1\alpha}$ (Fig. 7.12). Its formation by blood vessels may prevent undesirable deposition of platelets *in vivo*. Indeed, the capacity of vessel walls to synthesize prostacyclin *in vivo* is less affected by aspirin than is the ability of platelets to make TXA_2, and so controlled doses of aspirin may be useful in the prevention of unwanted clot formation.

The prostaglandin cascade is yet another example of a controlled radical process put to a useful purpose; others may exist. For example, the formation of cyclic GMP from GTP by the enzyme guanylate cyclase appears to be stimulated by systems generating oxygen radicals, by prostaglandin *endo*-peroxides and fatty acid hydroperoxides, and to be inhibited by thiol compounds, but the physiological significance of these effects is not yet clear.

7.7. Lipoxygenase products

In Section 7.5 we described the importance of lipoxygenase enzymes as initiators of 'controlled lipid peroxidation' in the wound response of plant tissues. In 1974, it was discovered that a lipoxygenase enzyme is active in platelets, and similar enzymes have since been studied in many mamalian tissues. Like plant lipoxygenases, the animal enzymes are probably iron-dependent.

Platelet lipoxygenase acts upon arachidonic acid, its natural substrate, to form 12-hydroperoxy-5,8,11,14-eicosatetraenoic acid (12-HPETE for short!). This is unstable and is quickly reduced to the 12-hydroxy derivative (12-HETE) *in vivo*. Reduction appears to be achieved by platelet glutathione peroxidase activity in the presence of GSH, since 12-HETE formation is decreased in the platelets of selenium-deficient rats.

Lipoxygenases found in other tissues introduce oxygen atoms at different places in the carbon chain (Fig. 7.14). Formation of 12-HETE has been observed in skin, 11-HETE in lung and mast cells, and 5-HETE, 15-HETE, and 5,12-DHETE (5,12 dihydroxy-6,8,10,14-eicosatetraenoic acid) in rabbit and human neutrophils. Indeed, the lipoxygenase pathway for metabolism of arachidonic acid in neutrophils appears to be more important biologically than is the cyclooxygenase route. At concentrations effective against cyclooxygenase, aspirin and indomethacin do not inhibit lipoxygenase activity, and can sometimes 'divert' the metabolism of polyunsaturated fatty acids from cyclooxygenase into lipoxygenase routes *in vivo*. Anti-inflammatory steroids, by contrast, will stop both pathways by preventing formation of the fatty-acid substrate. Animal

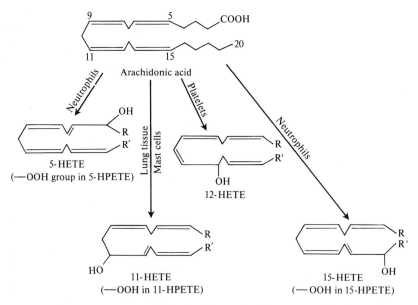

Fig. 7.14. Conversion of arachidonic acid catalysed by mammalian lipoxygenases.

lipoxygenases are inhibited by the antioxidant NGDA (Chapter 4) or by 5,8,11,14-eicosatetraynoic acid (ETYA).

The HPETE compounds are precursors of a range of chemicals with potent biological activity known as the leukotrienes. Figure 7.15 shows the products derived from 5-HPETE; but others arise from different HPETE isomers. Leukotriene A_4 is an unstable epoxide structure that can be reduced to leukotriene B_4. This pathway is especially important in neutrophils, LTB_4 being a powerful chemotactic agent for these cells that promotes aggregation, degranulation, and the respiratory burst. Human alveolar macrophages produce LTB_4 and other leukotrienes.

When lung and several other tissues, taken from humans allergic to a particular substance, is perfused with the substance, among the products formed are histamine and a 'slow-reacting substance' (SRS-A) that increases vascular permeability and contracts bronchial muscles. The contraction is slow and prolonged, unlike the rapid and short-acting action of histamine. Most SRS-A activity can be attributed to leukotrienes C_4 and D_4. LTC_4 is formed by the conjugation of LTA_4 with GSH in the presence of a glutathione transferase enzyme. Removal of a glutamic acid residue by glutamyltransferase activity yields LTD_4, which can be further degraded to LTE_4. This latter compound appears less biologically active.

In the next few years much more information about the chemical nature and biological properties of leukotrienes will surely become available. It is interesting to note that *in vitro*, both 15-HPETE and 12-HPETE are substrates for the peroxidase activity of cyclooxygenase and can lead to [O_x] formation, perhaps indicating an interaction between the two pathways.

Before leaving this area, we wish to point out that polyunsaturated fatty acids such as arachidonate are easily peroxidized by non-enzymic methods (Chapter 4), and commercially-available samples may be extensively peroxidized, thus causing variable results in, for example, cyclooxygenase assays. If such fatty acids are included in a system generating oxygen radicals, then products can be formed not only by cyclooxygenase and lipoxygenase pathways but also by the attack of species such as hydroxyl radical. This has already caused confusion in studies of arachidonate metabolism by activated neutrophils, which produce OH$\cdot$. Fortunately, products formed by enzyme-catalysed reactions are highly stereospecific, whereas those from non-enzymic reactions usually are not, and determinations of stereochemical purity should be made before assuming that a novel oxidized product has been produced by the action of an enzyme. The rat liver microsomal cytochrome P_{450} system has been claimed to hydroxylate arachidonic acid into 9-, 11-, 12-, and 15-HETEs. Hydroxyl radicals attack and destroy the biological activity of LTB_4, LTC_4, and LTD_4.

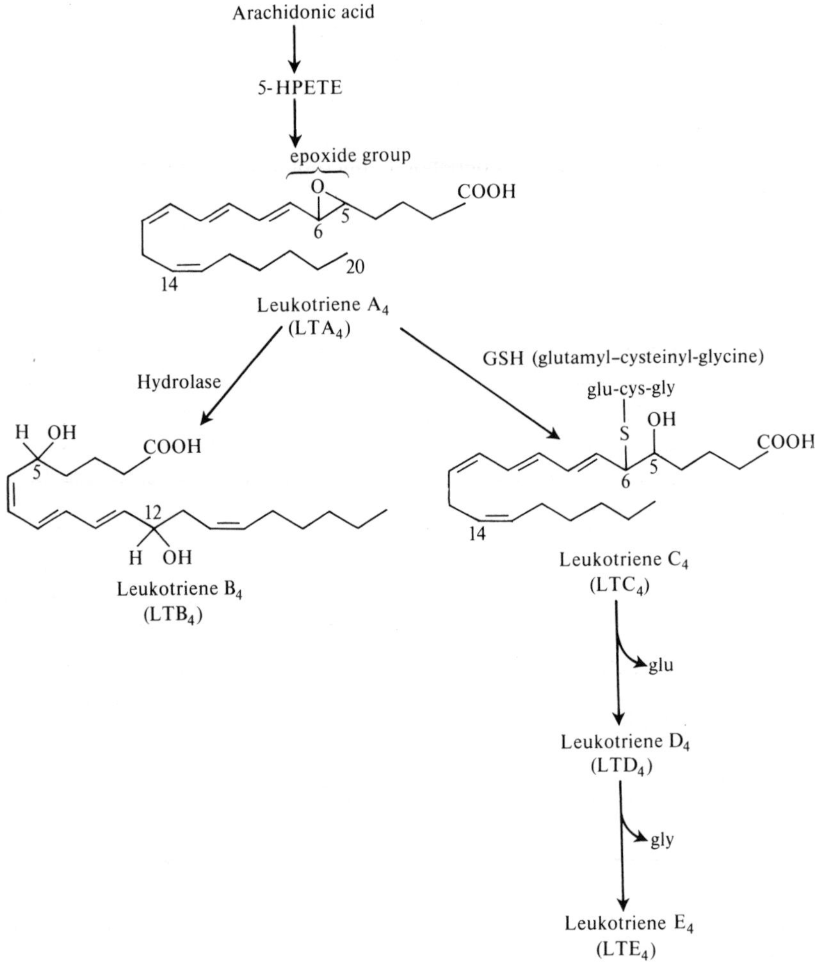

Fig. 7.15. The leukotrienes. Leukotriene A_4 is an epoxide (full name: 5,6-epoxy-7,9,11,14-eicosatetraenoic acid). It can be hydrolysed enzymatically into leukotriene B_4 (5,12-dihydroxy-6,*cis*-8,*trans*-10,*trans*-14,*cis*-eicosatetraenoic acid) or non-enzymically to other 5,12- and 5,6-dihydroxyacids. Only LTB_4 is shown for simplicity. LTA_4 is also converted by conjugation with GSH into LTC_4, which can be converted into LTD_4 and LTE_4 by successive removal of amino acids. LTC_4 and LTD_4 are responsible for the biological activity of most preparations of slow-reacting substance of anaphylaxis (SRS-A). Another group of leukotrienes can be formed by initial oxygenation of arachidonic acid at C_{15}. 'SRS-A' is a substance produced during allergic reactions that causes increased vascular permeability and, in lung, prolonged spasm of smooth muscle. Other hydroxylated compounds such as trihydroxy(20-OH-LTB_4) and dicarboxylate (20-COOH-LTB_4) derivatives have also been observed to be formed in neutrophils.

7.8. Further reading

Albich, J. M., McCarthy, C. A., and Hurst, J. K. (1981). Biological reactivity of hypochlorous acid. Implications for microbicidal mechanisms of leukocyte myeloperoxidase. *Proc. natn. Acad. Sci. U.S.A.* **78,** 210.

Babior, B. M. (1978). Oxygen-dependent microbial killing by phagocytes *New Engl. J. Med.* **298,** 721 and 645.

Badwey, J. A., and Karnovsky, M. L., (1980). Active oxygen species and the functions of phagocytic leukocytes. *Annu. Rev. Biochem.* **49,** 695.

Borgeat, P., De Laclos, B. F., and Maclouf, J. (1983). New concepts in the modulation of leukotriene synthesis. *Biochem. Pharmac.* **32,** 381.

Brennan, R., and Frenkel, C., (1977). Involvement of hydrogen peroxide in the regulation of senescence in pears. *Plant physiol.* **59,** 411.

Carpenter, M. P. (1981). Antioxidant effects on the prostaglandin endoperoxide synthetase product profile. *Fed. Proc.* **40,** 189.

De Chatelet, L. R., *et al.* (1977). Oxidative metabolism of the human eosinophil. *Blood* **50,** 525.

Del Maestro, R. F., Björk, J., and Arfors, K. E. (1981). Increases in microvascular permeability induced by enzymatically generated free radicals. *Microvasc. Res.* **22,** 239.

Edwards, S. W., Hallett, M. B., and Campbell, A. K. (1984). Oxygen-radical formation during inflammation may be limited by oxygen concentration. *Biochem. J.* **217,** 851.

Esnouf, M. P., *et al.* (1980). The production and utilisation of superoxide during the carboxylation of glutamyl residues in preprothrombin. In *Biological and clinical aspects of superoxide and SOD* (eds W. H. Bannister and J. V. Bannister) p. 72. Elsevier/North Holland, New York.

Föerder, C. A., Klebanoff, S. J., and Shapiro, B. M. (1978). Hydrogen peroxide production, chemiluminescence and the respiratory burst of fertilisation. Inter-related events in early sea urchin development. *Proc. natn. Acad. Sci. U.S.A.* **75,** 3183.

Foote, C. S., Goyne, T. E., and Lehrer, R. I. (1983), Assessment of chlorination by human neutrophils *Nature* **301,** 715.

Forney, L. J., Reddy, C. A., Tien, M., and Aust, S. D. (1982). The involvement of hydroxyl radical derived from hydrogen peroxide in lignin degradation by the white-rot fungus *Phanerochaete chrysosporium*. *J. biol. Chem.* **257,** 11455.

Galliard, T. (1978). Lipolytic and lipoxygenase enzymes in plants and their action in wounded tissue. In *Biochemistry of wounded plant tissues* (ed. G. Kahl). Walter de Gruyter and Co, Berlin.

Goldenberg, H. (1982). Plasma membrane redox activities. *Biochim. Biophys. Acta* **694,** 203.

Gross, G. G., Janse, C., and Elstner, E. F. (1977). Involvement of malate, monophenols and the superoxide radical in hydrogen peroxide formation by isolated cell walls from horseradish. *Planta* **136,** 271.

Gutteridge, J. M. C., and Kerry, P. J. (1982). Detection by fluorescence of peroxides and carbonyls in samples of arachidonic acid. *Br. J. Pharmac.* **76,** 459.

Halliwell, B. (1977). Superoxide and hydroxylation reactions. In *Superoxide and Superoxide dismutases* (Eds A. M. Michelson, J. M. McCord and I. Fridovich) p. 335. Academic Press, London.

Halliwell, B. (1978). Lignin synthesis: the generation of hydrogen peroxide and superoxide by horseradish peroxidase and its stimulation by manganese(II) and phenols. *Planta* **140**, 81.

Harrison, J. E., Watson, B. D., and Schultz, J., (1978). Myeloperoxidase and singlet oxygen: a reappraisal. *FEBS Lett.* **92**, 327.

Henry, J. P., Monny, C., and Michelson, A. M., (1975). Characterisation and properties of *Pholas* luciferase as a metalloglycoprotein. *Biochemistry* **14**, 3458.

Hocking, W. G., and Golde, D. W., (1979). The pulmonary-alveolar macrophage. *New Engl. J. Med.* **301**, 580 and 639.

Holme, E., *et al.* (1982). Does superoxide anion participate in 2-oxoglutarate-dependent hydroxylation? *Biochem. J.* **205**, 339.

Johnston Jr., R. B., *et al.* (1975). The role of superoxide anion generation in phagocyte bactericidal activity. Studies with normal and chronic granulomatous disease leukocytes. *J. Clin. Invest.* **55**, 1357.

Konze, J. R., and Elstner, E. F. (1978). Ethane and ethylene formation by mitochondria as indication of aerobic lipid degradation in response to wounding of plant tissues. *Biochim. Biophys. Acta* **528**, 312.

Muchmore, D. B., Little, S. A., and de Haen, C. (1982). Counter-regulatory control of intracellular hydrogen peroxide production by insulin and lipolytic hormones in isolated rat epididymal fat cells: a role of free fatty acids. *Biochem.* **21**, 3886.

Mukherjee, S. P., and Mukherjee, C. (1982). Similar activities of nerve growth factor and its homologue proinsulin in intracellular hydrogen peroxide production and metabolism in adipocytes. *Biochem. Pharmac.* **31**, 3163.

Newburger, P. E., and Tauber, A. I., (1982). Heterogeneous pathways of oxygen radical production in human neutrophils and the HL-60 cell line. *Pediatr. Res.* **16**, 856.

O'Flaherty, J. T. (1982). Biology of disease. Lipid mediators of inflammation and allergy. *Lab. Invest.* **47**, 314.

Repine, J. E. (1982). Neutrophils and lung oedema. State of the art. *Chest* **81S**, 475.

Sbarra, A. J., and Straus, R. R. (eds) (1980). *The reticuloendothelial system. A comprehensive treatise*, Vol. 2: *Biochemistry and metabolism*. Plenum Press, New York.

Segal, A. W., *et al.* (1983). Absence of cytochrome b in chronic granulomatous disease. *New Engl. J. Med.* **308**, 245.

Shimizu, T., Kondo, K., and Hayaishi, O., (1981). Role of prostaglandin endoperoxides in the serum thiobarbituric acid reaction. *Archs Biochem. Biophys.* **206**, 271.

Thelander, L., and Reichard, P. (1979). Reduction of ribonucleotides. *Ann. Rev. Biochem.* **48**, 133.

Till, G. O., *et al.* (1982). Intravascular activation of complement and acute lung injury. Dependency on neutrophils and toxic oxygen metabolites. *J. Clin. Invest.* **69**, 1126.

Weiss, J., Victor, M., Stendhal, O., and Elsbach, P., (1982). Killing of gram-negative bacteria by polymorphonuclear leukocytes. Role of an O_2-independent bactericidal system. *J. Clin. Invest.* **69**, 959.

Weissman, G., *et al.* (eds). *Advances in inflammation research*, Vols 1–5. Raven Press, New York.

Zimmerman, G. A., Renzetti, A. D., and Hill, H. R. (1983). Functional and metabolic activity of granulocytes from patients with adult respiratory distress syndrome. *Am. Rev. respir. Dis.* **127**, 290.

8 Free radicals, ageing, and disease

We have seen that free radicals are of great importance in the mode of action of several toxic substances (Chapter 6) and in the process of inflammation, in which both oxygen radicals and products derived from the action of lipoxygenase and cyclooxygenase enzymes play major roles. Radicals have been suggested to be involved in ischaemia and in degenerative arterial disease (Chapters 3 and 4), and it has recently been reported that neutrophils from patients with elevated concentrations of lipoprotein in the blood produce more O_2^- than normal upon activation. Two other major problems in Western societies are cancer and ageing populations, so it is perhaps appropriate to examine the role of radical reactions in these processes. Before doing so, however, we shall discuss the autoimmune diseases.

8.1. Chronic inflammation and the autoimmune diseases

We saw in Chapter 7 that the acute inflammatory response is beneficial to the organism in that it deals with unwanted and potentially dangerous foreign particles such as bacteria. Inflammation is normally a self-limiting event. However, anything causing abnormal activation of phagocytes has the potential to provoke a devastating response. For example, the major biochemical feature of *gout* is an elevated concentration of uric acid in the blood. Inflammation of joints is triggered by the deposition within them of sodium urate crystals. These crystals can provoke inflammation by a variety of mechanisms, including the stimulation of a respiratory burst in neutrophils and production of leukotriene B_4, which will attract more neutrophils (Chapter 7). Gout can be treated with *allopurinol*, an inhibitor of the xanthine dehydrogenase activity that converts hypoxanthine and xanthine into urate *in vivo* (Chapter 3). Perhaps the most striking consequences of abnormal phagocyte actions, however, are seen in the *autoimmune diseases*.

The body has mechanisms to prevent formation of antibodies against its own components. Any failure of these mechanisms allows formation of *autoantibodies* that can bind to normal body-components, and provoke attack by phagocytic cells. In some autoimmune diseases, only a single tissue is attacked. For example, in *Hashimoto's thyroiditis* infiltration of the thyroid gland by phagocytes is accompanied by tissue changes and fibrosis, and the presence of circulating antibodies against certain thyroid constituents, such as thyroglobulin. In *myasthenia gravis,* a neuromuscular disorder characterized by weakness and fatigue of voluntary muscles, antibodies against the transmitter acetylcholine are present. *Chronic autoimmune gastritis,* in which antibodies to gastric parietal cells are present, is a third example.

More serious than these are the autoimmune diseases in which lesions are widespread and autoantibodies are present against many tissues. These diseases include *systemic lupus erythematosus, dermatomyossitis,* and *autoimmune vasculitis.* To take one example, systemic lupus affects mainly young women, and produces a wide variety of lesions involving the skin, kidneys, muscles, joints, heart, and blood vessels. A wide range of autoantibodies is present, including circulating antibodies to DNA and RNA, anti-erythrocyte antibodies, and antibodies to subcellular organelles and plasma proteins. The kidney lesions are probably due to deposition of immune complexes on the basement membranes of renal glomeruli, followed by complement activation. Phagocytic cells have been identified in rat glomeruli, and respond to activation by mounting a respiratory burst. *Rheumatoid arthritis,* a disease characterized by chronic joint inflammation, especially in the hands and legs, has many features of an autoimmune disease, although the exact cause is unknown. The blood serum and joint fluid of rheumatoid patients often contains an antibody (rheumatoid factor) that binds to immunoglobulin G. The onset of the disease is usually slow, but may sometimes be rapid. The synovium of the joints becomes swollen and damaged, and there is an increased deposition of the iron proteins, ferritin and haemosiderin (Chapter 2) within it. The disease often proceeds to erosion of the joint cartilage. Production of synovial fluid, the natural joint lubricant, is increased, but its viscosity is much below normal. This decrease in viscosity is due to the breakdown of the high-molecular-weight polymer *hyaluronic acid* (Fig. 8.1) into smaller pieces. The cartilage wear particles, produced by increased friction in the joints, can activate neutrophils and make matters worse. Joint inflammation often accompanies other autoimmune diseases, such as systemic lupus.

Autoantibodies are present in the nonspecific inflammatory bowel diseases such as *Crohn's disease* and *ulcerative colitis.* The former is a recurrent inflammation and ulceration of the whole digestive tract, although it is often most severe in the lower part of the ileum, and in the colon and rectum, whereas in the latter disease the ulceration and inflammation affect the colon and rectum only. In both conditions autoantibodies to bowel components can be found in the blood serum. *Multiple sclerosis* has some features of an autoimmune disease.

Autoimmune diseases generally have active and quiescent phases,

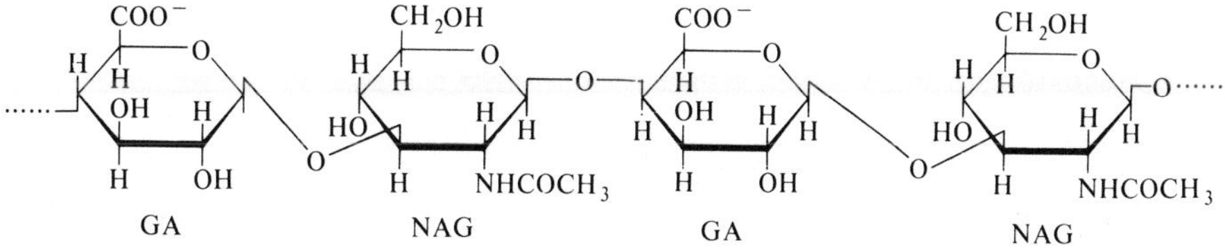

Fig. 8.1. Structure of hyaluronic acid. Hyaluronic acid is a long polymer formed by joining together alternately two different sugars: glucuronic acid (GA) and N-acetylglucosamine (NAG). The negative charge on the carboxyl groups of GA at physiological pH means that these subunits repel each other, so that the molecule extends out in solution. Hence solutions of hyaluronic acid are extremely viscous.

which makes the evaluation of medical treatment especially difficult. This should be borne in mind when assessing the effectiveness of any therapy, such as the use of superoxide dismutase injections, in other than a properly controlled clinical trial over a long period of time. How autoimmune diseases arise is not known, although there is an inherited predisposition to them; and viral infections have often been suggested to be involved. Certain drugs can induce a condition resembling systemic lupus erythematosus: most significantly, hydralazine, isoniazid, chlorpromazine, and procainamide. A few other drugs have been shown to induce conditions resembling lupus, but much less frequently; these include penicillamine, α-methyldopa, and diphenylhydantoin. It has been suggested that reaction of drug-derived metabolites with normal tissues can produce products that behave as 'foreign antigens'. At least in the cases of hydralazine, α-methyldopa, and isoniazid (Chapter 6), this binding may involve radical reactions. Autoantibodies are often demonstrable in small amounts in a few 'normal' members of the population, and their incidence increases with age. Post-mortem examination of apparently healthy people with such antibodies often shows minor lesions, too small to produce symptoms. Perhaps oxidative damage to tissues during the ageing process (Section 8.2) creates new antigens.

8.1.1. Autoimmune diseases and phagocytic action

Oxygen radicals, prostaglandins, lipoxygenase-derived products and hydrolytic enzymes produced by macrophages, monocytes, and neutrophils are important in mediating inflammation. Indeed, anti-inflammatory steroids and cyclooxygenase inhibitors are frequently employed in the control of autoimmune diseases. Superoxide dismutase, and other scavengers of oxygen radicals, have been observed to suppress inflammation in some animal model systems, such as the reversed passive

Arthus reaction in skin. (The 'Arthus reaction' is the name given to a local inflammation that results when an antigen is injected into the skin of an animal that has a high level of circulating antibody against that antigen. It is largely mediated by neutrophils. It can also be observed if the antibody is injected into the animal's bloodstream rather than being formed by the animal itself; this is the passive Arthus reaction. In the reversed passive reaction, the antigen is injected intravenously, and the antibody locally.)

For example, injection of human serum into the bloodstream of rats, followed by injection of an antibody against it into the skin, causes swelling and heat, which is largely mediated by neutrophils. McCord's group in the USA showed that intravenous injection of SOD into the animals had little anti-inflammatory effect in this system, because SOD is cleared from the circulation within minutes by the kidneys. However, if clearance was prevented by binding the SOD to a high-molecular-weight polymer (e.g. Ficoll), there was a marked anti-inflammatory effect. Ficoll-bound SOD also had an inhibitory effect against inflammation induced by injecting carrageenan (an irritating substance derived from seaweed) into the feet of rats. In neither case did Ficoll-bound catalase have an anti-inflammatory effect. Native SOD was observed to protect against glomerulonephritis induced by the intravenous injection of preformed antigen–antibody complexes into mice. These become deposited in the kidneys and provoke inflammation. The effectiveness of native SOD in this system may be due to its rapid accumulation in the kidneys prior to removal from the body. McCord attributed the above results to the action of SOD in removing O_2^-, so preventing O_2^--dependent formation of a factor chemotactic for neutrophils. This explanation has been challenged, however. The lack of inhibition by Ficoll-bound catalase suggested that hydroxyl radical formation was not important. Some other workers have confirmed these anti-inflammatory effects of SOD in related systems, but have often also found protective effects of catalase and other antioxidants, including the protein caeruloplasmin (Chapter 4). For example, both SOD and catalase showed inhibitory effects against inflammation induced by the implantation of carrageenan-soaked sponges beneath the skin of rats. Injection of xanthine oxidase into the hind-feet of rats produced a swelling that could be partially inhibited by SOD, catalase, or the hydroxyl radical scavenger, mannitol. By contrast, instillation of a mixture of xanthine and xanthine oxidase into the lungs of rats produced oedema that could be decreased by SOD but not by catalase. In all such experiments, it is essential to perform controls with heat-denatured (or metal-free) SOD and catalase (or other inactive proteins) in order to rule out 'non-specific' effects. In collaboration with D. R. Blake, the authors have shown that large doses of the iron-chelator desferrioxamine, which prevents formation of hydroxyl radicals from O_2^- and hydrogen peroxide in the presence of iron salts (the iron-catalysed Haber–Weiss reaction, Chapter 3) have an anti-inflammatory effect in three animal model systems. Small doses have a pro-inflammatory effect however, indicative of the complexity of the systems! Ward *et al.* in the USA found that desferrioxamine decreases complement- and neutrophil-mediated lung injury in rats caused by injection of a factor from cobra venom.

It would therefore seem that oxygen radicals and released enzymes, such as myeloperoxidase and hydrolases, might play an important role in the autoimmune diseases, and that therapy directed against them might prove clinically useful. Perhaps the best evidence consistent with this suggestion has been provided by Emerit and Michelson in France. They have shown that the serum of patients with lupus erythematosus contains a 'clastogenic factor' that induces chromosome damage when it is added to cultures of lymphocytes isolated from normal human donors. The chemical identity of clastogenic factor is unknown, but they have reported that inclusion of SOD in the culture medium prevents its effect. Lupus patients are light-sensitive and in about forty per cent of cases develop a characteristic 'butterfly' rash across the cheeks and the bridge of the nose when exposed to sunlight. Lymphocytes from lupus patients are damaged by exposure to light, in the wavelength range 360–400 nm, in the presence of clastogenic factor: an effect again prevented by SOD. Neutrophils from children with rheumatoid arthritis have been reported to contain less SOD activity than normal and to release more O_2^- during the respiratory burst.

Other evidence for the importance of oxygen radical reactions in autoimmune disease has been provided by Harman in the USA. He has studied New Zealand Black (NZB) mice, a strain that spontaneously develops an autoimmune disease that has some of the features of human systemic lupus. He finds that addition of antioxidants such as tocopheryl acetate or Santoquin (ethoxyquin, Table 4.5, Chapter 4) to their diet decreases the manifestations of this disease, and prolongs the average lifespan of the mice. The serum of NZB mice also contains a clastogenic factor whose action is decreased by SOD.

A few individual case-reports have appeared as to the benefit of oxygen radical scavengers in the treatment of human autoimmune disease, although proper controlled trials have not been done in most cases. SOD and simple copper chelates that can react with O_2^- (Chapter 3) have been reported as helpful in lupus and in dermatomyositis, and SOD encapsulated in liposomes (Chapter 4) has been used in the treatment of Crohn's disease. Large doses of vitamin E have sometimes been reported as helpful in lupus cases. Direct injection of SOD into inflamed joints

Table 8.1. Clinical testing of superoxide dismutase (Orgotein®) in human patients suffering from rheumatoid arthritis

(a) Double-blind trials were performed (i.e. neither the patient nor the patient's doctor knew whether the SOD or a placebo was being administered, so preventing bias in the results) to compare the effect of Orgotein® on a group of patients with rheumatoid arthritis already undergoing treatment with aspirin and/or corticosteroids. Improvement was found to be statistically significant in all the parameters listed below. 40 injections of 8 mg SOD were given deep-subcutaneously over a period of 8 weeks.

Parameter	Response
Pain	less pain, fewer analgesics used.
Circumference of joint	smaller, less swollen.
Grip strength	improved.
Morning stiffness	less marked.
Patient's assessment of disease	improved.

(b) 12 mg doses of Orgotein® were injected three times per week for 4 weeks, twice per week for the next 12 weeks, and then once a week for the last 10 weeks. A comparison of this regimen with gold therapy was conducted by clinicians not aware of the regimen used. Lower doses of Orgotein® were less effective. Gold dosing was 75 mg per week after an initial loading dose of 50 mg. It can be seen that SOD and the gold treatment are of comparable effectiveness.

Parameter	Medication	Start of treatment (mean ± SE)	End of treatment (mean ± SE)
Erythrocyte	SOD	47.2 ± 6.2	30.3 ± 6.4
sedimentation rate	Gold	61.8 ± 8.0	35.5 ± 9.8
Morning stiffness	SOD	98.2 ± 25.0	40.8 ± 20.9
	Gold	171.0 ± 12.0	116.7 ± 54.6
Grip strength	SOD	16.5 ± 4.9	37.4 ± 5.3
	Gold	13.1 ± 2.1	28.2 ± 6.5
Pain	SOD	69.6 ± 4.6	25.0 ± 7.8
	Gold	61.8 ± 4.1	30.5 ± 8.3

Data were abstracted from the article by K. B. Menander-Huber in *Biological and clinical aspects of superoxide and superoxide dismutase* (eds W. H. Bannister and J. V. Bannister) pp 408–23. Elsevier/North Holland, 1980. Repeated injection of SOD into animals does provoke the formation of some antibodies, although it appears to be only weakly antigenic, and no side-effects were reported during the above trials.

has been reported to partially relieve the symptoms of rheumatoid arthritis. Although in the latter disease, double-blind clinical trials have been performed upon humans, and have shown a real anti-inflammatory effect of SOD (Table 8.1), control experiments with denatured enzyme or the apoenzyme have not been reported. It remains possible, therefore, that the effect of SOD is not due to removal of O_2^-, but to a nonspecific protein effect, perhaps by providing an alternative substrate for proteolytic enzymes released from activated phagocytes. Even if true, this would not detract from the clinical use of SOD as an anti-inflammatory agent however, and a preparation of this enzyme, under the name *Orgotein*®, is now approved for veterinary use in the USA. Indeed, synovial fluid from rheumatoid patients shows an accumulation of stable prostaglandins, thiobarbituric-acid-reactive material (Chapter 4), and fluorescent products of lipid peroxidation (Chapter 4), indicative of the effects produced by neutrophils becoming activated when they contact immune complexes. Other products of neutrophils are also present in abnormally high amounts including lactoferrin and lysozyme.

Like most extracellular fluids of the body, synovial fluid contains little, if any, SOD or catalase activities, and so O_2^- and hydrogen peroxide generated during the respiratory burst can interact to form hydroxyl radical, $OH^{\bullet}$. Hydroxyl radical is known to attack and degrade hyaluronic acid (Fig. 8.1), which would account for the decrease in viscosity of rheumatoid synovial fluid as compared with normal fluid. Experiments by McCord and by Greenwald in the USA, and in the authors' laboratory, have shown that radical-generating systems such as activated neutrophils, or a mixture of xanthine and xanthine oxidase, can degrade hyaluronic acid *in vitro* in a way similar to that observed *in vivo*. The attacking species is known to be $OH^{\bullet}$, which requires traces of iron salts for its formation. The authors have shown that micromolar concentrations of non-protein-bound iron salts are present in human rheumatoid synovial fluid (Chapter 2). Caeruloplasmin is present, in synovial fluid and in serum from rheumatoid patients, in amounts greater than normal. However, the caeruloplasmin found in rheumatoid synovial fluid appears to have an unusually low stability upon storage, and other abnormalities of iron metabolism are seen in rheumatoid disease. A rapid fall in serum 'total iron' at the onset of inflammation is followed by a drop in haemoglobin concentration and increased deposition of iron proteins in the synovial membranes. Loss of iron from the blood has been observed in many other inflammatory conditions, in other forms of tissue injury, and during infections. It has often been thought to be an attempt by the body to 'withhold' iron from invading bacteria, and there is evidence consistent with this idea. For example, injection of iron salts can greatly increase the severity of a wide range of bacterial infections. However, in view of the importance of non-protein-bound iron salts, and possibly lactoferrin (Chapter 3), in promoting $OH^{\bullet}$-formation, and of the ability of iron bound to ferritin and haemosiderin to stimulate lipid peroxidation (Chapter 4), it is interesting to speculate that this redistribution of body iron potentiates damage by radical reactions. Indeed, several patients have been discovered who respond to oral doses of iron salts by a flare-up in their rheumatoid disease. Cases of arthritis associated with iron-overload conditions have been reported. It has recently been suggested that the risk of heart disease increases with high body iron stores. The enzyme elastase released by neutrophils appears to play a role in cartilage breakdown in the rheumatoid joint. Although the synovial fluid contains antiproteases, such as α-antitrypsin, the activity of these can be decreased by the action of oxygen radicals, and by myeloperoxidase (Chapter 6). Further, elastase can be released by neutrophils in contact with the cartilage where the antiproteases cannot reach it.

We conclude that oxygen radical reactions play an important role in chronic inflammation, and that further work is required to assess the clinical usefulness of therapy directed against such radicals.

8.2. Ageing

As an organism ages, its chance of death increases, so that all individuals of a given species are dead by some age, characteristic of that species. It seems likely that the maximum lifespan, probably around 90–100 years for humans, is encoded in DNA.

In primitive populations, few individuals reach their maximum lifespan because of infectious diseases or lack of an adequate food supply. In Western societies, such deaths are rare. Those few people who die under the age of 35 usually do so as a result of accidents, whereas older people more often die of diseases such as ischaemic damage to the heart (Chapter 3) or cancer (Section 8.3). Thus the average lifespan (or *mean lifespan*) of 'advanced' societies is greater than that in primitive ones, but the *maximum* lifespan is probably no different.

The nature of the ageing process has been the subject of considerable speculation. One suggested possibility is a progressive breakdown in the accuracy of protein synthesis, so that defective protein molecules accumulate and impair cell function: this is the *error-catastrophe* hypothesis. Another suggestion, *the free-radical theory of ageing*, is that progressive defects in protection against free-radical reactions allow tissue damage to occur. The effects of various toxins (Chapter 6) show that the antioxidant defences of many tissues are sufficient only to protect them against normal rates of radical generation; they cannot cope with increased rates. It seems reasonable to suggest that perhaps they can only cope with, say, 99.9 per cent of normal radical generation, so that there is a very slow progressive damage, undetectable in the short term in humans. In addition, failure of the normal 'self-recognition' mechanisms with increasing age may allow autoantibodies to form (Section 8.1). That radicals are somehow implicated in the ageing process is suggested by the accumulation of 'age pigments' (Section 8.2.1), and by the observation that certain antioxidants prolong the mean lifespan of small animals and some lower life forms (see below). On the other hand, evidence for a failure of radical protective mechanisms is less clearcut. Gershon in Israel observed that superoxide dismutase purified from the livers of old rats is less stable to heating, and of lower activity, than the enzyme from younger rats. The SOD activities of mouse brain, rat lens, and rat erythrocytes have been observed to decrease somewhat with age, although no such change has been observed in several other systems, including human cells in culture or human erythrocytes. Further, the SOD activities of tissues in several animal species showed no correlation to the lifespan of the species; and two different strains of *Drosophila* whose lifespans differed by 40 per cent had the same mitochondrial and total SOD activities. However, SOD is only one factor in protection against oxygen toxicity (Chapter 3) and it is interesting to note that the concentration of glutathione, and the glutathione reductase activities of several tissues, are decreased in old mice.

While maximum potential lifespan may well be genetically determined, the lifespan actually achieved by various organisms can be altered by environmental conditions. Restriction of food intake during the early growth phase of life has been shown to produce statistically significant increases in average lifespan in several species including rats, *Daphnia*, rotifers, and *Drosophila*. The results are complicated and the age at which underfeeding is introduced is critical. The composition of the diet is another important factor.

Lifespan can be affected by temperature. 'Cold-blooded' animals, such as insects and reptiles, live longer at lower temperatures. For example, *Drosophila* has an average lifespan of 120 days at 10 °C, but only 14 days at 30 °C. It seems likely that this effect is due to greater metabolic activity and oxygen consumption at the higher temperatures. Indeed, there is some evidence for an inverse correlation between the basal metabolic rates of animals and their lifespan in that, in general, larger animals consume less oxygen per unit of body mass than do smaller ones, and they live longer. The more oxygen is consumed, the more oxygen radicals will be made *in vivo*. We saw in Chapter 1 that cold-blooded animals are much more resistant to the toxic effects of elevated oxygen concentrations at lower environmental temperatures.

Insects consume much more oxygen when flying than when they are at rest. Prevention of houseflies from flying by removing their wings, or by confining them in small bottles, has been shown by Sohal in the USA to produce a marked increase in lifespan (Fig. 8.2), consistent with the above ideas.

Attempts have been made to test the role of free radicals in the ageing process by supplying antioxidants to various animal species, and examining their effect on longevity. Thiol compounds, such as glutathione and mercaptoethylamine, butylated hydroxytoluene, α-tocopherol, and

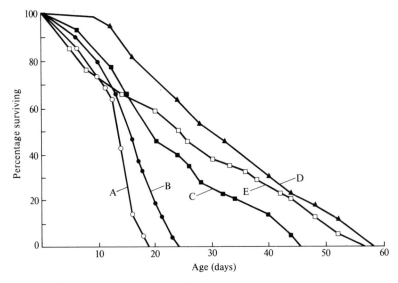

Fig. 8.2. Effect of activity on the lifespan of the housefly, *Musca domestica*. The survival curves of male houseflies under different experimental conditions. A: 50 flies in a large cage. B: 50 de-winged flies in a large cage. C: Each fly in a large cage. (The absence of other flies means that the insect is less disturbed and flies less.) D: Each normal fly in a small bottle to prevent flying. E: Each de-winged fly in a small bottle. Data by courtesy of Dr R. S. Sohal.

Santoquin (ethoxyquin) have been used, and, in several cases, a statistically significant effect has been seen. For example, addition of mercaptoethylamine to the diet of two-month-old LAF_1 mice raised the mean lifespan from 24.5 ± 6.8 months to 27.6 ± 4.1 months (0.5 per cent antioxidant in diet) or to 31.6 ± 6.8 months (1 per cent antioxidant in diet). Adding 0.5 per cent Santoquin to the diet increased the mean lifespan of male CH3 mice by 18 per cent. Vitamin E, by contrast, has no significant effect on lifespan in mice, although it has been reported to prevent the decline in immune response with age in one strain of female mice. However, vitamin E does prolong the average lifespan of several more primitive organisms such as *Drosophila*, nematode worms, and the rotifer *Philodina* (Fig. 8.3). NDGA (Chapter 4) increases the mean lifespan of *Drosophila* by 20 per cent, whereas the singlet oxygen scavengers, DABCO (Chapter 3) and β-carotene, have no effect on the lifespan of this insect. A recent study by Harman in the USA showed that the lifespan of mice can be extended by addition of antioxidants to the diet of the mother during pregnancy.

The effects of antioxidants on the lifespan of mammals are small. It therefore seems unlikely that ageing is due to a genetically-predetermined increase in oxygen radical reactions with time; much larger protective

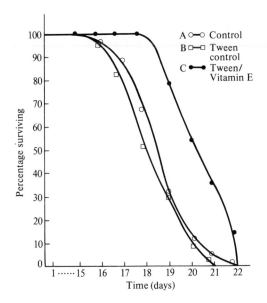

Fig. 8.3. Survival of the rotifer *Philodina* in the presence of vitamin E. Three groups, each of 32 rotifers, were used. Group A is the control group. In group C, vitamin E dissolved in the detergent Tween was added to the growth medium. Group B received Tween alone. Rotifers were counted as dead if they did not move when prodded with a pipette. Data from the article by H. Enesco and C. Verdone-Smith (1980) in *Exp. Gerontol.* **15,** 335–8, with permission.

effects would be expected. The antioxidants may well be acting to diminish tissue damage caused by radical reactions induced by toxins in the food supply, excessive amounts of polyunsaturated fatty acids, cosmic radiation, other ionizing radiation (such as X-rays and gamma-rays), and by exposure to the oxygen in the air, especially if respiratory activity is high. For example, formation of a hydroxyl radical by ionizing radiation (Chapter 2) can initiate lipid peroxidation (Chapter 4), so leading to membrane damage and formation of a range of cytotoxic aldehydes. A lipid-soluble antioxidant, by inhibiting peroxidation, would decrease these 'side-effects' of the hydroxyl radical, and so only the original lipid molecule that it attacks would suffer.

Lipid peroxidation is known to increase in older tissues. For example, the membranes of the oldest erythrocytes in the bloodstream show decreased fluidity and increased cross-linking of membrane proteins. Older rats exhale more hydrocarbon gases (Chapter 4) than do younger ones. The evidence quoted most often is the accumulation of 'age pigment'. Interest has also been expressed in the protein amyloid as an index of ageing.

8.2.1. Age pigments

The first description of the intracellular pigment, now known as *lipofuscin* or *age pigment*, was made in 1842 by Hanover, who reported its presence in the neurones of a number of animal species. The fluorescent properties of the pigment were described in 1911, and it was found to accumulate in amounts increasing with age in many tissues in humans and a wide variety of animals, including rats, nematodes, *Drosophila*, and houseflies, and also in several fungi and in cells in culture. In general, the most metabolically active tissues show most lipofuscin deposition. For example, there is little or no lipofuscin in the human heart muscle up to the second decade of life, but it then accumulates at about 0.3 per cent of the total heart volume per further decade of life. On the other hand, lipofuscin appears earlier in some tissues, such as the spinal cord. In parts of the nervous system of animals and humans, lipofuscin may be impregnated with the pigment melanin to give *neuromelanin*.

The colour of lipofuscin varies from red, through yellow, to dark-brown, and it occurs intracellularly as granules bounded by a single membrane, their diameters being in the range 1–5 µm (Fig. 8.4). Both the

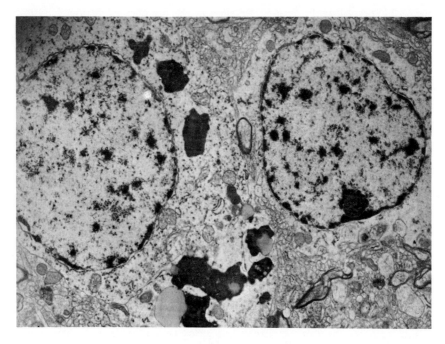

Fig. 8.4. Lipofuscin granules. The dark 'lumps' are lipofuscin granules in the brain tissue of a 53-year-old male. (Magnification ×11 500.) Photograph by courtesy of Prof. D. Armstrong.

number of granules, and their size, increase with age. Extraction of lipofuscin granules with a mixture of organic solvents (chloroform plus methanol) solubilizes part of the material, and the solution so obtained shows fluorescence characteristics quite similar to those of the conjugated Schiff bases formed during lipid peroxidation (Chapter 4). Fluorescent polymers of malonaldehyde (polyMDA) (Chapter 4) have been identified in the age pigments from the fungus *N. crassa*, and may be important in other places. It is therefore thought that lipofuscin represents the end-product of the oxidative destruction of lipids, and their cross-linking with proteins and other compounds bearing amino groups. Many different types of lipid are present in the pigments, including triglyceride, phospholipid, and cholesterol, and an equally wide variety of proteins as judged from the amino-acid composition of hydrolysed pigments. Lipofuscin contains a high concentration of metal ions such as zinc, copper, and especially iron.

Further evidence for the formation of lipofuscin as being due to lipid peroxidation comes from the observation that the amount of this pigment produced in animal tissues is greatly increased if they are fed on diets deficient in vitamin E. The term *ceroid* is sometimes used to refer to this rapidly accumulating age-pigment, which is also seen in some disease states (Section 8.2.3), but there is no real chemical difference between ceroid and lipofuscin (except perhaps that the former is not as extensively cross-linked). Increased lipofuscin (or ceroid) deposition can also be achieved by feeding animals on diets abnormally rich in polyunsaturated fatty acids. Such feeding induces a 'yellow fat disease' in, for example, pigs and minks that can be prevented by feeding excess vitamin E or other antioxidants such as *N,N'*-diphenyl-*p*-phenylenediamine (Chapter 4). In houseflies, the lipofuscin content of the muscles increases with flight activity: flies that are more active live for shorter times (Fig. 8.2) and accumulate pigment faster.

Deposition of age pigment is not an irreversible process. For example, the drug centrophenoxine, also known as meclophenoxate, lucidril, or dimethylaminoethyl-*p*-chlorophenoxyacetate, has been observed to decrease the number of age-pigment granules in the central nervous system of animals *in vivo*. Its mode of action is unclear, although it has sometimes been prescribed for elderly patients as a tonic.

Although there seems little doubt that lipofuscin and ceroid represent products of lipid peroxidation, it is not clear exactly how the granules are formed. Histochemical studies strongly suggest that they are derived from lysosomes. The cellular lysosomes are continually digesting parts of the cell cytoplasm, a process known as 'autophagy'. Proteins and lipids are taken into the lysosomes, which are bounded by a single membrane, and degraded. It is possible that lysosomes have a special affinity for peroxidized lipids and so gradually accumulate them; but an alternative, and perhaps more likely, explanation would be that lipids taken into the lysosomes are peroxidized more rapidly than normal. Disruption of membrane structure by lysosomal hydrolytic enzymes should facilitate lipid peroxidation, as would the high internal concentrations of copper and iron salts within lysosomes. These metal ions are probably derived from ingested metalloproteins, including transferrin (Chapter 2). The association of age pigments with lysosomes is seen especially in *Chédiak–Higashi syndrome*, a rare human disease, recessively inherited, that is characterized by the presence of abnormally large cytoplasmic granules in several tissues. These granules are of lysosomal origin, and frequently enclose large amounts of pigmented lipid material. Their presence in neutrophils is correlated with impaired bactericidal capacity, causing recurrent infections. Other symptoms include defective skin pigmentation and nervous disorders. The primary cause of Chédiak–Higashi syndrome is unknown.

Lipofuscin (ceroid) deposition is promoted by a number of abnormalities of fat metabolism, including *abetalipoproteinaemia*. In this disease, dietary fat is digested and absorbed, but held up in the intestinal mucosal cells. Presumably more lipid than usual is degraded within lysosomes in such diseases. Collins in Sweden exposed human glial cells in culture to rat liver mitochondria, some of which could be taken up by phagocytosis. The result was an increased accumulation of age pigment in the glial cells, presumably as the mitochondrial lipids were degraded within lysosomes.

8.2.2. Amyloid

Amyloid is a predominantly extracellular fibrous material, composed mainly of protein. It is deposited in tissue in a number of diseases and to a much smaller extent in normal tissues during the ageing process (e.g. it is found in the seminal vesicles of 21 per cent of men over the age of 75). Diseases associated with amyloid deposition include tuberculosis, syphilis, rheumatoid arthritis, and multiple myeloma. Several, rare, inherited types of severe amyloid deposition have been described, such as familial Mediterranean fever, found principally in Mediterranean Jews and Armenians. This disease is eventually fatal, often because of kidney damage.

Amyloid deposition secondary to illness is called *secondary amyloidosis,* whereas *primary amyloidosis* refers to amyloid deposition in the absence of a known predisposing factor, such as during ageing, or in inherited amyloid deposition. Organs severely affected by amyloid deposition are enlarged, abnormally firm, and their cut surfaces are waxy in appearance. Severe amyloid deposition interferes with the normal passage of water and solutes across the walls of blood vessels. In amyloidosis secondary to multiple myeloma, the amyloid deposit appears to be partially derived from antibody molecules, but this is not the case in other diseases. Serum amyloid A, an acute-phase protein synthesized in increased amounts during tissue injury (Chapter 4), may be an important component of other amyloid deposits but different proteins have been found in some deposits. Harman, in the USA, observed that Santoquin, which increases the mean lifespan of LAF_1 mice (see above), completely prevented spontaneous amyloidosis in these mice. The mechanism of this action is unknown.

8.2.3. 'Premature ageing' disorders

Neuronal ceroid lipofuscinosis (NCL)

'Neuronal ceroid lipofuscinosis' refers to a series of recessively inherited disorders that occur world-wide in about 1 per 100 000 people, but more frequently than this in Scandinavia. Some patients develop symptoms during infancy (infantile NCL), others in early childhood (late infantile NCL), others in late childhood (juvenile NCL), and yet others after adolescence (adult NCL). Juvenile NCL is sometimes known as *Batten's Disease*.

The onset of the disease is marked by behavioural abnormalities which worsen to include disturbances of vision and speech, as well as muscular and mental deterioration, associated with seizures. In the terminal stages the brain is severely damaged and patients assume a contracted position (Fig. 8.5). A similar disease has been observed in English red setter dogs. There is a rapid accumulation of age pigments (Fig. 8.6), not only in the brain but also in other tissues. Hence the 'neuronal' part of the name is slightly misleading. The cause of NCL disease is unknown. No significant decreases in catalase, superoxide dismutase, or glutathione peroxidase activity have been observed in any of the body tissues studied. In collaboration with Dr. T. Westermarck in Finland, the authors have observed an increased content of non-protein-bound iron salts in the cerebrospinal fluids of patients with juvenile NCL, which should promote lipid peroxidation *in vivo*. This is associated with a decreased antioxidant protection against iron-dependent hydroxyl radical formation when these fluids are examined *in vitro*. Whether the rise in non-protein-bound iron is a cause or a consequence of the disease remains to be established. In a regime pioneered by Dr. Westermarck, many of these patients respond to treatment with selenium. It is possible that the defect is either one in lipid metabolism or in iron metabolism, and the accumulation of age pigment is secondary to this.

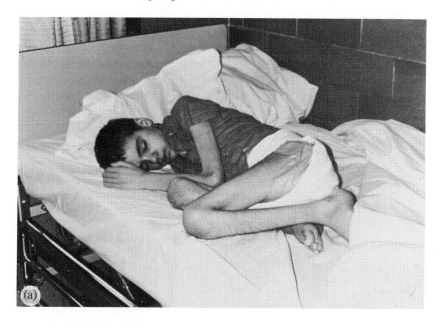

Fig. 8.5. Neuronal ceroid lipofuscinoses. (a) Shows the terminal clinical condition of a patient with juvenile NCL. (b) Shows a similar condition in an English setter dog. The dog, having been placed in such an awkward position, will not move, indicating the extensive damage suffered by the brain. Photograph by courtesy of Prof. D. Armstrong.

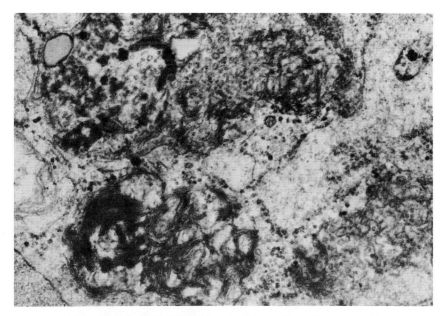

Fig. 8.6. Neuronal ceroid lipofuscinosis. The figure shows lipofuscin (ceroid) arranged in a 'fingerprint' manner in brain tissue from a 7-year-old child with juvenile form of the disease. Photograph by courtesy of Prof. D. Armstrong.

Senile dementia and Alzheimers disease

Some shrinkage in the size and weight of the brain (*atrophy*) is normal in elderly adult humans. Severe atrophy is usually associated with loss of mental function, the condition being termed *senile dementia*. It is increasingly common in people aged over 70, and is becoming a serious problem as the average lifespan increases. In *Alzheimers disease*, the pathology is similar, but the changes occur at lower ages. In both cases the pathological findings resemble a gross exaggeration of those seen in the normal ageing process of the brain. There is a marked loss of tissue, especially in the part of the brain known as the 'nucleus basilis', and the deposition of large numbers of 'senile plaques' in the cortical grey matter, but a greater than normal accumulation of lipofuscin has not yet been rigorously demonstrated. The plaques often have a core of amyloid material. The ability of the brain to synthesize acetylcholine is greatly diminished. The basic cause of the conditions is unknown, nor is it clear if antioxidants would be helpful in prevention or treatment. In *Huntington's chorea*, a disease inherited dominantly, that appears in middle age, there is progressive mental and physical degeneration, leading to death, and a massive accumulation of age pigment in the brain. For example, the content of age pigment in the brain of one such patient was equivalent to the level found in human brain extrapolated to 195 years of age. Again, the cause of this unpleasant disease is unknown.

Model systems

Munkres in the USA has described a 'rapidly-ageing' mutant of the fungus *Neurospora crassa*. The mutant ages, and accumulates age-pigment much more rapidly than does the wild-type strain. However, accumulation of lipofuscin is decreased, and lifespan increased, by including in the culture medium antioxidants such as NDGA, thiol compounds, or vitamin E. This mutant has increased activities of superoxide dismutase, catalase, and glutathione peroxidase relative to the wild-type fungus. Perhaps the primary defect of the mutant is an increased rate of radical generation, which stimulates the synthesis of the above protective enzymes, although not to a sufficiently great extent. Munkres has suggested that the mutant is unable to 'organize' its cell-membranes properly, leading to an increased rate of lipid peroxidation. Indeed, a *N. crassa* mutant that was unable to synthesize inositol (Chapter 4) for incorporation into its membrane lipids was observed to show increased lipid peroxidation when placed on a medium deficient in this substance. Antioxidants offered some protection against the damage.

8.3. Cancer

A tumour may be defined as an abnormal lump or mass of tissue, the growth of which exceeds, and is uncoordinated with, that of the normal tissue, continuing after the stimuli that initiated it have ceased. Most tumours form discrete masses, but in the *leukaemias* (tumours of myeloid or lymphoid cells) the tumour cells are spread through the bone-marrow or lymphoid tissues, and also circulate in the blood. Tumours vary widely in their growth rates. The most important classification of tumours is that of benign or malignant. The cells of benign tumours remain at the site of origin, forming a cell mass. When growing in a solid tissue, they usually become enclosed in a layer of fibrous material, the capsule, formed by compression of the surrounding tissues. Benign tumours rarely kill, unless they press on a vital structure or secrete abnormal amounts of hormones.

Most fatal tumours are *malignant*, or *cancerous*. The cells of malignant tumours invade locally, and also pass through the bloodstream and lymphatic system to form secondary tumours (*metastases*) at other sites. The rate of growth, and metastasis formation, differs from tumour to tumour. For example, breast cancers often grow very slowly, as do many rodent ulcers. Rapidly growing malignant tumours usually lose their histogical resemblance to their tissue of origin. Basal cell cancer of the epidermis rarely metastasizes, whereas melanoma frequently does.

The process of conversion of a normal cell to the malignant state is called *carcinogenesis,* and agents that induce it are called *carcinogens.* Carcinogenesis is a very complicated process: factors predisposing to malignancy include both inherited traits and environmental factors. Some people are much more sensitive to environmental carcinogens than are others, for example. A low dose of a carcinogen, itself too small to induce a tumour, can be effective if supplied together with certain non-carcinogenic substances known as *promoters.* For example, croton oil, a non-carcinogen, promotes the development of cancer by sub-carcinogenic doses of the hydrocarbon methylcholanthrene, but only if the oil is given simultaneously with, or after, the methylcholanthrene. Croton oil is obtained from seeds of the plant *Croton tiglium.* Fractionation of the oil has shown that the most powerful tumour promoter present is phorbol myristate acetate (PMA, Fig. 8.7), a compound that induces a respiratory burst in phagocytic cells (Chapter 7). It is interesting to note in this context than many tumour promoters, including PMA, are potent inducers of inflammation.

Dietary factors exist that depress the action of certain carcinogens. These include the synthetic antioxidants BHA, BHT, and ethoxyquin (Chapter 4), which probably act by stimulating the synthesis of enzymes that convert ingested carcinogens into harmless products. Other such dietary agents include vitamin A, β-carotene, and selenium. Selenium-deficient animals seem more sensitive to some carcinogens, especially if they are fed a diet rich in polyunsaturated fats. Indeed, there have been suggestions that excessive consumption of polyunsaturated fats, by humans and animals, predisposes to cancer, although they are not well supported by the available evidence. Lipid peroxides, and several carbonyl compounds derived from them (Chapter 4), have been reported to be carcinogenic in some animals, although claims that malonaldehyde is a carcinogen have been disputed (Chapter 4). The organic peroxide

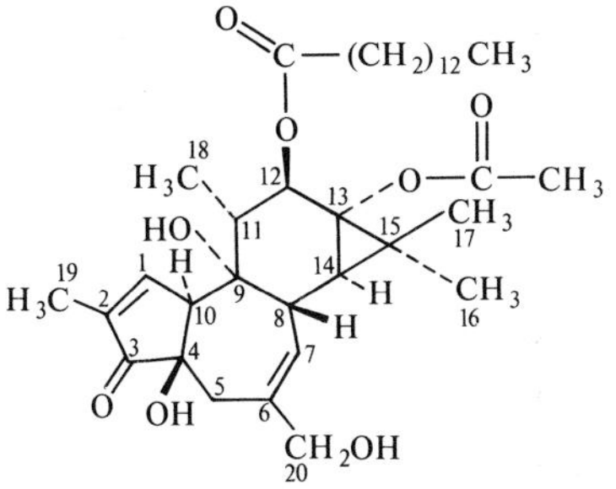

Fig. 8.7. Structure of PMA (phorbol myristate acetate). The proper chemical name of this compound is 12-*O*-tetradecanoylphorbol-13-acetate.

benzoylperoxide has been reported to act as a tumour-promoter in some experiments. Claims that ascorbic acid prolongs the survival time of human cancer patients have not been rigorously supported by experimental evidence, however, although ascorbate does decrease the action of several carcinogens (see below). On the other hand, diets *deficient* in vitamin B_2 (riboflavin) have been reported to slow the growth of tumours in experimental animals.

Free-radical reactions have been suggested to be involved in the initiation of cancer, in 'immune surveillance', in the mode of action of some carcinogens, and in the action of antitumour drugs. Let us now look at the evidence available for the role of radical reactions in these processes. Their importance, in cancer therapy using ionizing radiation and hypoxic cell-sensitizers, has already been discussed (Chapter 6).

8.3.1. Free radicals and initiation of cancer

The interplay of a number of elements, including genetic factors, is responsible for the development of malignant disease. Viruses have been observed to cause leukaemia in a number of animal species (e.g. feline leukaemia virus in cats), and their possible role in human cancer is being intensively investigated. It has been found that a number of human and animal cancer cells contain genes (*oncogenes*) that can be transferred into cultured cells which then undergo cancerous transformation. For one of these genes, found in a line of human bladder cancer cells, the difference between the oncogene and the normal gene is a single nucleotide change, replacement of a guanine by a thymine base. How this change causes transformation is unclear. A few human tumours develop mainly, or wholly, as a result of a genetic anomaly which makes it almost certain that affected individuals will develop a particular malignancy. *Familial polyposis coli,* in which colon cancer almost invariably develops, is an example. *Xeroderma pigmentosum*, an inborn defect in the mechanism that repairs damaged DNA, manifests itself as severe skin damage and development of skin cancer on exposure to sunlight or other sources of ultra-violet light. Certain families have a higher than average incidence of cancer.

In view of the importance of genetic factors in the origin of cancer, it seems likely that any substance capable of reacting with DNA, and producing chemical changes that might lead to expression of oncogenes, would be potentially carcinogenic. The same comment would apply to ionizing radiation. Hence all mutagens should be thought of as carcinogens unless proved otherwise. Indeed, many of the drugs used in cancer chemotherapy (Section 8.3.4) are powerful mutagens, and may thus be carcinogenic. A mutagen has even been detected in the faeces of 3 per cent of the North American population.

Exposure of DNA to γ-irradiation not only damages it but also produces a series of mutagenic products, apparently derived from the attack of hydroxyl radicals upon the sugar deoxyribose. The ability of oxygen radicals, such as OH·, to damage DNA fits in well with many observations on bacteria, and upon animal and plant cells in culture, that show a mutagenic effect of elevated oxygen concentrations. Indeed, exposure of bacteria or cultured mammalian cells to activated human neutrophils is known to produce DNA damage, and the DNA within the neutrophils themselves is fragmented during phagocytosis, presumably as O_2^- and hydrogen peroxide interact *in vivo* to form hydroxyl radicals. This may tend to promote the development of cancer in chronically inflamed tissues.

Several rare diseases associated with chromosome abnormalities, such as Down's syndrome (mongolism), Bloom's syndrome, and Fanconi's syndrome, carry an increased risk of cancer. The number of chromosomal aberrations in lymphocytes, taken from patients with the latter syndrome, is strikingly decreased by lowering the oxygen concentration in the culture medium, which suggests a possible defect in the protection of these cells against oxygen radical damage. Indeed, the SOD activity of the red blood cells (and possibly other cells as well) is decreased by 30–40 per cent in Fanconi patients, whereas the activities of catalase and glutathione peroxidase are normal. The erythrocyte SOD present has normal specific activity and mobility on electrophoresis, so there is no evidence for an inborn error in the SOD gene. Cultures of fibroblasts from patients with Bloom's syndrome have been reported to release a clastogenic factor that can induce chromosome breaks in normal lymphocytes, an effect that is suppressed by including SOD in the culture medium. Similar clastogenic factors have been described in some autoimmune diseases (Section 8.1). On the other hand, in Down's syndrome, the SOD content of erythrocytes, leukocytes, fibroblasts, platelets and probably other tissues, is increased by about 50 per cent. In this condition, three number-21 chromosomes are present instead of the normal pair, although it is not clear how this leads to the physical and mental defects observed. The increased content of SOD is accounted for by the fact that the gene for the copper–zinc enzyme is located on chromosome-21. The activity of mitochondrial (manganese) SOD in platelets is decreased by 30 per cent, and there are no changes in the activities of catalase, glutathione reductase, and the pentose phosphate pathway enzymes in erythrocytes, although there may be some increase in erythrocyte glutathione peroxidase activity. It would be interesting to know if manganese SOD activities are decreased in tissues other than platelets in Down's syndrome patients.

Mitochondria from several malignant animal tumours, and from tumour cells in culture, are deficient in manganese SOD activity when

Table 8.2. Activities of 'antioxidant' enzymes and GSH concentrations in some tumour cell lines

Cell line	Catalase (units mg^{-1} protein)	SOD (μg mg^{-1} protein)	Glutathione peroxidase (units mg^{-1} protein)	Glutathione reductase (units mg^{-1} protein)	G6PDH[1] (units mg^{-1} protein)	GSH (μg mg^{-1} protein)
A	<0.02	0.25	11	0.04	0.05	6.6
B	<0.02	0.53	36	0.18	0.11	18.1
C	<0.02	0.50	13	0.06	0.16	24.4

[1] Glucose-6-phosphate dehydrogenase, the first enzyme of the pentose phosphate pathway.

Data were abstracted from Bozzi *et al.* (1979) *Cancer Biochem. Biophys.* **3**, 135–41. Line A was Ehrlich ascites cells; B was Yoshida ascites cells; C, Novikoff ascites cells.

compared to mitochondria from normal tissues. Lowered activities of copper–zinc SOD are often seen in tumours, but not in all cases. These observations led Oberley and Buettner, in the USA, to propose that decreased antioxidant protection, together with increased radical generation, might explain many of the properties of cancer cells. Many animal tumour cell lines are low in catalase activity also (Table 8.2). Oberley further observed that growth of a tumour model system in mice was decreased by injection of a low-molecular-weight copper chelate that can scavenge O_2^-, copper(II)(3,5-diisopropylsalicylate)$_2$. This compound has also been reported to decrease the tumour-promoting effect of a phorbol ester in mice. In mice bearing transplantable Ehrlich ascites tumours, tissues other than the tumour, such as liver, spleen, and lung, have decreased manganese SOD activity. This implies an effect of the tumour on these other tissues, perhaps mediated by a circulating factor.

However, studies carried out on human malignant tumours biopsied during surgery have shown no significant decrease in manganese SOD, and human tumour cell lines in culture are not generally deficient in copper–zinc SOD, manganese SOD, catalase, or glutathione peroxidase activities. It should be borne in mind that the centre of a large tumour mass often has a poor blood supply, and is thus anoxic, which may decrease its SOD activity (Chapter 3). For example, Petkau *et al.*, in Canada, showed that the SOD content of a rat breast cancer was 54 ($\pm$10) μg g^{-1} of tissue at the centre of the tumour, and 117 ($\pm$38) μg g^{-1} at the periphery. Exposure of the tumour-bearing rats to elevated oxygen concentrations raised these values to 162 ($\pm$73) and 286 ($\pm$103) μg g^{-1} respectively.

The role of transition-metal ions, such as iron, in the conversion of O_2^- into more reactive radicals such as OH$^•$ (Chapter 3) is of particular interest because malignant disease, like chronic inflammation, produces changes in body iron distribution. Iron is lost from the blood (e.g. the percentage iron saturation of circulating transferrin drops markedly) and it accumulates in liver, spleen, and bone-marrow. Perhaps the organism is attempting to withhold iron from the tumour and so slow its growth since, in general, tumours seem to contain less 'total iron', and have a lower degree of iron-saturation in ferritin than do normal tissues. This is not always the case, however, since human breast tumours appear to accumulate iron, and in Hodgkin's disease, heavy deposits of iron and ferritin are seen surrounding the tumour nodules. Many cases of liver cancer have been documented in iron-overload patients. In some cancers, including Hodgkin's disease, breast cancer, and leukaemia, the normally low concentrations of ferritin present in the blood are greatly increased. A number of metals, such as nickel salts, can damage DNA if supplied to cells in culture; and Shires, in the USA, observed that incubation of Fe^{2+} salts with isolated rat liver nuclei causes extensive oxygen-dependent DNA fragmentation, an observation perhaps relevant to iron-overload conditions (Chapter 2).

Several groups have attempted to detect radical reactions by subjecting tissue samples to electron spin resonance studies (Chapter 2). For example, Slater in England, and Benedetto *et al.* in Italy, observed that frozen samples of normal human cervix and endometrium give strong ESR signals, which are much less marked in cancerous tumours of these tissues. The signal has been attributed to an organic peroxy radical (Chapter 4); but this is not rigorously proved. ESR studies on freeze-dried blood samples seem to detect the semidehydroascorbate radical (Chapter 3), the amounts of which change during leukaemia.

In general, the results summarized above do not support 'increased radical reactions' or defects in antioxidant protection as a general mechanism for the origin of cancer, although increased oxygen-radical generation can lead to DNA damage, and is thus potentially carcinogenic.

8.3.2. Immune surveillance

There is considerable evidence that the immune response of the host influences the development of malignant tumours. For example, there is an increased risk of malignancy in patients undergoing immunosuppressive therapy. The overall increased risk of skin cancer in kidney-transplant patients, treated with immunosuppressive drugs to prevent rejection of the 'foreign' kidney, has been estimated as over seven-fold. Certain lifestyles, such as promiscuous homosexuality, can occasionally lead to deficiencies in the immune system, as manifested by life-threatening infections with organisms that are normally easily dealt with (such as *Pneumocystis carinii*, which causes a form of pneumonia) and by predisposition to certain unusual cancers such as Kaposi's sarcoma.

Burnet originally suggested that one of the functions of the immune system is to detect and destroy abnormal cells, a concept referred to as *immune surveillance.* Although such surveillance is by no means efficient against all tumours, in that many cancers develop without any apparent defect in the immune system, there is no doubt that cells of the immune system can sometimes attack cancer cells. In 1973, Edelson and Cohn, in the USA, demonstrated that the myeloperoxidase–H_2O_2–halide system can kill a number of tumour cells, and monocytes, neutrophils, and macrophages have all been shown to destroy tumour cells during the respiratory burst *in vitro.* Macrophages are especially effective. The killing mechanism probably involves released enzymes and oxygen-derived species such as hydrogen peroxide and O_2^-. Many animal tumour cells are especially sensitive to hydrogen peroxide because they contain little, if any, catalase activity and often only low activities of enzymes associated with glutathione metabolism (Table 8.2). Depletion of the GSH content of a tumour cell line by chemical manipulation has been shown to increase the rate at which the cells can be lysed by activated neutrophils or macrophages.

It is, of course, necessary to activate the phagocyte respiratory burst, either by coating the tumour cells with antibody that the phagocyte can recognize, or by supplying another inducer. Hence the efficiency of tumour cell destruction *in vivo* will depend upon the circumstances, not least the extent to which the tumour cells have foreign antigens on their surface that cause antibody production. Merely exposing tumour cells to neutrophils or macrophages does not induce killing, i.e. cancer cells themselves do not provoke a respiratory burst.

Another cell apparently involved in attack against tumour cells is a type of lymphocyte known as the *natural killer* (NK) cell. Human NK cells are large lymphocytes with a single cytoplasmic granule. They can attack a wide variety of cells, including certain types of cell in the normal bone-marrow and thymus, cells infected by viruses or by certain parasites, and some tumour cells. No preexposure of the target cells to antibody is necessary. Little is known as yet about the mechanisms by which NK cells recognize their targets. Claims that O_2^- or hydroxyl radicals produced by NK cells are responsible for target-cell killing remain to be substantiated. NK cells additionally contain various proteolytic enzymes that might be involved in killing. The number of tumour cells that are destroyed *in vivo* by NK cells has yet to be established.

8.3.3. Carcinogens

About one-third of all cancer cases, in Europe and North America, can be related to the presence of carcinogens in cigarettes and other tobacco

Table 8.3. Some chemical carcinogens

Industrial chemicals

Chemical	Site of cancer	Chemical	Site of cancer
Aromatic amines	Bladder	Arsenic	Skin, bronchus
Asbestos	Bronchus, pleura	Benzene	Bone-marrow
Tars, oils	Skin, lungs	Vinyl chloride	Liver
Diethylstilboestrol	Vagina	Aromatic hydrocarbons	Lung

Naturally occurring chemicals

Chemical	Site of occurrence	Site of cancer
Cyasin	Cycad plants	Liver, kidney, intestine (rats)
Not yet purified	Bracken fern	Bladder, intestine (rats, cows)
Not yet purified	Betel nut	Mouth (betel nut and leaves are often chewed in Southeast Asian countries)
Safrole	Oil of sassafras	Liver (rats)
Aflatoxins (several known, most active is B_1)	*Aspergillus flavus* (fungus)	Liver (rats)

products (Chapter 6). A vast range of naturally occurring and synthetic chemicals have been shown to be carcinogenic; and Table 8.3 lists a few of them.

Some chemical carcinogens, such as nitrosoureas, nitrosoguanidine, or nitrogen mustard, are highly reactive, and can attack and modify DNA directly. Most, however, have to undergo metabolism before they become active. The necessary metabolism is sometimes carried out by gut bacteria. For example, cycasin (Table 8.3) is hydrolysed in the gut to give the true carcinogen. More often, host enzymes are involved. The first example of metabolic activation to be discovered was the conversion of the aromatic amine 2-acetylaminofluorene (Fig. 8.8) to an N-hydroxylated derivative. This type of metabolism occurs with several aromatic amines including 2-naphthylamine, which induces cancer of the bladder in humans. The N–OH product can then undergo several reactions, including combination with sulphate to give an N-O-sulphate ester that is probably the ultimate carcinogen. It can also be oxidized to an N—O$^{\cdot}$ radical by peroxidase enzymes. The N-hydroxylation is catalysed by the cytochrome P_{450} system. Cytochrome P_{450} is also involved in conversion of the fungal product aflatoxin B_1 (Table 8.3) into the true carcinogen, an epoxide (Fig. 8.8) that combines with DNA. P_{450} converts vinyl chloride into a carcinogenic epoxide, and it acts upon dimethylnitrosamine

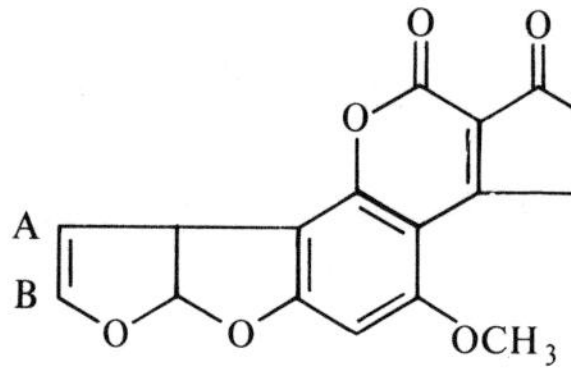

Aflatoxin B$_1$

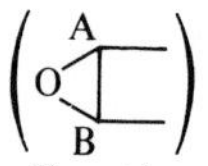

Epoxide

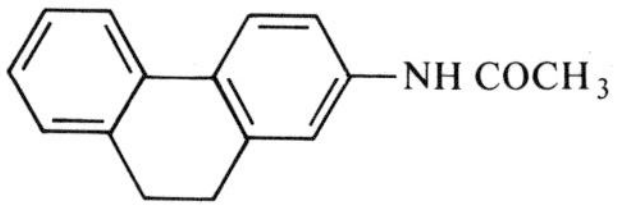

2-Acetylaminofluorene   —NH COCH$_3$

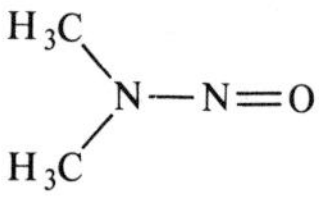

Dimethylnitrosamine

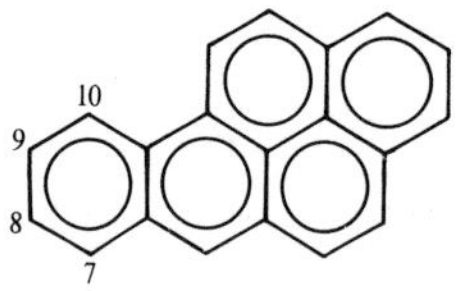

Benzpyrene

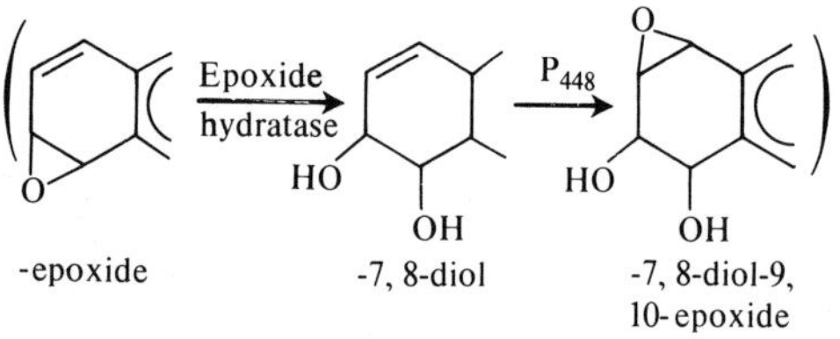

Epoxide hydratase   →   P$_{448}$ →

-epoxide   -7, 8-diol   -7, 8-diol-9, 10-epoxide

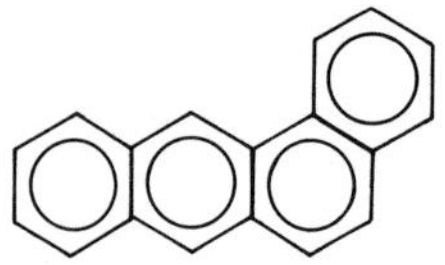

Benzanthracene

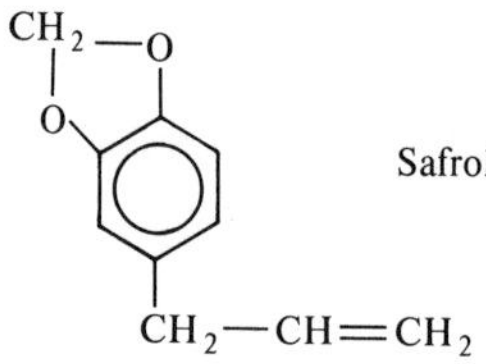

Safrole

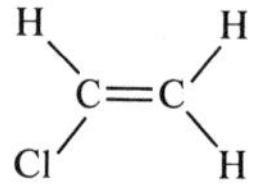

Vinyl chloride

CH$_3$—NH—NH—CH$_3$ 1,2-Dimethylhydrazine

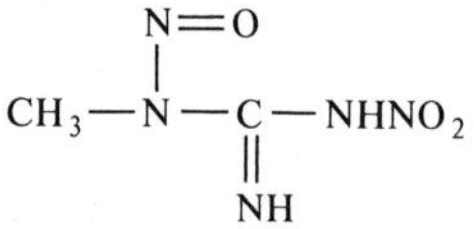

CH$_3$—N—C—NHNO$_2$ N-Methyl-N'-Nitronitrosoguanidine

Fig. 8.8. Structure of some carcinogens.

(Fig. 8.8) to form methyl ions (CH_3^+) that combine with guanine residues in DNA (Chapter 3).

Perhaps the most studied reaction of the cytochrome P_{450} system is its ability to metabolize benzpyrene, benzanthracene (Fig. 8.8), and other carcinogenic hydrocarbons into the 'true' carcinogens. These, and several other hydrocarbons, are generated on combustion of organic materials, e.g. petrol in car engines, and are thought to be responsible, at least in part, for the carcinogenicity of coal-tar and cigarette tar. *In vivo*, benzpyrene is converted into a wide variety of metabolites including a 6-hydroxy derivative that rapidly oxidizes into a mixture of benzpyrene quinones. These quinones can be reduced by GSH, NADH, or NADPH into diols that can then re-oxidize, producing O_2^- and hydrogen peroxide that might contribute to DNA damage.

However, the most carcinogenic metabolite of benzpyrene is the 7,8-diol-9,10-epoxide (Fig. 8.8). Cytochrome P_{450}, or probably one (or more) special form(s) of it, known collectively as cytochrome P_{448}, converts benzpyrene into a 7,8-epoxide. This is acted upon by the enzyme *epoxide hydratase*, to form the 7,8-diol that is itself a substrate for further epoxidation (Fig. 8.8). Epoxidation of the benzpyrene molecule at other positions occurs as well, although the products formed are less carcinogenic. The peroxidase activity of prostaglandin synthetase (Chapter 7) is capable of hydroxylating benzpyrene to 6-hydroxybenzpyrene, and converting the 7,8-diol into the 7,8-diol-9,10-epoxide *in vitro*, probably by the action of $[O]_x$. Peroxidizing microsomes have also been reported to epoxidize the 7,8-diol. The roles played by lipid peroxidation and prostaglandin synthetase in metabolizing benzpyrene in different tissues, as compared with metabolism by 'cytochrome P_{448}', have not been established. Prostaglandin synthetase can also co-oxidize benzidine, 2-aminofluorene, 2-naphthylamine, and 2,5-diaminoanisole to mutagenic products.

As well as forming diols, epoxides are substrates for the glutathione-S-transferase enzymes (Chapter 3), being converted into glutathione derivatives than can be excreted. This, and several other metabolic pathways, represent detoxification mechanisms, and so the overall effect of benzpyrene and other carcinogens *in vivo* probably depends on the balance between activation and detoxification mechanisms. A number of antioxidants, such as BHA and BHT (Chapter 4), decrease the carcinogenicity of benzpyrene and other metabolites by stimulating detoxification pathways. Ascorbic acid decreases the carcinogenicity of nitroso-compounds by reducing the nitroso group (Fig. 8.8) into an inactive product. GSH, and possibly other thiol compounds, can combine directly with reactive intermediates, GSH also being a substrate for transferase enzymes.

8.3.4. Anti-tumour drugs

The object of cancer chemotherapy is to kill cancer cells, with as little damage as possible to normal cells. In cancerous tumours, many cells are dividing, and so most drugs used are designed to interfere with cell growth and division by blocking synthesis of DNA, RNA, or protein. Examples of such drugs include methotrexate, which interferes with synthesis of DNA precursors, and cytosine arabinoside, which contains an arabinose sugar instead of ribose, and which blocks DNA polymerase activity. Alkylating agents such as nitrogen mustard and cyclophosphamide chemically modify DNA and RNA, interfering with replication, transcription, and translation. *cis*-Platinum (*cis*-diamminedichloroplatinum(II), see Fig. 8.10) probably acts by cross-linking guanine residues in DNA. Hydroxyurea inhibits ribonucleoside diphosphate reductase (Chapter 8). Vincristine and vinblastine interfere with spindle formation during mitosis. Of course, any normal cells undergoing division will also be damaged, as in the intestinal epithelium, hair follicles, and bone marrow. These side-effects limit the dose frequency, and size of dose that can be administered. Most of the agents that chemically modify DNA are themselves mutagenic, and thus can be regarded as potentially carcinogenic, meaning that they must be handled with care.

A number of anti-tumour antibiotics occur naturally and can be harvested after microbiological fermentation. Usually they act by binding to DNA and interfering with replication and/or transcription or causing strand breakage. The anti-tumour antibiotics can be grouped under several distinct chemical types. These include the anthracylines and other quinone-containing drugs (such as mitomycins, streptonigrin, daunomycin, and doxorubicin, otherwise known as adriamycin), the metal-chelators (such as tallysomycin and the bleomycins), the protein anti-tumour antibiotics (such as macromomycin and neocarzinostatin), and aureolic-acid-based antibiotics (such as mithramycin, chromomycins, and olivomycins). It is in the mode of action of anti-tumour antibiotics that free radical reactions have been suggested to be most important. Let us look at a few examples in detail.

Bleomycin

Bleomycin is a glycopeptide antibiotic produced by *Streptomyces verticillus*. It was first isolated by Umezawa in Japan. The clinical preparation commonly used, Blenoxane, is a mixture of several bleomycins that differ slightly in structure, although bleomycin A_2 is the major component (Fig. 8.9). Bleomycins are active against several human cancers, including Hodgkin's disease and cancer of the testes.

Bleomycins produce their action by binding to DNA, especially

Bleomycin A$_2$

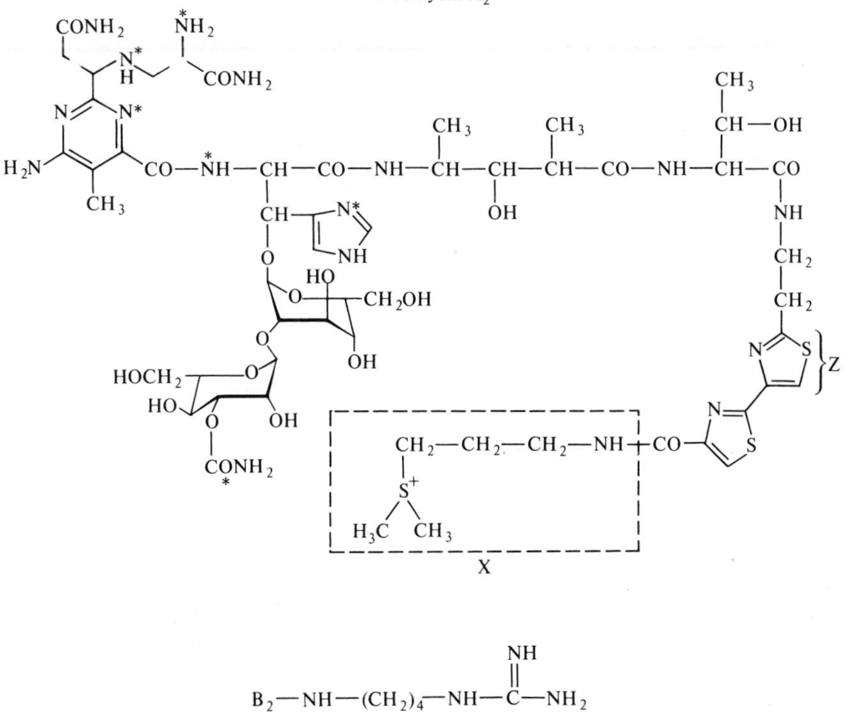

Fig. 8.9. Structure of bleomycin A$_2$. In bleomycin B$_2$ the substituent shown replaces the terminal group (X) of bleomycin A$_2$. The asterisks denote the atoms that probably interact with bound transition-metal ions. The *phleomycin* antibiotics differ from the bleomycins only in the absence of one of the double bonds in the ring marked Z. Phleomycins have not so far been used clinically because they cause kidney damage. Bleomycin is usually supplied by the manufacturers as a sulphate salt.

adjacent to guanine residues. They cause single-strand, and some double-strand, breaks and degradation of the deoxyribose sugar to form products that react with thiobarbituric acid to give a pink colour (Chapter 4). Bleomycins are powerful chelators of transition-metal ions, such as Cu^{2+}, Co^{2+}, Zn^{2+}, Fe^{2+}, Fe^{3+}, Ni^{2+}, by donation of electrons from nitrogen atoms and from a $>$C$=$O group to the metal ion (Fig. 8.9).

In vitro and presumably *in vivo* as well, only the bleomycin–Fe^{2+} complex is capable of degrading DNA. DNA-degradation additionally requires oxygen, and so bleomycin-toxicity is reduced if cells are kept under hypoxic conditions. The bleomycin–Fe^{2+} complex may be generated by adding Fe(II) salts to bleomycin or by reducing a bleomycin–Fe^{3+} complex with such biological reducing agents such as ascorbate, GSH, or a system generating superoxide radicals (O_2^-). Complexes of bleomycin with metals other than iron are ineffective in degrading DNA.

Incubation of bleomycin with an Fe(II) salt in aqueous solution causes formation of superoxide (O_2^-) and hydroxyl ($OH^\bullet$) radicals, detected by spin-trapping experiments (Chapter 2). In such incubations, the drug itself undergoes chemical modification that destroys its activity, but it can be protected by addition of DNA. It is tempting to suggest that binding of a bleomycin–Fe^{2+}–O_2 complex to the DNA causes formation of $OH^\bullet$ that reacts immediately with the DNA molecule, although this is difficult to prove. Radical scavengers such as superoxide dismutase, low-molecular-weight scavengers of O_2^- (Chapter 3), catalase, caeruloplasmin, or the $OH^\bullet$ scavengers methanol, thiourea, mannitol, or dimethylsulphoxide offer little, if any, protection against bleomycin damage to DNA. Indeed, propyl gallate and other phenolic antioxidants with reducing properties make the damage much worse! Removal of iron from the complex by chelating agents such as EDTA, DETAPAC, and desferrioxamine (Chapter 2) prevents DNA degradation completely, however. Indeed, the authors have used the iron-requirement of bleomycin as a means of measuring the micromolar iron content of body fluids (Chapter 2). As an explanation of the lack of effect of $OH^\bullet$ scavengers, it could be argued that $OH^\bullet$ is formed so close to the DNA that it cannot be effectively scavenged. Alternatively, the damaging species might not be $OH^\bullet$ but perhaps a ferryl complex, or something similar (Chapters 3 and 4).

A major side-effect of therapy with bleomycins is lung damage. The lung appears to lack an enzyme named 'bleomycin hydrolase' which breaks down the bleomycin molecule into an inactive form. Lung damage may again involve radical reactions, since microsomal fractions from rat lungs can reduce Fe^{3+}–bleomycin in the presence of NADPH, forming Fe^{2+}–bleomycin and hence oxygen radicals. Consistent with this, bleomycin-induced lung damage in hamsters is increased by exposure to elevated oxygen concentrations. The toxicity of bleomycin to some strains of bacteria might also involve oxygen radicals, since Umezawa has reported that strains of *E. coli* K12 with increased SOD and catalase activities due to previous exposure to paraquat (Chapter 3) are more resistant to the toxic effects of bleomycin.

Quinone anti-tumour antibiotics

Streptonigrin
Streptonigrin (Fig. 8.10), an antibiotic produced by *Streptomyces flocculus,* was one of the first quinone antibiotics whose action was shown to involve oxygen radicals. It does have anti-tumour effects in Man, but it is not generally used clinically because of side-effects. The toxic action of

streptonigrin on *E. coli* requires the presence of oxygen and is decreased if the superoxide dismutase activity of the cells is raised (Chapter 3). Increased uptake of iron salts by *E. coli*, following the addition of citrate to the growth medium, enhances the bacteriocidal action of streptonigrin. This effect can be inhibited by desferrioxamine, suggesting a role for iron in the oxygen radical reactions leading to bacterial damage. It seems that the quinone part of the molecule (Fig. 8.10) can be reduced by bacterial enzymes to a semiquinone form that can then reduce oxygen to form O_2^-. Whether or not oxygen radicals are responsible for the observed DNA-degrading effects of streptonigrin in cancer cells is hard to determine, since the semiquinone itself is a reactive and damaging molecule. However, the toxicity of streptonigrin to mouse mammary tumour cells is decreased under hypoxic conditions.

Actinomycin D
Actinomycin D is a peptide antibiotic produced by *Streptomyces* species (Fig. 8.10). Its anti-tumour activity is generally attributed to its binding (intercalation) to DNA, so preventing RNA-synthesis. Intercalation occurs adjacent to guanine bases. This may not be the whole story, however, since actinomycin D, incubated with microsomal fractions in the presence of NADPH, is slowly reduced into a radical intermediate that can convert oxygen into O_2^-. DNA-bound actinomycin D cannot be reduced, however. Oxygen radical production may be relevant *in vivo* since the toxicity of actinomycin D to mouse mammary tumour cells is decreased at low oxygen concentrations.

Mitomycin C
Mitomycin C, isolated from *Streptomyces caespitosus,* is used in the palliative treatment of a large number of advanced human cancers. *In vitro*, it can be reduced by microsomal or nuclear electron-transport chains at the expense of NADPH. Reduction forms a semiquinone (Fig. 8.10) that can react with oxygen to generate O_2^-. The P_{450} system also converts mitomycin C into products that attack DNA directly, and produce cross-linking of the strands. Usually the toxicity of mitomycin C is greater under conditions of low oxygen concentration.

The anthracycline antibiotics
Anthracyclines are tetracyclic (i.e. have four joined ring structures) antibiotics produced by various *Streptomyces* strains. They are widely used in the treatment of acute leukaemia, breast cancer, Hodgkin's disease, and sarcomas. The best known are daunorubicin (sometimes called 'daunomycin') and doxorubicin (often called 'adriamycin'), although several others are now in use. Like all anti-tumour drugs, the anthracyclines produce a number of side effects, the most serious being

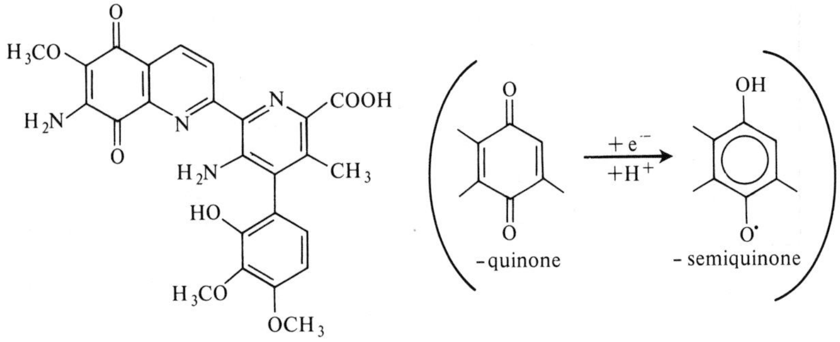

(a)

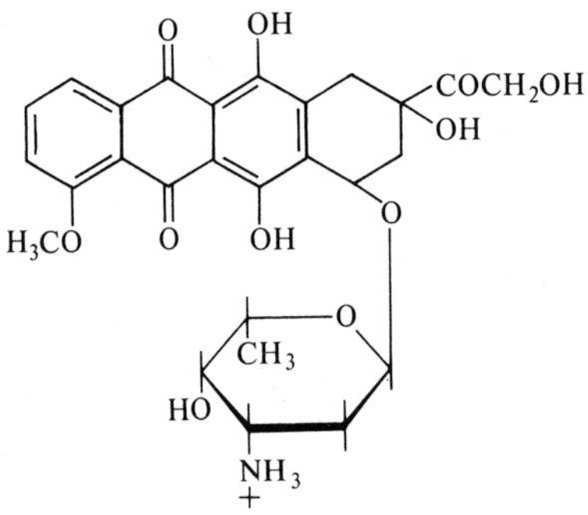

(b)

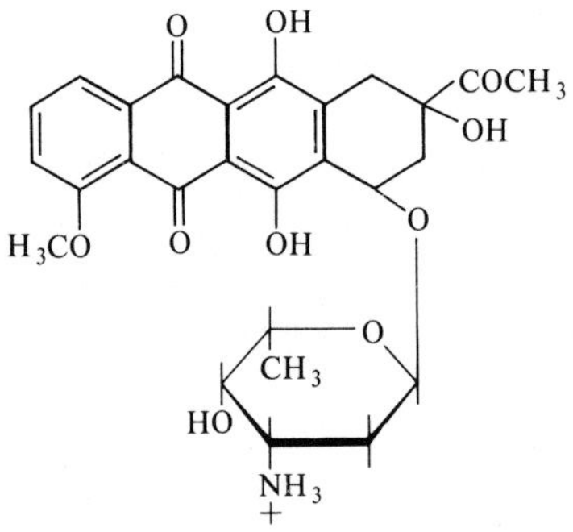

(c)

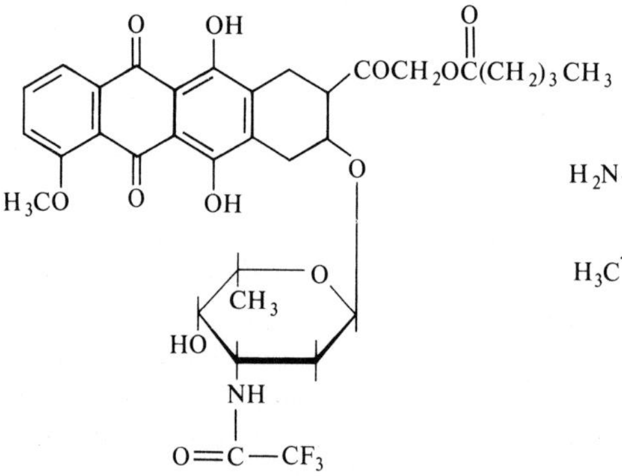

(d)

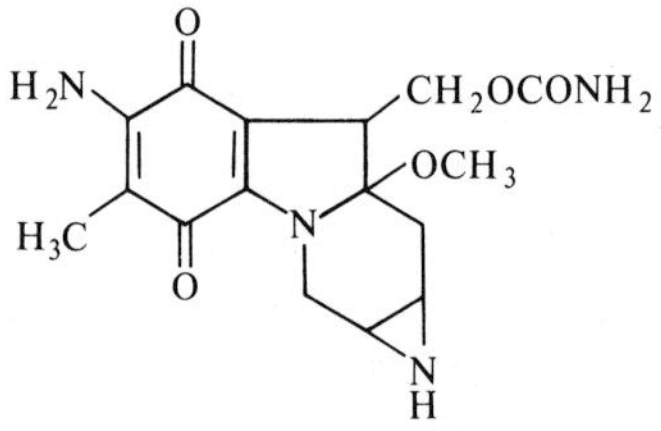

(e)

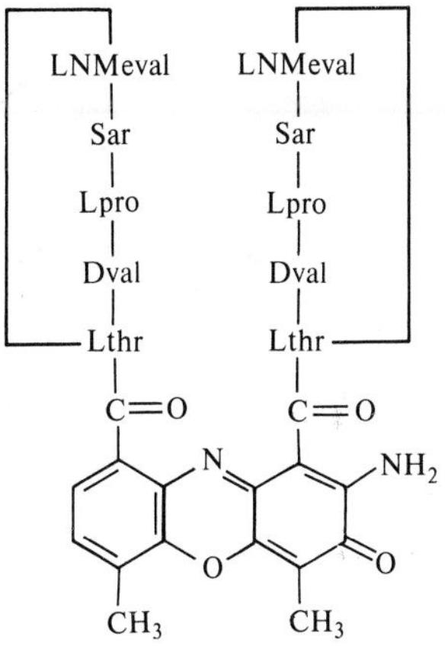

Fig. 8.10. Structures of some antitumour drugs. (a), Streptonigrin; (b) adriamycin; (c), daunomycin; (d), AD32; (e), mitomycin C; (f) actinomycin D (thr, threonine; val, valine; pro, proline; sar, sarcosine; NMeval, *N*-methylvaline), (g) *cis*-diamminedichloroplatinum(II). (*Cis* means that similar groups are next to each other, i.e. the two —NH₂ groups and the two —Cl groups are adjacent.) The *trans*-isomer [in brackets] is not an effective anti-tumour drug.

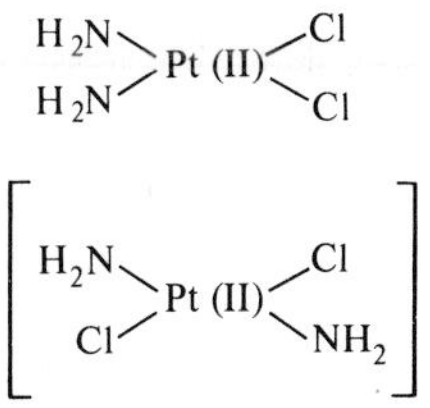

damage to the heart. It is this cardiotoxicity that limits the doses of adriamycin and daunorubicin that can safely be given to patients.

The mechanism of action of anthracyclines is not entirely clear, and they probably have multiple effects. Adriamycin and daunorubicin bind strongly to DNA, and so interfere with replication and RNA synthesis. *In vivo* they produce DNA strand breaks, and possibly interstrand cross-links. However, adriamycin is toxic to at least one tumour cell line, even if it is prevented from entering the cells. It binds to cell membranes and alters their permeability to ions such as Ca^{2+}. Indeed the adriamycin derivative AD32 (*N*-trifluoroacetyladriamycin-14-valerate; see Fig. 8.10) will not bind to DNA, and yet it is still an effective anti-tumour agent. Increased membrane permeability to Ca^{2+} ions may in part explain the cardiotoxic effects of adriamycin. This drug also interferes with electron transport in cardiac mitochondria and administration of ubiquinone (Chapter 3) has been reported to protect experimental animals against adriamycin-induced heart damage.

The pioneering work of Handa and Sato, in Japan, showed that quinone drugs can be reduced by the microsomal electron-transport chain to form reactive semiquinones that can combine directly with cellular components, or react with oxygen to form the superoxide radical. Reduction of anthracycline drugs is similarly catalysed by several enzymes, such as xanthine oxidase or ferredoxin reductase, and by the electron-transport chain located in the nuclear envelope. Binding of anthracyclines to DNA prevents reduction but reduction of unbound drug could be important in other aspects of anthracycline action. Formation of O_2^- and its conversion to more reactive radicals might account for some of the membrane-damaging effects of anthracyclines, and could be involved in the cardiotoxicity. Treatment of various animals with ascorbic acid, vitamin E, or the thiol compound, *N*-acetylcysteine has shown limited protective effects against cardiotoxicity. Of course, these scavengers might react directly with the semiquinones. Heart tissue is low in catalase activity, and its glutathione peroxidase activity decreases after adriamycin treatment. Mixtures of vitamin E and dimethylsulphoxide have been observed to offer some protection against ulceration induced by the direct application of adriamycin to animal tissues, but none of these scavengers has yet been shown to protect against the side-effects of adriamycin in human patients.

The toxicity of the anthracyclines to a number of tumour cells is decreased under hypoxic conditions, again suggesting a role for oxygen radicals *in vivo*. This is not always the case, however. For example, the toxicity of adriamycin to mouse mammary tumour is *greater* under hypoxic conditions.

Any damage done by an increased rate of O_2^--formation *in vivo* would presumably result from conversion of this radical into more reactive species, such as OH·, in the presence of transition-metal ions. It is therefore interesting to note that adriamycin and daunorubicin form complexes with a number of metal ions including Fe^{2+}, Fe^{3+}, and Cu^{2+}. Complexes of adriamycin and Fe^{2+} can oxidize in air to generate superoxide radicals. When adriamycin complexes with ferric ions, the complex greatly stimulates lipid peroxidation *in vitro*. It has recently been proposed by Nakano, in Japan, that the Fe(III) is reduced to Fe(II) during such complexing, and that this accounts for the greatly stimulated lipid peroxidation (Chapter 4). The metal-ion-chelating agent ICRF-187, a derivative of EDTA, has been reported to decrease the cardiotoxicity of adriamycin in rabbits and dogs.

Protein antitumour drugs

Neocarzinostatin

Neocarzinostatin is a protein, 109 amino acids long, secreted by a mutant strain of *Streptomyces carzinostaticus*. It has actions against a number of tumour cells *in vitro*. The protein has, bound to it, a small fluorescent molecule that is responsible for its biological activity. This fluorescent chromophore binds to the DNA of the target cells, mainly adjacent to adenine and thymine bases. It promotes strand-scission and damage to deoxyribose that results in formation of TBA-reactive material. DNA damage *in vitro* requires the presence of both oxygen and a reducing agent, such as a thiol compound. Indeed, incubation of the neocarzinostatin chromophore with thiols inactivates it, perhaps by forming a reactive radical species that can also attack DNA. DNA degradation by neocarzinostatin is not prevented by SOD, catalase, scavengers of hydroxyl radical, or metal-ion-chelating agents, but α-tocopherol does offer some protection. Elucidation of the mechanism of the damage must await the final determination of the structure of the chromophore.

The protein part of neocarzinostatin is itself inactive, and probably serves merely as a vehicle for transporting the chromophore into cells.

Other protein drugs

Macromomycin, a protein 112 amino acids long, isolated from *Streptomyces macromomyceticus,* has some anti-tumour effects, and induces DNA strand breakage. During purification of macromomycin, a small fluorescent molecule is released. Re-addition of it greatly increases the antibacterial and antitumour activities of the protein, the complex of macromomycin and the fluorescent chromophore being known as auromomycin. This chromophore itself can induce DNA damage *in vitro,* although a reducing agent is not required.

8.4. Further reading

Autor, A. P. (ed.) (1982). *Pathology of oxygen.* Academic Press, New York.

Blake, D. R., *et al.* (1981). The importance of iron in rheumatoid disease. *Lancet* **ii,** 1142.

Blake, D. R., *et al.* (1983). Effect of a specific iron chelating agent on animal models of inflammation. *Ann. Rheumatic Dis.* **42,** 89.

Bragt, P. C., Bansberg, J. I., and Bonta, I. L. (1980). Antiinflammatory effects of free radical scavengers and antioxidants. *Inflammation* **4,** 289.

Burton, G. W., Cheesman, K. H., Ingold, K. U., and Slater, T. F. (1983). Lipid antioxidants and products of lipid peroxidation as potential tumour protective agents. *Biochem. Soc. Trans.* **11,** 261.

Dingle, J. T., and Gordon, J. L. (eds) (1981). *Cellular Interactions.* Research Monographs in cell and tissue physiology, Volume 6. Elsevier/North Holland, Amsterdam.

Editorial (1981). Immunocompromised homosexuals. *Lancet* **ii,** 1325; *and* Gay compromise syndrome. *Lancet* **ii,** 1338.

Fauvadon, Y. (1982). On the mechanisms of reductive activation in the mode of action of some anticancer drugs. *Biochimie* **7,** 457.

Floyd, R. A. (1982). The role of free radicals in arylamine carcinogenesis. In *Free radicals and cancer* (ed. R. A. Floyd), p. 361. Marcel Dekker, New York.

Glass, G. A., and Gershon, D. (1981). Enzymatic changes in rat erythrocytes with increasing cell and donor age: loss of superoxide dismutase activity associated with increases in catalytically defective forms. *Biochem. biophys. Res. Commun.* **103,** 1245.

Greenwald, R. A., and Moy, W. W. (1980). Effect of oxygen-derived free radicals on hyaluronic acid. *Arthritis and Rheumatism* **23,** 455.

Halliwell, B. (1982). Production of superoxide, hydrogen peroxide and hydroxyl radicals by phagocytic cells. A cause of chronic inflammatory disease? *Cell Biol. Int. Reps* **6,** 529.

Harley, J. B., *et al.* (1982). Streptonigrin toxicity in *E. coli:* oxygen dependence and the role of the intracellular oxidation-reduction state. *Can. J. Microbiol.* **28,** 545.

Harman, D. (1982). The free-radical theory of ageing. In *Free radicals in biology* Vol. V (ed. W. A. Pryor) p. 255. Academic Press, London.

Hirschelmann R., and Bekemeier, H. (1981). Effects of catalase, peroxidase superoxide dismutase and 10 scavengers of oxygen radicals in carrageenin oedema and in adjuvant arthritis of rats. *Experientia* **37,** 1313.

Hodges, R. E. (1982). Vitamin C and cancer. *Nutr. Rev.* **40,** 289.

Husby, G., Marhaug, G., and Sletten, K. (1982). Amyloid A in systemic amyloidosis associated with cancer. *Cancer Res.* **42,** 1600.

Ioannides. C., and Parke, D. V. (1980). The metabolic activation and detoxication of mutagens and carcinogens. *Chem. Ind.* **November 1980,** 854.

Joenje, H., *et al.* (1981). Oxygen-dependence of chromosomal aberrations in Fanconi's anaemia. *Nature* **290,** 142.

Levin, D. E., *et al.* (1982). A new *Salmonella* tester strain (TA102) with A. T. base pairs at the site of mutation detects oxidative damage. *Proc. natn. Acad. Sci. U.S.A.* **79,** 7445.

Lippman, R. D. (1983). Lipid peroxidation and metabolism in aging: a biological, chemical and medical approach. *Rev. biol. Res. Aging* **1,** 315.

McBrien, D. C. H., and Slater, T. F. (eds) (1982). *Free Radicals, Lipid Peroxidation and Cancer.* Academic Press, London.

Marklund, S. L., Westman, N. G., Lundgren, E., and Roos, G. (1982). Copper and zinc-containing superoxide dismutase, catalase and glutathione peroxidase in normal and neoplastic human cell lines and normal tissue. *Cancer Res.* **42,** 1955.

Nathan, C., and Cohn, Z. (1980). Role of oxygen-dependent mechanisms in antibody-induced lysis of tumour cells by activated macrophages. *J. Exp. Med.* **152,** 198.

Oyanagui, Y. (1976). Participation of superoxide anions at the prostaglandin phase of carrageenan foot odema. *Biochem. Pharmac.* **25,** 1465.

Pras, M., *et al.* (1983). Primary structure of an amyloid prealbumin variant in familial polyneuropathy of Jewish origin. *Proc. natn. Acad. Sci. U.S.A.* **80,** 539.

Rister, M., and Bauermeister, K. (1982). Superoxid dismutase und superoxid-radikal-freisetzung bei juveniler rheumatoider Arthritis. *Klin. Wochenschr.* **60,** 561.

Robertson, M. (1983). Oncogenes and the origins of human cancer. *Br. Med. J.* **286,** 81.

Rothstein, M. (1982). *Biochemical approaches to aging.* Academic Press, New York.

Sausville, E. A., Peisach, J., and Horwitz, S. B. (1978). Effect of chelating agents and metal ions on the degradation of DNA by bleomycin. *Biochemistry* **17,** 2740.

Selkoe, D. J. (1982). Molecular pathology of the ageing human brain. *Trends in neurol. Sci.* **October 1982,** 32.

Sohal, R. S. (ed) (1981). *Age Pigments*. Elsevier/North Holland, Amsterdam.

Sohal, R. S., and Buchan, P. B. (1981). Relationship between physical activity and lifespan in the adult housefly *Musca domestica*. *Exp. Gerontol.* **16,** 157.

Sugioka, K., and Nakano, M. (1982). Mechanisms of phospholipid peroxidation induced by ferric ion-ADP-adriamycin-co-ordination complex. *Biochim. Biophys. Acta* **713,** 333.

Suthanthiran, M. *et al.* (1984). Hydroxyl radical scavengers inhibit human natural killer cell activity. *Nature* **307,** 276.

Theofilopoulos, A. N., and Dixon, F. B. (1982). Autoimmune diseases. Immuno-pathology and etiopathogenesis. *Am. J. Pathol.* **108,** 321.

Umezawa, H. (1983). Studies of microbial products in rising to the challenge of curing cancer. *Proc. R. Soc. (Lond.)* **B217,** 357.

Weinberg, E. D. (1978). Iron and infection. *Microbiol. Rev.* **42,** 45.

Weinberg, E. D. (1981). Iron and neoplasia. *Biochem. trace Elem. Res.* **3,** 55.

Wilmer, J., and Schubert, J. (1981). Mutagenicity of irradiated solutions of nucleic acid bases and nucleosides in *Salmonella typhimurium*. *Mutat. Res.* **88,** 337.

Wolf, G. (1982). Is dietary β-carotene an anti-cancer agent? *Nutr. Rev.* **40,** 257.

Young, R. C., Ozolo, R. F., and Myers, C. E. (1981). The anthracycline antineoplastic drugs *New Engl. J. Med.* **305,** 139.

Appendix: A consideration of atomic structure and bonding

A.1. Atomic structure

For the purposes of this book it will be sufficient to consider a simple model of atomic structure in which the atom consists of a positively charged nucleus that is surrounded by one or more negatively charged electrons. The nucleus contains two types of particle of approximately equal mass, the positively charged proton and the uncharged neutron. By comparison with these particles, the mass of the electron is negligible so that virtually all of the mass of the atom is contributed by its nucleus. The *atomic number* of an element is defined as the number of protons in its nucleus, the *mass number* as the number of protons plus neutrons. In the neutral atom, the atomic number also equals the number of electrons surrounding the nucleus. The simplest atom is that of the element hydrogen, containing one proton (atomic number equals one, mass number equals one) and one electron. All other elements contain neutrons in the nucleus.

Some elements exist as *isotopes*, in which the atoms contain the same number of protons and electrons, but different numbers of neutrons. These isotopes can be stable or unstable, the unstable ones undergoing radioactive decay at various rates. In this process, the nucleus of the radioactive isotope changes, and a new element is formed. For example, an isotope of the element uranium (atomic number 92) with a mass number of 238 undergoes nuclear disintegration to produce two fragments, one with two protons and two neutrons and the other with 90 protons and 144 neutrons, in fact an isotope of the element thorium. Fortunately the elements with which we are largely concerned in this book, carbon, hydrogen and oxygen, exist almost exclusively as one isotopic form in nature (Table A.1).

The electrons surrounding the atomic nucleus possess a negative charge. Since they do not spiral into the nucleus, they must possess energy to counteract the attractive electric force tending to pull them in. In 1900, Planck suggested that energy is quantized, i.e. that energy changes only occur in small, definite amounts known as 'quanta'. Application of Planck's quantum theory to the atom, by Bohr, produced a model in which the electrons exist in specific orbits, or 'electron shells', each associated with a particular energy level. The 'K'-shell electrons,

Table A.1. Isotopes of some common elements

Element	Isotope	Number of protons in nucleus	Number of neutrons in nucleus	Comments
Chlorine	$^{35}_{17}Cl$	17	18	Both isotopes are stable and occur naturally, ^{35}Cl being more abundant
	$^{37}_{17}Cl$	17	20	
Carbon	$^{12}_{6}C$	6	6	Over 90% of naturally occurring carbon is $^{12}_{6}C$. Small amounts of the radioactive isotope $^{14}_{6}C$ are formed by the bombardment of atmospheric CO_2 with cosmic rays (i.e. streams of neutrons arising from space). This isotope undergoes slow radioactive decay (50% decay after 5600 years)
	$^{13}_{6}C$	6	7	
	$^{14}_{6}C$	6	8	
Oxygen	$^{16}_{8}O$	8	8	Over 99% of naturally occurring oxygen is the isotope $^{16}_{8}O$
	$^{17}_{8}O$	8	9	
	$^{18}_{8}O$	8	10	
Hydrogen	$^{1}_{1}H$	1	0	Over 99% of hydrogen is $^{1}_{1}H$. Deuterium ($^{2}_{1}H$) is a stable isotope, whereas tritium ($^{3}_{1}H$) is radioactive. Deuterium oxide is known as 'heavy water', and is used in detecting the presence of singlet oxygen in biological systems (Chapter 2)
	$^{2}_{1}H$	1	1	
	$^{3}_{1}H$	1	2	

The superscript number on the left of the symbol for the element represents the mass number, and the subscript the atomic number. All atoms of a given element have the same number of protons, but sometimes have different numbers of neutrons, giving rise to isotopes.

lying closest to the nucleus, have the lowest energy, and the energy successively increases as one proceeds outwards to the so-called L-, M-, and N-shells. The K-shell can hold a maximum of two electrons, the L-shell, 8, M-shell, 18, and N-shell, 32. Table A.2 shows the location of electrons in each of these shells for the elements up to atomic number 36.

Subsequent developments of atomic theory have shown that an electron has some of the properties of a particle, and some of the properties of a wave motion. As a result, the position of an electron at a given time cannot be precisely located, but only the region of space where it is most likely to be. These regions are referred to as *orbitals*. Each electron in an atom has its energy defined by four so-called quantum numbers. The first, or *principal quantum number* (n) defines the main energy level the electron occupies. For the K-shell, $n = 1$; for L, $n = 2$; for M, $n = 3$; and for N, $n = 4$. The second, or *azimuthal quantum*

Table A.2. Location of electrons in shells for the elements with atomic numbers 1 to 36

Atomic number of element	Element	Symbol	Shell K	Shell L	Shell M	Shell N
1	Hydrogen	H	1			
2	Helium	He	2			
3	Lithium	Li	2	1		
4	Beryllium	Be	2	2		
5	Boron	B	2	3		
6	Carbon	C	2	4		
7	Nitrogen	N	2	5		
8	Oxygen	O	2	6		
9	Fluorine	F	2	7		
10	Neon	Ne	2	8		
11	Sodium	Na	2	8	1	
12	Magnesium	Mg	2	8	2	
13	Aluminium	Al	2	8	3	
14	Silicon	Si	2	8	4	
15	Phosphorus	P	2	8	5	
16	Sulphur	S	2	8	6	
17	Chlorine	Cl	2	8	7	
18	Argon	Ar	2	8	8	
19	Potassium	K	2	8	8	1
20	Calcium	Ca	2	8	8	2
21	Scandium	Sc	2	8	9	2
22	Titanium	Ti	2	8	10	2
23	Vanadium	V	2	8	11	2
24	Chromium	Cr	2	8	13	1
25	Manganese	Mn	2	8	13	2
26	Iron	Fe	2	8	14	2
27	Cobalt	Co	2	8	15	2
28	Nickel	Ni	2	8	16	2
29	Copper	Cu	2	8	18	1
30	Zinc	Zn	2	8	18	2
31	Gallium	Ga	2	8	18	3
32	Germanium	Ge	2	8	18	4
33	Arsenic	As	2	8	18	5
34	Selenium	Se	2	8	18	6
35	Bromine	Br	2	8	18	7
36	Krypton	Kr	2	8	18	8

number (l) governs the shape of the orbital and has values from zero up to $(n - 1)$. When $l = 0$, the electrons are called 's' electrons; when $l = 1$, they are 'p' electrons; $l = 2$, 'd' electrons; and $l = 3$ gives 'f' electrons. The third quantum number is the *magnetic quantum number* (m) and, for each value of l, m has values of $l, (l - 1), \ldots, 0, -1, \ldots, -l$. Finally, the fourth quantum number, or *spin quantum number*, can have values of either $\frac{1}{2}$ or $-\frac{1}{2}$ only. Table A.3 shows how various combinations of these four quantum numbers can fill the electron shells, and (hopefully!) makes the above explanation a bit clearer. *Pauli's principle* states that "no two electrons can have the same four quantum numbers". Since the spin

Table A.3. Orbitals available in the principal electron shells

Shell	Principal quantum number	Value of l (azimuthal quantum number)	Electron type	Value of m (magnetic quantum number)	Value of s (spin quantum number)	Maximum number of electrons in shell
K	1	0	s	0	$\pm\frac{1}{2}$	2 (1s-orbital)
L	2	0	s	0	$\pm\frac{1}{2}$	2 (2s-orbital)
		1	p	1, 0, −1	$\pm\frac{1}{2}$	3 × 2 (three 2p-orbitals)
M	3	0	s	0	$\pm\frac{1}{2}$	2 (3s-orbital)
		1	p	1, 0, −1	$\pm\frac{1}{2}$	3 × 2 (three 2p-orbitals)
		2	d	2, 1, 0, −1, −2	$\pm\frac{1}{2}$	5 × 2 (five 3d-orbitals)
N	4	0	s	0	$\pm\frac{1}{2}$	2 (4s-orbital)
		1	p	1, 0, −1	$\pm\frac{1}{2}$	3 × 2 (three 4p-orbitals)
		2	d	2, 1, 0, −1, −2	$\pm\frac{1}{2}$	5 × 2 (five 4d-orbitals)
		3	f	3, 2, 1, 0, −1, −2, −3	$\pm\frac{1}{2}$	7 × 2 (seven 4f-orbitals)

L shell total: 8
M shell total: 18
N shell total: 32

quantum number has only two possible values ($\pm\frac{1}{2}$), it follows that an orbital can hold only two electrons at most (Table A.3).

In filling the available orbitals electrons will enter the orbitals with the lowest total energy content (*Aufbau principle*). The order of filling is:

1s 2s 2p 3s 3p 4s 3d 4p 5s 4d 5p 6s 4f 5d 6p 7s 5f

lowest energy   increasing energy   highest energy

Table A.4 gives the electronic energy configurations of the elements with atomic numbers from 1 to 30. When the elements are arranged in the *Periodic Table* (Fig. A.1), elements with similar electronic arrangements fall into similar 'groups' (vertical rows), e.g. the group II elements all have two electrons in their outermost electron shell, and the group IV elements have four. Since the 4s-orbital is of lower energy than the

Table A.4. Electronic configuration of the elements

Element	Atomic number	Symbol	Configuration	Place in periodic table
Hydrogen	1	H	$1s^1$	uncertain
Helium	2	He	$1s^2$	Group 0 (inert gases)
Lithium	3	Li	$1s^2 2s^1$	Group I (alkali metals)
Beryllium	4	Be	$1s^2 2s^2$	Group II (alkaline-earth metals)
Boron	5	B	$1s^2 2s^2 2p^1$	Group III
Carbon	6	C	$1s^2 2s^2 2p^2$	Group IV
Nitrogen	7	N	$1s^2 2s^2 2p^3$	Group V
Oxygen	8	O	$1s^2 2s^2 2p^4$	Group VI
Fluorine	9	F	$1s^2 2s^2 2p^5$	Group VII (halogen elements)
Neon	10	Ne	$1s^2 2s^2 2p^6$	Group 0
Sodium	11	Na	$1s^2 2s^2 2p^6 3s^1$	Group I
Magnesium	12	Mg	$1s^2 2s^2 2p^6 3s^2$	Group II
Aluminium	13	Al	$1s^2 2s^2 2p^6 3s^2 3p^1$	Group III
Silicon	14	Si	$1s^2 2s^2 2p^6 3s^2 3p^2$	Group IV
Phosphorus	15	P	$1s^2 2s^2 2p^6 3s^2 3p^3$	Group V
Sulphur	16	S	$1s^2 2s^2 2p^6 3s^2 3p^4$	Group VI
Chlorine	17	Cl	$1s^2 2s^2 2p^6 3s^2 3p^5$	Group VII
Argon	18	Ar	$1s^2 2s^2 2p^6 3s^2 3p^6$	Group 0
Potassium	19	K	$1s^2 2s^2 2p^6 3s^2 3p^6 4s^1$	Group I
Calcium	20	Ca	$1s^2 2s^2 2p^6 3s^2 3p^6 4s^2$	Group II
Scandium	21	Sc	$1s^2 2s^2 2p^6 3s^2 3p^6 4s^2 3d^1$	d-block
Titanium	22	Ti	$1s^2 2s^2 2p^6 3s^2 3p^6 4s^2 3d^2$	d-block
Vanadium	23	V	$1s^2 2s^2 2p^6 3s^2 3p^6 4s^2 3d^3$	d-block
Chromium	24	Cr	$1s^2 2s^2 2p^6 3s^2 3p^6 4s^1 3d^5$	d-block
Manganese	25	Mn	$1s^2 2s^2 2p^6 3s^2 3p^6 4s^2 3d^5$	d-block
Iron	26	Fe	$1s^2 2s^2 2p^6 3s^2 3p^6 4s^2 3d^6$	d-block
Cobalt	27	Co	$1s^2 2s^2 2p^6 3s^2 3p^6 4s^2 3d^7$	d-block
Nickel	28	Ni	$1s^2 2s^2 2p^6 3s^2 3p^6 4s^2 3d^8$	d-block
Copper	29	Cu	$1s^2 2s^2 2p^6 3s^2 3p^6 4s^1 3d^{10}$	d-block
Zinc	30	Zn	$1s^2 2s^2 2p^6 3s^2 3p^6 4s^2 3d^{10}$	d-block
Gallium	31	Ga	$1s^2 2s^2 2p^6 3s^2 3p^6 4s^2 3d^{10} 4p^1$	Group III
Germanium	32	Ge	$1s^2 2s^2 2p^6 3s^2 3p^6 4s^2 3d^{10} 4p^2$	Group IV

Groups

s – block

I	II
1 H	4 Be
3 Li	12 Mg
11 Na	20 Ca
19 K	38 Sr
37 Rb	56 Ba
55 Cs	88 Ra
87 Fr	

d – block

21 Sc	22 Ti	23 V	24 Cr	25 Mn	26 Fe	27 Co	28 Ni	29 Cu	30 Zn
39 Y	40 Zr	41 Nb	42 Mo	43 Tc	44 Ru	45 Rh	46 Pd	47 Ag	48 Cd
57 La	72 Hf	73 Ta	74 W	75 Re	76 Os	77 Ir	78 Pt	79 Au	80 Hg
89 Ac									

Groups

III	IV	V	VI	VII	0
				1 H	2 He
5 B	6 C	7 N	8 O	9 F	10 Ne
13 Al	14 Si	15 P	16 S	17 Cl	18 Ar
31 Ga	32 Ge	33 As	34 Se	35 Br	36 Kr
49 In	50 Sn	51 Sb	52 Te	53 I	54 Xe
81 Ti	82 Pb	83 Bi	84 Po	85 At	86 Rn

p – block

f – block

Lanthanides

58 Ce	59 Pr	60 Nd	61 Pm	62 Sm	63 Eu	64 Gd	65 Tb	66 Dy	67 Ho	68 Er	69 Tm	70 Yb	71 Lu

Actinides

90 Th	91 Pa	92 U	93 Np	94 Pu	95 Am	96 Cm	97 Bk	98 Cf	99 Es	100 Fm	101 Md	102 No	103 Lr

Fig. A.1. Periodic table of the elements. Hydrogen has features of the group I and group VII elements, so its position is uncertain. Group I: alkali metals—highly electropositive elements. Group II: alkaline-earth metals—electropositive. Group VII: halogen elements—highly electronegative. Group 0: inert gases. In general, metallic character, and electropositivity, increases down a group, e.g. group IV begins with carbon (a non-metal), and ends with lead (a metal).

3d-orbitals, these latter orbitals remain empty until the 4s-orbital is filled (e.g. see the elements potassium and calcium in Table A.4). In subsequent elements the five 3d-orbitals receive electrons, creating the first row of the so-called *d-block* in the Periodic Table (Fig. A.1). Some of these d-block elements are called *transition elements*, meaning elements in which an inner shell of electrons is incomplete (in this case these are electrons in the fourth shell, but all the d-orbitals of the third shell are not yet full). The term transition element, as defined above, applies to scandium and subsequent elements as far as nickel, although it is often extended to include the whole of the first row of the d-block.

If orbitals of equal energy are available, e.g. the three 2p-orbitals in the L-shell, or the five 3d orbitals in the M-shell (Table A.3), each is filled with one electron before any receives two (*Hund's rule*). Hence one can further break down the electronic configurations shown in Table A.4. The element boron has two 1s, two 2s, and one 2p electrons. Three 2p-orbitals of equal energy are available (Table A.3), and they are often written as $2p_x$, $2p_y$, and $2p_z$. If we represent each orbital as a square box and an electron as an arrow

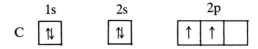

For the next element, carbon, the extra electron enters another 2p-orbital in obedience to Hund's rule

And for nitrogen we have

Further electrons will now begin to 'pair-up' to fill the 2p-orbitals, e.g. for the oxygen atom

Table A.5. Electronic configuration of the elements scandium to zinc in the first row of the d-block of the Periodic Table

Element	Electron configuration	3d					4s
Scandium	Ar	↑					↑↓
Titanium	Ar	↑	↑				↑↓
Vanadium	Ar	↑	↑	↑			↑↓
Chromium	Ar	↑	↑	↑	↑	↑	↑
Manganese	Ar	↑	↑	↑	↑	↑	↑↓
Iron	Ar	↑↓	↑	↑	↑	↑	↑↓
Cobalt	Ar	↑↓	↑↓	↑	↑	↑	↑↓
Nickel	Ar	↑↓	↑↓	↑↓	↑	↑	↑↓
Copper	Ar	↑↓	↑↓	↑↓	↑↓	↑↓	↑
Zinc	Ar	↑↓	↑↓	↑↓	↑↓	↑↓	↑↓

'Ar' is used as an abbreviation for the argon configuration, $1s^2 2s^2 2p^6 3s^2 3p^6$, to simplify the table, i.e. each element begins with the argon configuration. The 'unusual' electronic configurations of chromium and copper seem to be due to the increased relative stability of atoms in which each 3d-orbital contains either one or two electrons.

Hund's rule is particularly important in the d-block elements, e.g. Table A.5 uses the same 'electrons-in-boxes' notation for the elements in the first row of this block. Each of the five 3d-orbitals receives one electron, before any receives two.

We shall now consider how atoms join together to form molecules in chemical reactions.

A.2. Bonding between atoms

A.2.1. Ionic bonding

As with our consideration of atomic structure, the account of chemical bonding that follows is the simplest possible model consistent with the requirements of this book.

Essentially two types of chemical bond can be distinguished. The first is called *ionic bonding*. This tends to happen when so-called electropositive elements combine with electronegative ones. Electropositive elements, such as those in groups I and II of the Periodic Table (Fig. A.1), tend to lose their outermost electrons easily, whereas electronegative elements (group VII, and oxygen and sulphur in group VI) tend to accept extra electrons. By doing so, they gain the electronic configuration of the nearest inert gases, which seems to be a particularly stable configuration in view of the relative lack of reactivity of these elements! Consider, for example, the combination of an atom of sodium with one of chlorine. Sodium, an electropositive group I element, has the electronic configuration $1s^2 2s^2 2p^6 3s^1$. If a sodium atom loses one electron, it then has the configuration $1s^2 2s^2 2p^6$, that of the inert gas, neon. It is still the element sodium because its nucleus is unchanged, but the loss of one electron leaves the atom with a positive charge, forming an ion or, more specifically, a *cation* (positively charged ion). For chlorine, configuration $1s^2 2s^2 2p^6 3s^2 3p^5$, acceptance of an electron gives the argon-electron-configuration $1s^2 2s^2 2p^6 3s^2 3p^6$, and produces a negatively charged ion (*anion*) Cl^-.

In the case of a group II element such as magnesium, it must lose two electrons to gain an inert-gas electron-configuration. Thus one atom of magnesium can provide electrons for acceptance by two chlorine atoms, giving magnesium chloride a formula $MgCl_2$ i.e.

$$Mg \longrightarrow Mg^{2+} + 2e^-.$$

$1s^2 2s^2 2p^6 3s^2$ $1s^2 2s^2 2p^6$
(neon configuration)

An atom of oxygen, however, can accept two electrons and combine with magnesium to form an oxide MgO:

$$O + 2e^- \rightarrow O^{2-}$$

$1s^2 2s^2 2p^4$ $1s^2 2s^2 2p^6$
(neon configuration)

Once formed, anions and cations are held together by the electric attraction of their opposite charges. Each ion will exert an effect on each other ion in its vicinity, and these effects cause the ions to pack together into an *ionic crystal lattice*, as shown in Fig. A.2 for NaCl. Each Na^+ ion is surrounded by six Cl^- ions, and vice versa. Once the lattice has formed, it cannot be said that any one Na^+ ion 'belongs' to any one Cl^- ion, nor can 'molecules' of sodium chloride be said to exist in the solid. The formula of an ionic compound merely indicates the combining-ratio of the elements involved. A considerable amount of energy is required to disrupt all the electrostatic forces between the many millions of ions in a crystal of an ionic compound, so such compounds are usually solids with high melting-points. Ionic compounds are mostly soluble in water, and

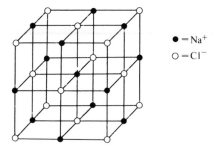

Fig. A.2. Crystal structure of sodium chloride. The exact type of lattice formed by an ionic compound depends on the relative sizes of the ions. NaCl forms a cubic lattice, as shown.

the solutions conduct electricity because of the presence of ions to carry the current. The properties of an ionic compound are those of its constituent ions.

A.2.2. Covalent bonding

The covalent bond involves a sharing of a pair of electrons between the two bonded atoms. In 'normal' covalent bonding, each atom contributes one electron to the shared pair; but in *dative covalent bonding*, one atom contributes both of the shared electrons. The element hydrogen is usually found in nature as *diatomic* molecules, H_2. Two hydrogen atoms are sharing a pair of electrons. If we represent the electron of each hydrogen atom by a cross ($\times$) we can write

$$H \times + {}^{\times}H \rightarrow H \underset{\times}{\overset{\times}{}} H$$

where $\underset{\times}{\overset{\times}{}}$ is the shared pair of electrons. Many other gaseous elements, including oxygen and chlorine, exist as covalently bonded diatomic molecules.

When chlorine combines with hydrogen, the covalent compound, hydrogen chloride, is formed. If we represent the outermost electrons of the chlorine atom as circles, we can write

$$\underset{\text{oo}}{\overset{\text{oo}}{}} \text{Cl}^{\text{o}} + H \times \longrightarrow H \underset{\text{oo}}{\overset{\text{oo}}{}} \text{Cl} \underset{\text{oo}}{\overset{}{}}$$
$$\text{hydrogen chloride}$$

where $\overset{\times}{\text{o}}$ is the shared pair of electrons. In reality, of course, one electron is the same as any other, in that once the bond is formed, the electron originating from the chlorine cannot be distinguished from that which came from the hydrogen. Similarly for the covalent compound ammonia, NH_3

$$\overset{\text{H}}{\underset{\text{H}}{8\text{N} + 3\text{H} \times \longrightarrow \text{N} \text{H}}}$$

In all the above cases, each atom has contributed one electron to the covalent bond. The compound ammonia also undergoes dative covalent bonding using the spare pair ('*lone-pair*') of electrons on the nitrogen atom. For example, it forms a covalent bond with a proton (H^+), formed by the loss of one electron from a hydrogen atom and thus possessing no electrons of its own.

$$\overset{\text{H}}{\underset{\text{H}}{\text{N}}} + \text{H}^+ \longrightarrow \left[\overset{\text{H}}{\underset{\text{H}}{\text{H} \text{N} \text{H}}} \right]^+$$

ammonia   ammonium ion, NH_4^+

Once formed, each of the four covalent bonds in NH_4^+ is indistinguishable from the others.

Covalent compounds do not conduct electricity and are usually gases, liquids, or low-melting-point solids at room temperature, because the forces of interaction between the molecules are weak (by contrast, covalent bonds themselves are usually very strong). Covalent bonds, unlike ionic bonds, have definite directions in space, and so their length, and the angles between them, can be measured and quoted.

The orbital theory applied to atomic structure (Section A.1) can also be applied to covalent compounds, the bonding electrons being considered as occupying *molecular orbitals* formed by interaction of the atomic orbitals in which the electrons were originally located. The various possible interactions produce molecular orbitals of different energy levels, each of which can hold a maximum of two electrons with opposite values of the spin quantum number (i.e. Pauli's principle is obeyed). In the simplest case, the hydrogen molecule, there are two possible molecular orbitals formed by interaction of the 1s atomic orbitals. The lowest energy orbital is the *bonding molecular orbital* (often written as σ1s) in which the electron is most likely to be found between the two nuclei. There is also an *antibonding molecular orbital* (written as σ*1s) of higher energy in which there is little chance of finding an electron between the two nuclei. A bonding molecular orbital is more stable than the atomic orbitals that might give rise to it, whereas an antibonding molecular orbital is less stable. The two electrons in the hydrogen molecule have opposite spin, and both occupy the bonding molecular orbital. Hence H_2 is much more stable than are the isolated H atoms. By contrast, helium atoms have the electron configuration $1s^2$, and if they combined to give He_2, both the bonding and antibonding molecular orbitals would contain two electrons, and there would be no effective gain in stability. Hence He_2 does not form.

The combination of p-type atomic orbitals can produce two types of molecular orbital by overlapping in different ways. These are known as σ and π. Hence, for one of the 2p-orbitals (say $2p_x$) combining with another such orbital, there will be two bonding molecular orbitals, $\sigma 2p_x$ and $\pi 2p_x$, and two antibonding molecular orbitals, $\sigma^* 2p_x$ and $\pi^* 2p_x$. Energy increases in the order

$$\sigma 2p_x \quad < \quad \pi 2p_x \quad < \quad \pi^* 2p_x \quad < \quad \sigma^* 2p_x.$$

With this in mind, we can consider bonding in three more complicated cases: the gases, nitrogen, oxygen, and fluorine. The nitrogen atom has the configuration $1s^2 2s^2 2p^3$. If two atoms join together to form a diatomic molecule N_2, the four 1s-electrons (two from each atom) fully occupy both a $\sigma 1s$ bonding and a $\sigma^* 1s$ antibonding orbital, and so there is no net bonding. The four 2s-electrons similarly occupy $\sigma 2s$ and $\sigma^* 2s$ molecular orbitals, and again no net bond results. Six electrons are left, located in two $2p_x$, two $2p_y$, and two $2p_z$ atomic orbitals. If the axis of the bond between the atoms is taken to be that of the $2p_x$ orbitals, they can overlap along this axis to produce a bonding $\sigma 2p_x$ molecular orbital that can hold both electrons. The $2p_y$ and $2p_z$ atomic orbitals cannot overlap along their axes, but they can overlap laterally to give bonding $\pi 2p_y$ and $\pi 2p_z$ molecular orbitals, each of which holds two electrons with different values of the spin quantum number. The 2p antibonding orbitals are not occupied; and the net result is a triple covalent bond N≡N, i.e. one σ covalent bond and two π covalent bonds. N_2 is thus far more stable than are the individual N-atoms.

The oxygen atom (configuration, $1s^2 2s^2 2p^4$) has one extra electron, and so when O_2 is formed there are two more electrons to consider. These must occupy the next highest molecular orbital in terms of energy. In fact, there are two such orbitals of equal energy, $\pi^* 2p_y$ and $\pi^* 2p_z$. By Hund's rule, each must receive one electron. Since the presence of these electrons in antibonding orbitals energetically cancels out one of the $\pi 2p$ bonding orbitals, the two oxygen atoms are effectively joined by a double bond, i.e. O=O.

The fluorine molecule contains two more electrons than does O_2, and so the $\pi^* 2p_y$ and $\pi^* 2p_z$ orbitals are both full. Since three bonding and two antibonding molecular orbitals are occupied, the fluorine molecule effectively contains a single bond, F—F.

A.2.3. Non-ideal character of bonds

The discussion so far has implied an equal sharing of the bonding electrons between two atoms joined by a covalent bond. However, this is only the case when both atoms have a similar attraction for the electrons, i.e. are equally electronegative. This is often not the case. Consider, for example, the water molecule, which contains two oxygen–hydrogen covalent bonds:

$$2H\times + \overset{\circ\circ}{\underset{\circ\circ}{\circ O}} \longrightarrow \overset{H}{\underset{\circ\circ}{\circ O \times}} H$$

Oxygen is more electronegative than hydrogen, and so takes a slightly greater 'share' of the bonding electrons than it should, giving it a slight negative charge (written as δ^-). The hydrogen similarly has a slight positive charge i.e.

$$\overset{\delta^+}{H}\diagdown \quad \diagup \overset{\delta^+}{H}$$
$$\underset{\delta^-}{O}$$

where the dash between the atoms represents a covalent bond.

   The existence of these charges gives water many of its properties. They attract water molecules to each other, so raising the boiling point to 100 °C at normal atmospheric pressure e.g.

weak electrostatic bond

These weak electrostatic bonds are called *hydrogen bonds*. The small charges also allow water to hydrate ions; water molecules cluster around ions and help to stabilize them, e.g. for A^+ and B^- ions

The energy obtained when ions become hydrated is what provides the energy to disrupt the crystal lattice when ionic compounds dissolve in water. In those cases where the energy of hydration would be much smaller than the energy needed to disrupt the lattice, then the ionic compound will not dissolve in water.

A.2.4. Hydrocarbons and electron delocalization

The element carbon has four electrons in its outermost shell (Table A.4), and normally forms four covalent bonds. Carbon atoms can covalently bond to each other to form long chains. For example, the compound butane, used as a fuel in cigarette lighters, has the structure

$$H-\underset{\underset{H}{|}}{\overset{\overset{H}{|}}{C}}-\underset{\underset{H}{|}}{\overset{\overset{H}{|}}{C}}-\underset{\underset{H}{|}}{\overset{\overset{H}{|}}{C}}-\underset{\underset{H}{|}}{\overset{\overset{H}{|}}{C}}-H \quad (C_4H_{10}),$$

each dash (—) representing a covalent bond. Butane is referred to as a *hydrocarbon*, since the molecule contains carbon and hydrogen only. Two other hydrocarbon gases, ethane and pentane, are released during the peroxidation of membrane lipids (Chapter 4). They have the structures

$$H-\underset{\underset{H}{|}}{\overset{\overset{H}{|}}{C}}-\underset{\underset{H}{|}}{\overset{\overset{H}{|}}{C}}-H \text{ (ethane)}, \quad H-\underset{\underset{H}{|}}{\overset{\overset{H}{|}}{C}}-\underset{\underset{H}{|}}{\overset{\overset{H}{|}}{C}}-\underset{\underset{H}{|}}{\overset{\overset{H}{|}}{C}}-\underset{\underset{H}{|}}{\overset{\overset{H}{|}}{C}}-\underset{\underset{H}{|}}{\overset{\overset{H}{|}}{C}}-H \text{ (pentane)}.$$

Carbon atoms can also form double covalent bonds (written as $>C=C<$) and triple covalent bonds (—C≡C—) with each other. A double bond consists of four shared electrons (two pairs), and a triple bond has six shared electrons (three pairs). The simplest hydrocarbon containing a double bond is the gas *ethene*, otherwise known as ethylene.

It has the structure
$$\underset{H}{\overset{H}{\diagdown}}C=C\underset{H}{\overset{H}{\diagup}}$$
. Ethene is produced in several assay for the detection of hydroxyl radicals (Chapter 2).

Ethyne, otherwise known as acetylene, contains a triple bond and has the structure H—C≡C—H.

Organic compounds containing carbon–carbon double or triple bonds are often said to be *unsaturated*. Many constituents of membrane lipids are of this type (see Chapter 4).

The organic liquid *benzene* has the overall formula C_6H_6. Given that carbon forms four covalent bonds, the structure of benzene might be drawn as containing three carbon–carbon single bonds, and three double bonds, i.e.

This structure cannot be correct, however, since benzene does not show the characteristic chemical reactions of compounds containing double bonds. A carbon–carbon single bond is normally 0.154 nm long (one nanometre, nm, is 10^{-9} metre), and a carbon–carbon double bond, 0.134 nm; yet all the bond lengths between the carbon atoms in benzene are equal at 0.139 nm, i.e. intermediate between the double and single bond lengths. The six electrons, that should have formed three double bonds, appear to be 'spread around' all six bonds. This is often drawn as

or, in abbreviated form,

This abbreviated form is used in this book. Compounds containing the benzene ring are called *aromatic compounds*. This delocalization of electrons over several bonds greatly increases the stability of a molecule. Other examples can be seen in haem rings (Chapter 2), which show extensive delocalization of electrons, and in several ions such as nitrate (NO_3^-) and carbonate (CO_3^{2-}). In each case the negative charge is spread between each of the bonds, i.e.

(each O has, on average, one-third of the negative charge)

$$\left[\begin{array}{c} O \\ \vdots \\ C \\ O \quad \quad O \end{array} \right]^{2-}$$

(each O has, on average, two-thirds of a negative charge)

A.3. Further reading

Harrison, P. M., and Hoare, R. J. (1980). *Metals in biochemistry.* Chapman and Hall, London.

Hughes, M. N. (1981). *The inorganic chemistry of biological processes.* Wiley and Sons, London.

Liptrot, G. F. (1978). *Modern inorganic chemistry.* Mills and Boon, London.

Index

Where a topic is discussed in several places in the text, those pages containing the most detailed account of it are printed in **bold type.** The term *def* in parentheses after a page number indicates that that page contains a definition of the term indexed.